Simplified TV Trouble Diagnosis

By Robert Goodman

FIRST EDITION

FIRST PRINTING—MARCH 1973

Printed in the United States
of America

Hardbound Edition: International Standard Book No. 0-8306-3633-1

Paperbound Edition: International Standard Book No. 0-8306-2633-6

Library of Congress Card Number: 72-94810

This book covers some of the older color TV tube-type circuits and many of the modern solid-state chassis systems now in use or just coming off the production line. Thus, it is to your advantage to be well informed about how these newer solid-state circuits operate. Try not to be frustrated by this progress and feel that it is a threat to your electronics career, but think of it as a challenge. The past of TV electronics has been great; let's now prepare for its fantastic future.

Always follow a logical troubleshooting procedure for tracking down electronic circuit faults. This includes evaluating all symptoms and making an impartial judgment as to where or in which section(s) of the chassis the fault(s) have occurred. Then make an "eyeball" inspection of the chassis for any obvious defects and then check out all tubes or solid-state devices in the suspected section(s).

It is beyond the realm of this book to point out each and every circuit detail troubleshooting procedure. Many circuit faults differ and it would be impractical to go into great detail for every possible fault. However, use this book as your guide in order to solve the problem in the minimum amount of time.

I cannot stress this point enough — always use your oscilloscope to its full advantage and look at the circuit waveforms in order to solve most circuit defects in short order.

And always remember, don't overlook any circuit malfunction possibility or take any component for granted.

A block diagram of an RCA CTC-40 color TV chassis is shown in the frontispiece. As you know, the modern color TV receiver is a very complicated electronic instrument and can develop many problems which the TV service technician must solve. Serious study of this drawing even for a few minutes as a memory refresher will be useful as you progress through this book.

And that is the prime mission of this book, to take the mystery out of these many blocks and help improve your troubleshooting speed. You will find between these covers many TV circuit functions explained, troubleshooting procedures, and actual "case-history" problems that have occurred in these circuit systems. If you really want to probe these color TV circuits and see what makes them tick, then you should find this book "ver-ry in-terrr-esting."

I want to thank the many companies and persons listed below for their information and assistance in making this book possible. I greatly appreciate the material, information, and photos they have furnished.

Castle TV Tuner Service, Inc.—Cedric Western
Philco-Ford Corp.—John R. Krawczyk, Fred Fischer
Zenith Radio Corp.—Frank Hadrick, Ed Kobe, Joe Barrett, Ed Krol, Leo Smith, Brian Marohmic
RCA Sales Corp.—George F. Corne, H.C. Horton
General Electric Co.—Frank Boston, Bob Movak
Sony Corp.—G. P. McGinty
Motorola Inc.
The Magnavox Company
Texas Instruments, Inc.
B & K—Carl Korn
Rauland Corp.
Tektronix
Sencore— Norm Pedersen
RCA Commercial Engineering
Geo. H. Lehleitner Co., Inc., New Orleans, La.—Charles (Pinky) Osborne

Robert L. Goodman

CONTENTS

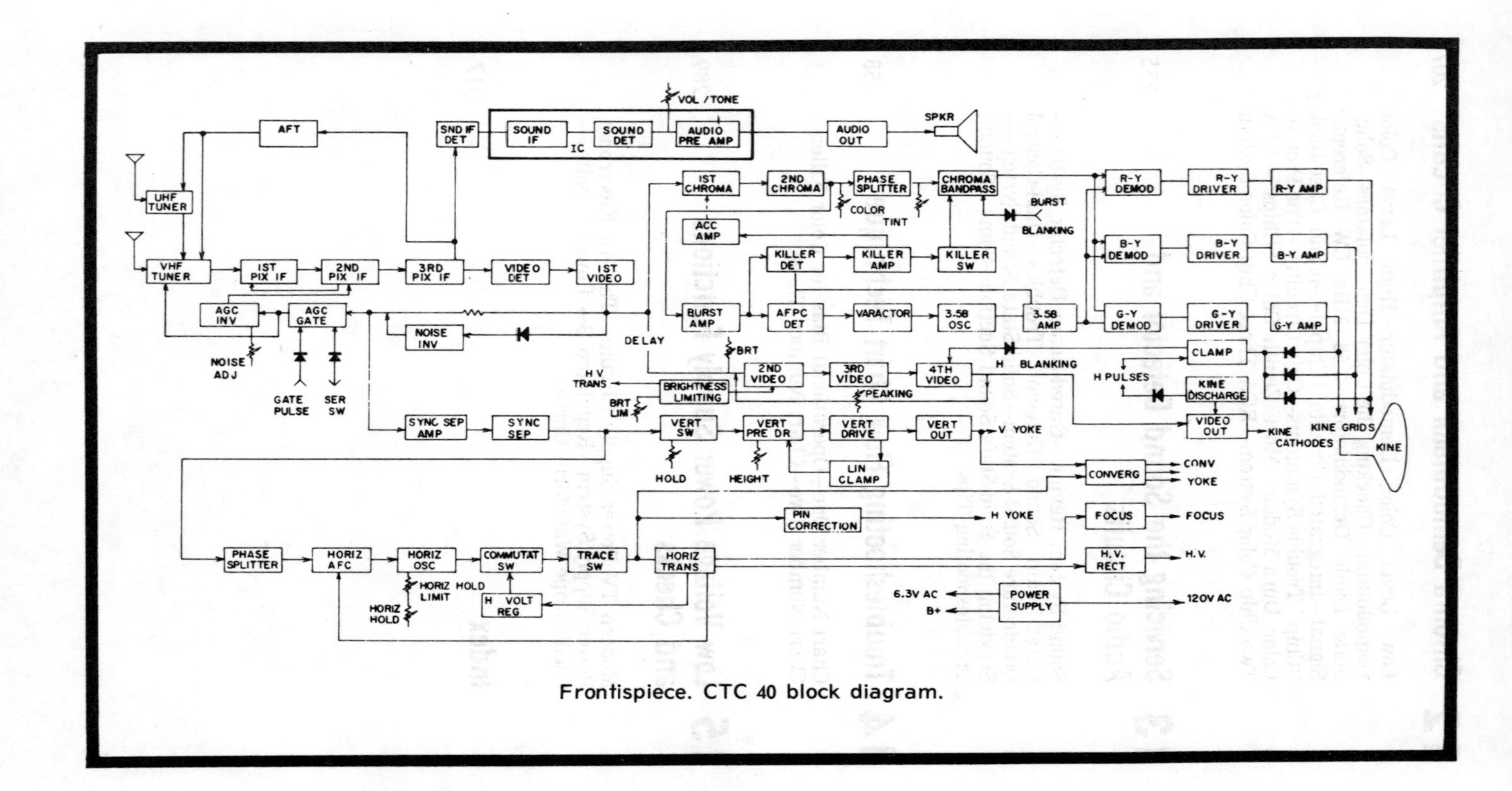

Frontispiece. CTC 40 block diagram.

Vertical Sweep Circuits

There are some differences between the vertical sweep systems found in black and white as compared with those in color TV receivers. One reason for differences is that the color CRT is not as efficient as the B and W picture tube, thus more deflection power and higher voltage are needed. Also the yoke is less efficient. These requirements call for higher-power output tube stages, a larger sweep output transformer, and a beefed-up HV system. Also the sweep tolerance is more stringent for a color set. Which reminds us that the vertical sweep rate for color reception is 59.94 Hz.

Another difference in the vertical sweep system is that color receivers, because of additional requirements, need some additional circuitry. Dynamic convergence signals must be taken from the vertical sweep circuits. On the rectangular color tube receivers, in order to improve picture geometry, pincushion correction circuits are needed to keep the edges of the raster straight. This pincushion distortion is due partly to yoke construction and will cause a straight line to curve inward at the tube center. See Fig. 1-1

VISUAL INDICATIONS OF VERTICAL SWEEP FAULTS

- *Raster of picture has vertical foldover
- *Raster has insufficient height
- *Bright horizontal line across screen
- *Picture nonlinear vertically

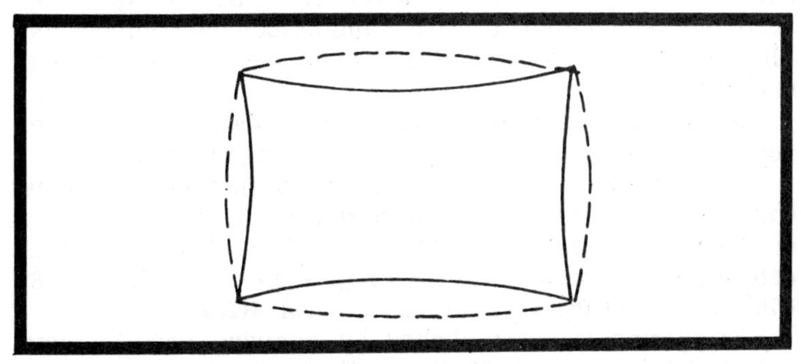

Fig. 1-1. Pincushion distortion.

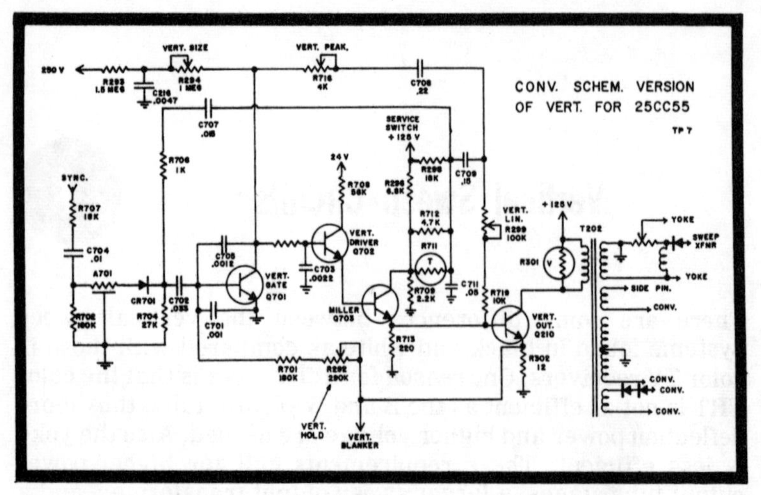

Fig. 1-2. Vertical sweep circuit schematic.

* Top of picture stretched out or compressed
* Bottom of picture stretched out or compressed
* Loss of vertical frequency or picture roll
* Vertical keystoning
* Intermittent condition of any symptoms that have been listed above.

Also the following symptoms can be caused by trouble in the vertical sweep section:

* Loss of, or poor vertical convergence
* Excessive pincushion distortion

In the following discussion we will see how some of the modern color TV vertical sweep circuits operate.

SOLID-STATE VERTICAL SWEEP (MILLER) CIRCUIT

This vertical circuit (Fig. 1-2) is composed of Q701 vertical gate, Q702 vertical driver, Q703 vertical Miller and Q210, the vertical sweep output transistor and associated components. These circuits are all on a plug-in module except the vertical output stage.

Adjustment controls include vertical hold, vertical size, vertical linearity, and the vertical peaking control.

The circuit is of a multivibrator design in which the collector output of the Miller amplifier is fed back to the base of the vertical gate transistor. The rate of oscillation is determined by the RC network composed of C701, C702, C708, C709, R704 and the vertical hold control, R292.

Triggering is accomplished by applying vertical sync pulses to the base of Q701, the vertical gate transistor. Sync from the video processor is integrated and shaped by R707, C704, R702, integrator A701, and CR701.

The vertical gate conducts only during retrace time, and is cut off during scan time. All other transistors in the vertical circuit are in conduction during vertical scan time.

For a better understanding of how this circuit operates, refer to Fig. 1-3. In its simplest form the circuit consists of a switch, a capacitor, and a transistor. When the switch is closed, the capacitor charges to the B+ voltage. When the switch is open, the capacitor is discharged through and by the transistor at a linear rate.

To provide sufficient current gain in this type of circuit, a transistor can be added between the switch and the original transistor (Fig. 1-4).

Basic circuit action is the same. By selection of the proper size capacitor and operation of the switch at a chosen frequency, a sawtooth waveform can be generated by this circuit.

Of course, this type of circuit would not be practical. If we replace the switch electronically, add appropriate feedback to operate the switch electrically, and add feedback for tuning, height and linearity controls, the circuit would appear as in Fig. 1-5.

When final refinements are added the completed vertical circuit is as shown in Fig. 1-2. The Miller feedback capacitor is now seen as two capacitors in series (C708, C709). Part of the signal at the base of Q210, vertical output transistor, is coupled through the vertical linearity control to shape the discharge waveform. Additional shaping is accomplished by the vertical peaking control. This affects the top one inch of the raster. The vertical size control functions by varying the voltage applied to a portion of the vertical circuit. The vertical hold control

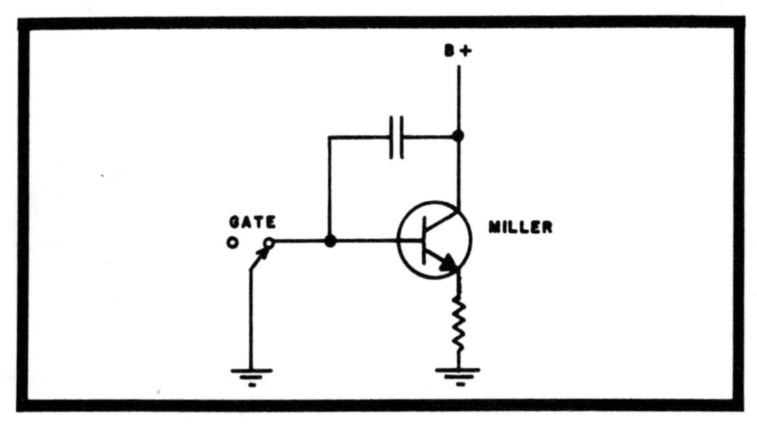

Fig. 1-3. Basic Miller circuit.

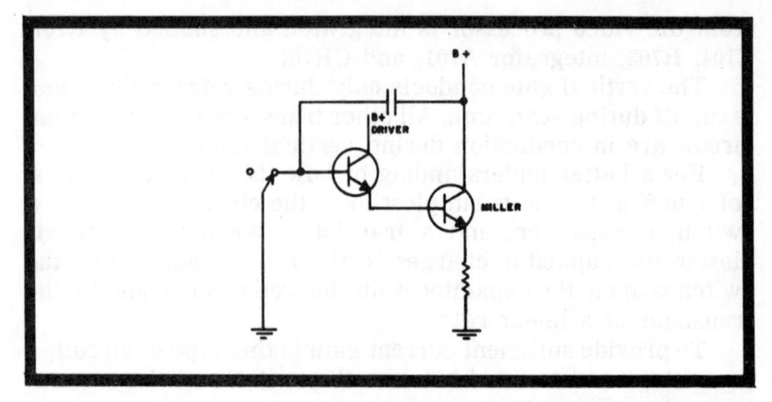

Fig. 1-4. Adding a transistor to the basic circuit, for current gain.

handles a major portion of the feedback to sustain oscillation and maintain proper frequency. In practice, vertical hold is adjusted so that the oscillator runs slightly high in frequency and the sync pulses trigger the oscillator at the TV station frequency. Thermister R711 is used to maintain vertical size with any temperature change. Vertical collapse, for CRT setup purposes, is accomplished by the setup switch which removes the +125 volts to Q703 and the Miller capacitors.

When the set is first turned on, capacitors C702 and C707 are charged by the B+ which is coupled through R296, R711, and R298. This positive voltage drives the gate into conduction. Miller capacitors C708 and C709 are charged during the retrace interval as follows: The high B+ voltage at the collector of Q703 is applied to the capacitors through the peaking control and Q701 (in conduction) to ground. As the charge at C702 bleeds off, Q701 stops conducting. This allows

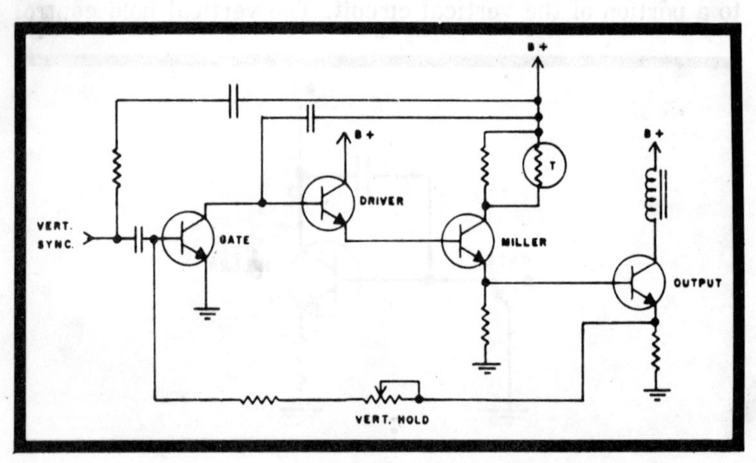

Fig. 1-5. A practical vertical circuit.

capacitors C708 and C709 to discharge (during the trace interval) through the driver to the base of Q703. The resulting positive-going sawtooth drives Q703 into conduction, causing the Miller capacitors to keep on discharging. A positive sawtooth also appears at the emitter of Q703. The sawtooth is reflected back to the base of Q701. This, along with the discharge of C702, causes the cycle to start again.

Since the vertical peaking control is in series with this circuit, adjustment of this control will affect the discharge waveform.

The waveform coupled from the emitter of Q704 into the base of Q210 (vertical output transistor, shown later) is amplified and applied through the vertical output transformer. The vertical section of the deflection yoke as well as the convergence panel is supplied with this sawtooth waveform.

VERTICAL SWEEP CIRCUIT POINTS OF SERVICE

Servicing this type circuitry is always complicated because there are several multiple feedback paths. If the basic reasons for these paths are understood, troubleshooting can be simplified. Refer to the module circuit in Fig. 1-6 for the following service procedures.

1- If the feedback path including R292, the vertical hold control, or its circuit connections becomes open, oscillation will cease.

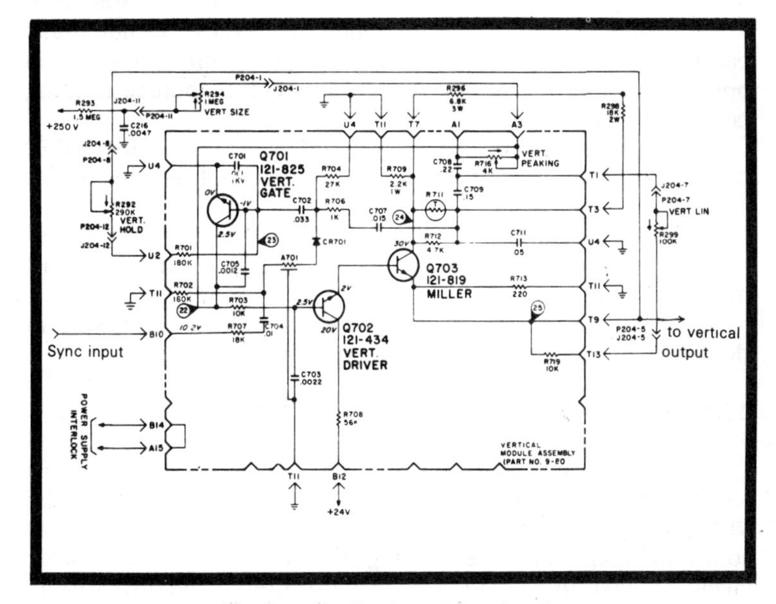

Fig. 1-6. Vertical module circuit.

2 - The feedback loop consisting of C707 and R706 is also necessary to sustain oscillation.

3 - The feedback through C709, C708, and the vertical peaking control primarily affects the frequency and linearity. An open circuit in this loop results in reduced vertical deflection, nonlinearity and foldover.

4 - Failure of any transistor in this circuit will result in complete vertical collapse. All transistors should be checked before operating the circuit or the DC coupling between the transistor stages may cause multiple failures.

5 - If C702 opens, the oscillator ceases to function as the proper AC reference for the vertical switch transistor is lost.

6 - Vertical sync defects will primarily involve components between edge connector B10 (R707) and the base of the vertical gate transistor.

ISOLATING A FAULT TO A STAGE

Use the following procedure to isolate the vertical deflection fault to a particular stage:

1 - Break the feedback loop at the vertical hold control by lifting one end of R701. Also lift off one end of C707.

2 - With the receiver "power off" connect one end of a .1 uF, 600-V capacitor from the green wire of the power transformer (dial lamp supply) with a clip lead to the base of the vertical output transistor. Full deflection (although nonlinear) will prove that the vertical output stage is working (Fig. 1-7).

Caution: For this and all following tests - Do not make connections to the vertical circuit with the receiver in operation. Remove power from the receiver before making connections to the circuit.

3 - Connect a 6.3-V AC signal to the base of Q703, the Miller transistor. Any screen deflection indicates that the Miller amplifier is operating.

4 - Connect signal to base of Q702, the vertical driver transistor. Deflection indicates proper operation of all previous stages.

5 - Continue back through the vertical gate transistor to determine location of any defective stage components.

TROUBLESHOOTING HYBRID VERTICAL SWEEP

The operation of this hybrid vertical sweep system is very similar to the previous solid-state circuit. So let's bypass the theory and go into a reliable method to pinpoint any faults that could develop. Troubleshooting the vertical multivibrator circuit, shown in Fig. 1-8, can be simplified if the difficulty is isolated to one stage.

No Vertical Sweep

1 - Break the multivibrator feedback loop by disconnecting one end of R254. Connect a lead to the 6.3 volt tap on the proper transformer (use pilot light lead). Connect a .1 uf, 600-volt capacitor to the other end of this lead.

2 - Connect the capacitor lead to the grid of the vertical output tube. If the stage is working, some deflection will be noted. The sweep on the screen will be distorted since the deflection is being generated by a sine wave rather than a sawtooth waveform. A normal operating output stage will

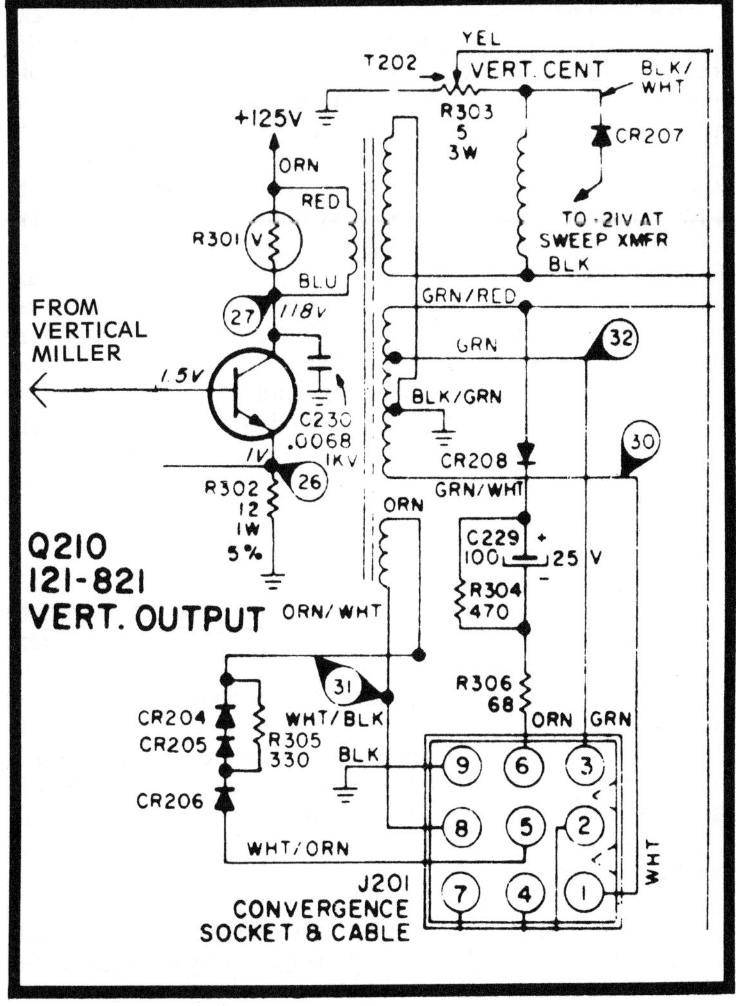

Fig. 1-7. Vertical output circuit.

produce about four inches of sweep with the vertical linearity control turned fully counterclockwise and about eight inches with the control turned fully clockwise. If some sweep was noted, proceed to the next step. If no sweep was produced by this test, make voltage measurements on the elements of the output tube. If the voltages are near normal, check for shorted windings and continuity of the output transformer, deflection yoke, and pincushion coils.

3 - Now inject the test signal by touching the capacitor to the collector of the oscillator. This is a quick check for coupling capacitor C227. Next, touch the capacitor to the base of the oscillator transistor. If the oscillator stage is operating correctly, about three or four inches of deflection will appear if the vertical size and linearity controls are turned fully counterclockwise. A severe overscan will be noted with both controls set fully clockwise. If the deflection at this point is sufficient, go on to the next step. If little or no sweep is obtained, measure the base voltage of the oscillator. When this stage is not oscillating, this voltage will measure +1.5 volts to +1.7 volts (variable with vertical hold control) if the transistor is conducting. A voltage reading of about +5 volts indicates either a defective transistor or an open diode in the emitter circuit. A shorted transistor is generally indicated by a low collector voltage reading. A low collector reading may also indicate a shorted regulator transistor.

4 - If everything has checked normal up to this point, the problem is in the multivibrator feedback loop. Check out capacitor C222 and C226 for an open condition. Check capacitor C221 for short or leakage.

Rapid Intermittent Partial Vertical Collapse

This symptom can be difficult to locate since the voltage fluctuations are measured all over the circuit. The problem can be isolated to one stage by applying the same technique as with no vertical sweep. Disconnect the feedback loop R254 and coupling capacitor C227. Connect the 6.3 volt AC test signal via the .1 uf capacitor to the vertical output grid. If the sweep is erratic, the trouble is in the output stage. If the sweep is steady, check the oscillator circuitry.

No Vertical Sync

If the horizontal sync locks in solid but the vertical sync is weak or nonexistent, the difficulty is generally localized to the circuitry in the emitter of the vertical oscillator. Check capacitor C209 for leakage. If this capacitor develops leakage,

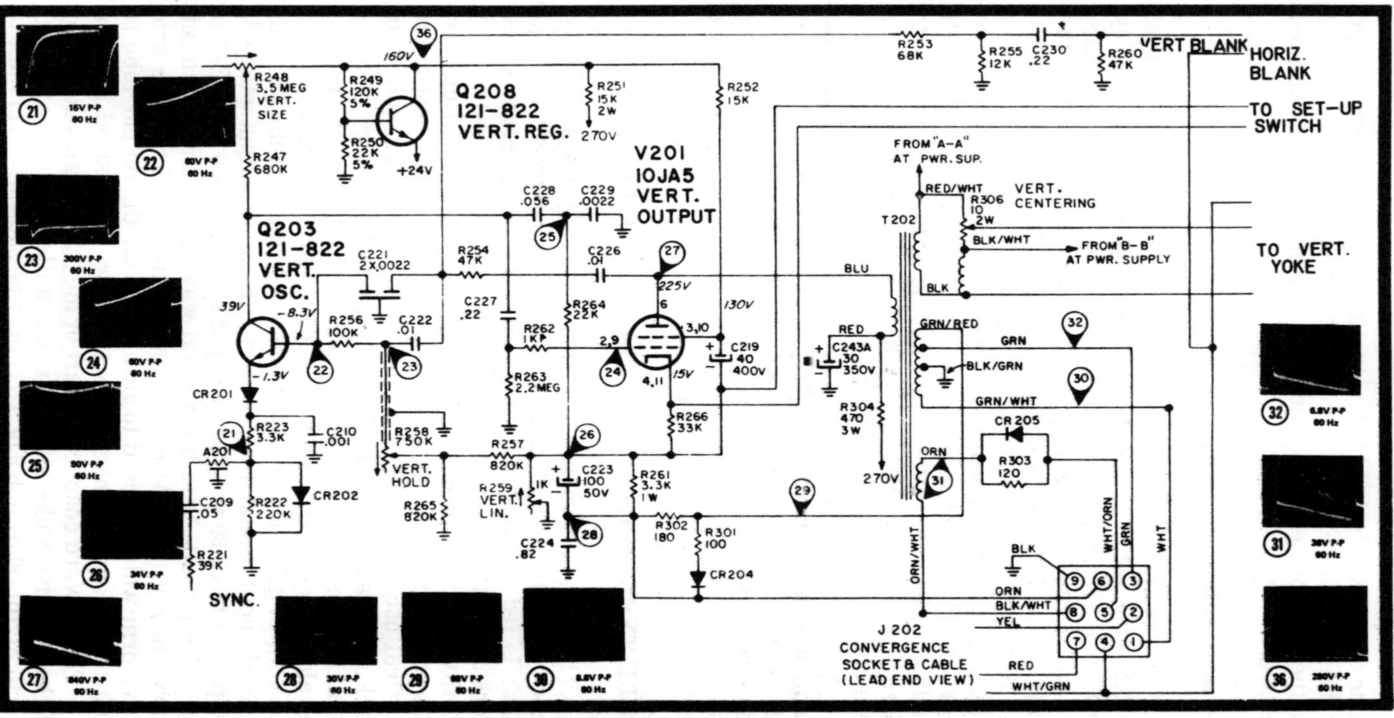

Fig. 1-8. Hybrid vertical sweep circuit.

diode CR202 will be held in conduction by the collector voltage at the sync limiter, thus presenting a short circuit to ground for the sync pulses. Also, check for leakage in CR202 and the vertical integrator, A201.

Vertical Jitter

This could be a symptom of vertical jitter and/or of raster interlace. This kind of difficulty is usually caused by horizontal sync information getting into the vertical circuitry. It can be caused by an open AGC bypass capacitor on the tuner, or the IF AGC bypass C215, or capacitor C207 located in the AGC section. Within the actual vertical circuitry, poor interlace may be caused by an open (capacitive section) integrator. It may also be caused by an open C221 dual unit feedback filter capacitor. The function of this capacitor is to filter out horizontal information that is reflected back from the deflection yoke and convergence system.

Another problem is where the vertical oscillator runs off-frequency and cannot be locked in by the hold control. This difficulty is generally associated with leakage or a change in value of capacitors C221, C222, and C226. This may also be caused by an off-value resistor or poor connection in the vertical hold control circuitry.

TUBE-TYPE VERTICAL DEFLECTION CIRCUIT

The vertical sweep system shown in Fig. 1-9 is a typical tube-type deflection circuit. A 3-volt P-P negative-going sync pulse is fed via coupling capacitor C104 and integrator network A1 to the cathode of V4B, the vertical oscillator. Any positive portion above 0.5 volt is clipped by diode X5. However, this pulse is not negative enough so that its effect is lost to the vertical oscillator where it coincides with the 140-volt feedback grid waveform. This pulse triggers V4B during vertical blanking and initiates the start of the sawtooth voltage developed across C110 and about 2.5 megohms of resistance in the control grid circuit of vertical sweep output tube V5. This typical input signal for the vertical sweep output stage is amplified from 80 volts peak-to-peak to over 1300 volts P-P in the plate circuit. Note the normal waveforms for this circuit shown in Fig. 1-10. Also, note the negative spike in the grid waveform which is used to firmly cut off the output tube so that the deflection yoke current follows the desired pattern.

The forward conduction part of the trace begins at the top of the negative spike and continues until the tube is at cutoff. Vertical roll (shown in Fig. 1-11) may be caused by a change in

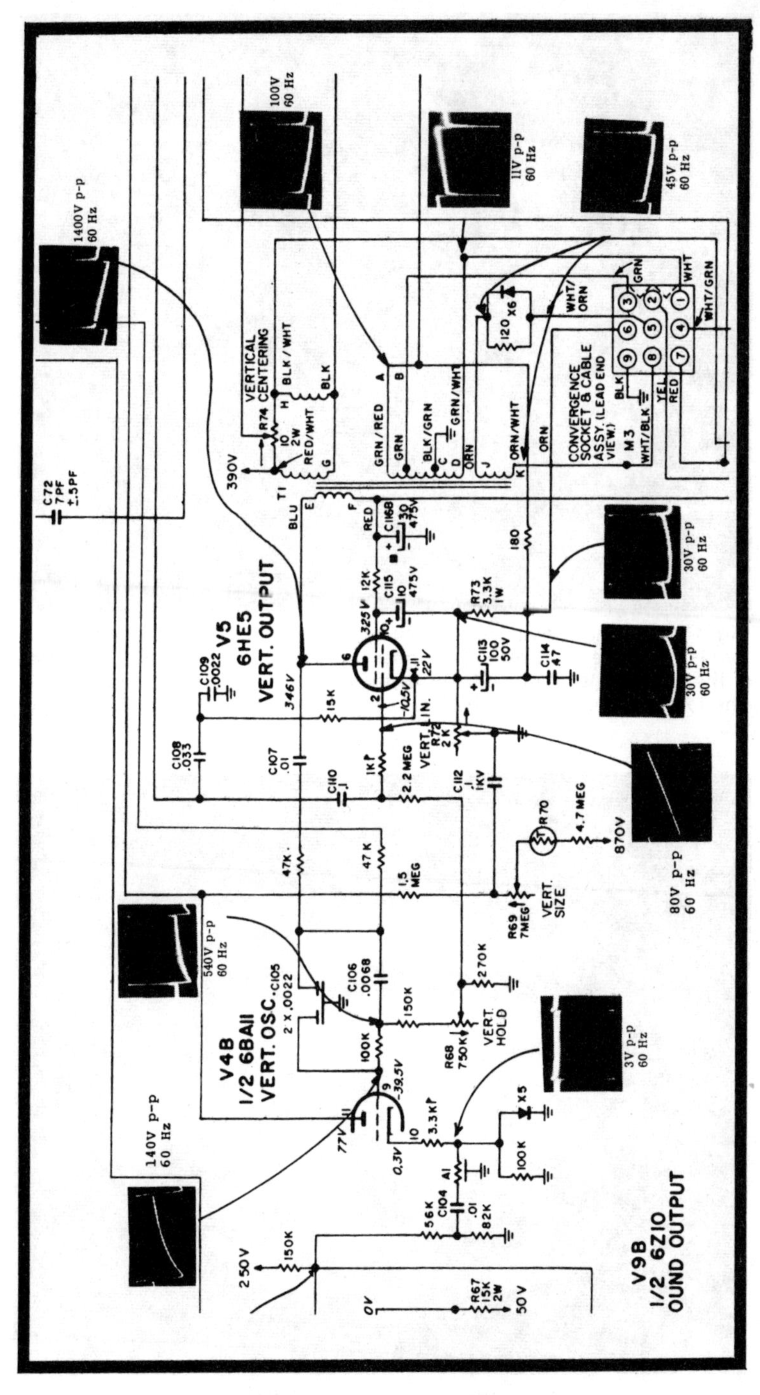

Fig. 1-9. Tube-type vertical deflection circuit.

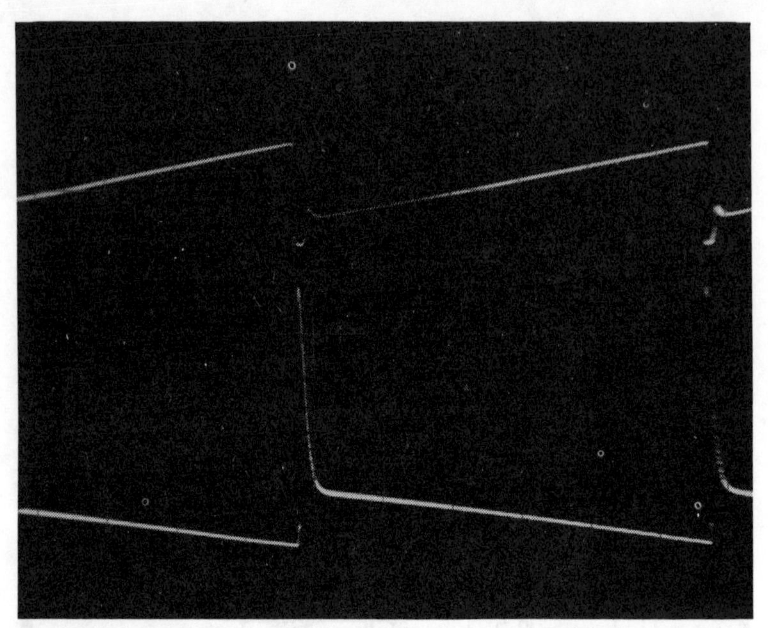

Fig. 1-10. Normal waveforms for circuit in Fig. 1-9. Scope is set at 30 Hz, that is, ½ the vertical rate. The upper trace is 80 V P-P taken at the plate of the vertical oscillator tube or at the grid of the vertical output stage. The lower trace is 1350 V P-P, taken at the plate of the vertical output tube.

Fig. 1-11. Vertical roll.

the oscillator grid bias resistor value or a leaky capacitor in the oscillator input grid circuit. Nonlinearity at center of the screen may be caused by an off-value replacement part or a leaky capacitor or resistor value change. Poor vertical linearity or insufficient height may be due to defective controls for linearity or height, or some leakage in the setup switch.

The vertical linearity and height (size) controls are in the cathode circuit of the vertical output and plate circuit of the oscillator. Feedback for the oscillator is taken from the plate of V5 by a 47K resistor, C107, and the 100K resistor to the oscillator grid. Dual capacitor C105 is a combination voltage divider and grid capacitor filter at the input of V4B.

Aside from providing the proper impedance match and linear current to the deflection yoke, the vertical output transformer performs other functions like providing voltages to the vertical dynamic convergence and pincushion correction system. A 100-volt peak-to-peak 60 Hz voltage is obtained from tap A of the vertical output transformer and an unfiltered 15,750 Hz parabolic waveform from a tap on the horizontal sweep transformer.

These two voltages, at the cathode and grid of the pincushion correction tube (not shown), key this stage into conduction during each switching interval and form the 150-volt "bow-tie" shaped waveform at the plate circuit. This waveform (which is adjustable by tuning the pincushion correction transformer core), is then applied to the deflection yoke and properly corrects the raster lines. Use a crosshatch generator in order to adjust the picture for straight raster lines.

If the vertical circuit shown in Fig. 1-9 has a vertical roll symptom (no locking action at all) check diode X5. This diode, located in the grid circuit of V4B, may be shorted and ground out all vertical sync information. Also check C104 for leakage.

Another vertical trouble symptom would be the picture bouncing up and down vertically like a rubber ball. This picture problem has also been referred to as yo-yoing up and down furiously. It's caused by a defective A1 integrator network.

A keystone picture is generally caused by a defective deflection yoke. However, a vertical sweep circuit fault can cause a slight keystone raster symptom. Here's a trouble history: The picture was about 3 inches wider at the top than the bottom. This trouble was caused by a faulty C116B waveshaping capacitor in the screen grid circuit of V5, vertical output stage. This capacitor may become open intermittently. A faulty C113 capacitor may give you a drifting convergence problem.

ONE TYPE OF PINCUSHION CORRECTION

Some pincushion circuits use a saturable coil that consists of specially shaped ferrox-cube E- cores with windings on all three limbs (or cores). Note transformer on left in Fig. 1-12. The center control winding is connected in series between the vertical deflection coils while the outer windings (being the controlled load impedance) are connected in series with the horizontal deflection coils and in series with each other. Note this circuit in Fig. 1-13. Connecting the two outer windings in series and in opposite direction minimizes interaction between the control and load windings. However, interaction only remains low as long as the core saturation is very low. At higher core saturations, interaction does occur and is used to obtain the required "top-bottom and side" pincushion corrections simultaneously.

The vertical deflection sawtooth current flowing through the center core winding produces a variable magnetic flux in the core. During one vertical scan, the permeability will vary from a low level at the start of the vertical scan, to a high level at mid-scan and again to a low level at the end of the vertical scan. This change in permeability has a parabolic type curve. The variation of permeability causes a change in the inductance value of the outer windings. Thus, the amplitude of the horizontal deflection current will vary in an approximately parabolic mode with vertical scan which will counteract pincushion distortion at the raster sides (Fig. 1-14).

During one horizontal scan period, the momentary value of the vertical deflection current can be considered to be constant. Also, at the time of a horizontal scan through picture center, the vertical deflection current is zero. The currents at horizontal frequency flowing through the outer windings of the saturable reactor produce a flux in the core that only passes through the outer cores but not the center core. However, this is true only if the outer cores are equally saturated. Due to differences in saturation of the outer cores, the flux at horizontal scan frequency will also partly pass through the center core. Note (Fig. 1-15) that the the intensity of this flux is dependent upon the intensity of the flux at the vertical frequency. The polarity of this flux is dependent on the polarity of the vertical deflection current. Thus, a flux at the horizontal frequency which is modulated in amplitude and phase when the vertical frequency passes through the center core induces a voltage in the center core which is an approximate sawtooth shaped waveform. See Fig. 1-16. Since this voltage acts in series with the vertical deflection coils, it causes currents of parabolic waveforms at horizontal frequency in the vertical deflection coils. The currents have

Fig. 1-12. Pincushion correction transformers.

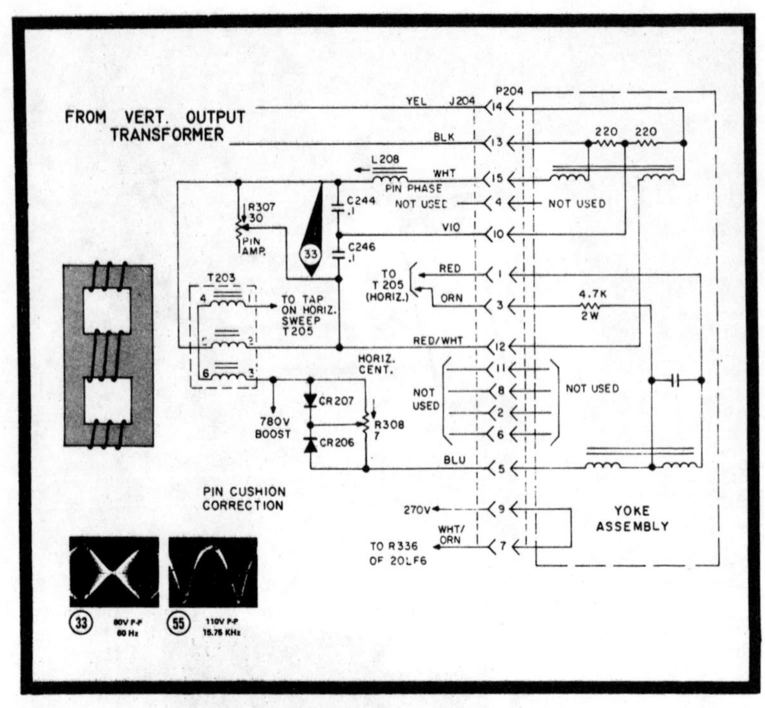

Fig. 1-13. Pincushion correction circuit.

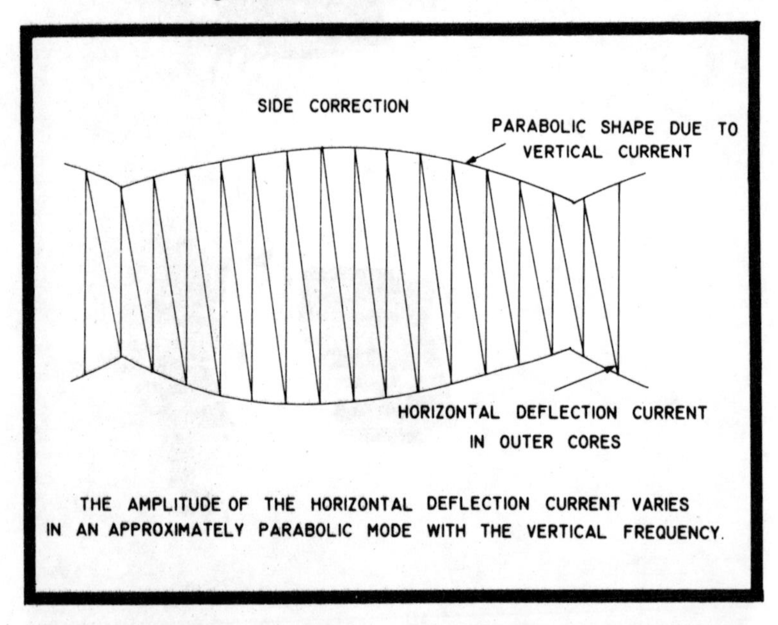

Fig. 1-14. Pincushion correction for the raster sides (expanded waveform).

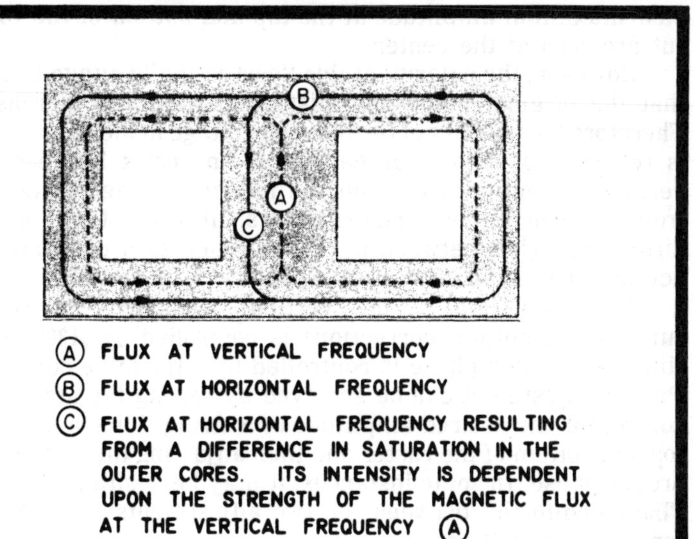

B
A
C
(A) FLUX AT VERTICAL FREQUENCY
(B) FLUX AT HORIZONTAL FREQUENCY
(C) FLUX AT HORIZONTAL FREQUENCY RESULTING FROM A DIFFERENCE IN SATURATION IN THE OUTER CORES. ITS INTENSITY IS DEPENDENT UPON THE STRENGTH OF THE MAGNETIC FLUX AT THE VERTICAL FREQUENCY (A)

Fig. 1-15. Flux paths in pincushion saturable reactor.

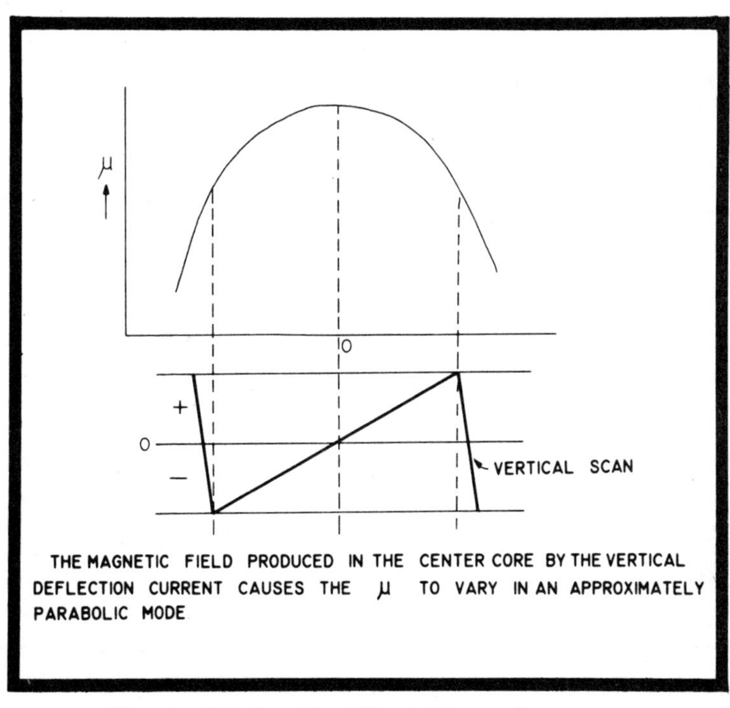

μ
O
+
O
—
VERTICAL SCAN
THE MAGNETIC FIELD PRODUCED IN THE CENTER CORE BY THE VERTICAL DEFLECTION CURRENT CAUSES THE μ TO VARY IN AN APPROXIMATELY PARABOLIC MODE

Fig. 1-16. Relationship of flux (u) to scan frequency.

their maximum amplitude at the top and bottom of the CRT, but are zero at the center.

However, the polarity of this "correcting" voltage is such that the original pincushion distortion would be increased. Therefore the polarity of the induced voltage in the center core is reversed by two .1 uf capacitors in series between the vertical deflection coils, and the induced voltage is tapped from between the coils and capacitors. Since a voltage (phase) difference exists between the voltage across a coil and one across a capacitor, the proper polarity is obtained.

The amplitude of the vertical frequency correction current (top-bottom correction) is controlled by R307, a 30 ohm control, and phase is controlled by adjustable coil L208. Proper adjustment can be achieved by setting the amplitude control for the desired amplitude of a straight line across the top and bottom of the raster and then adjusting the coil for the proper phase. In some instances, it may be desirable to work "back-and-forth" between the two adjustments a few times for optimum settings.

Horizontal Sweep and High Voltage Systems

The horizontal sweep section of a color TV receiver has more functions to perform than a black-and-white set. These additional functions include pulses for color killer action, automatic color control, burst separation, chroma section blanking, pincushion correction, and convergence circuitry.

Because the shadow mask color picture tube is not very efficient, a larger amount of power is required for the horizontal sweep and high voltage (HV) system. The color CRT needs 25 kV at the second anode that must be maintained at a constant level by a voltage regulator which also consumes more power. This is to prevent convergence shift and changes in picture size as beam current varies with changes in video modulation.

To maintain constant picture linearity and convergence, the yoke power in a color chassis is almost double that found in a B & W TV set. The result is greater power requirements which call for larger capacity output tubes and dampers, higher peak-to-peak drive voltage from the oscillator, and a larger horizontal sweep output transformer. Of course, with an increase in this power, more chassis heat is generated. And with all of these additional circuits, the chance of a system failure is also increased. So let's now look at some of these circuits and see how they operate and what problems may develop.

THE HORIZONTAL CONTROL CIRCUIT

A popular horizontal sweep system uses a 6U10 triple-triode tube to provide horizontal frequency oscillation and control. Refer to the schematic in Fig. 2-1. In addition, a pair of diodes is used as a phase detector which accepts a sawtooth waveform from the flyback circuit and compares its phase (timing) with the incoming horizontal sync pulse. The output from the diodes is a DC voltage which is fed to the control grid of one triode section. Any change in phase between the incoming sync pulse and the sawtooth waveform will be reflected as a change of the internal resistance for the control triode section. The cathode-plate resistance of the control

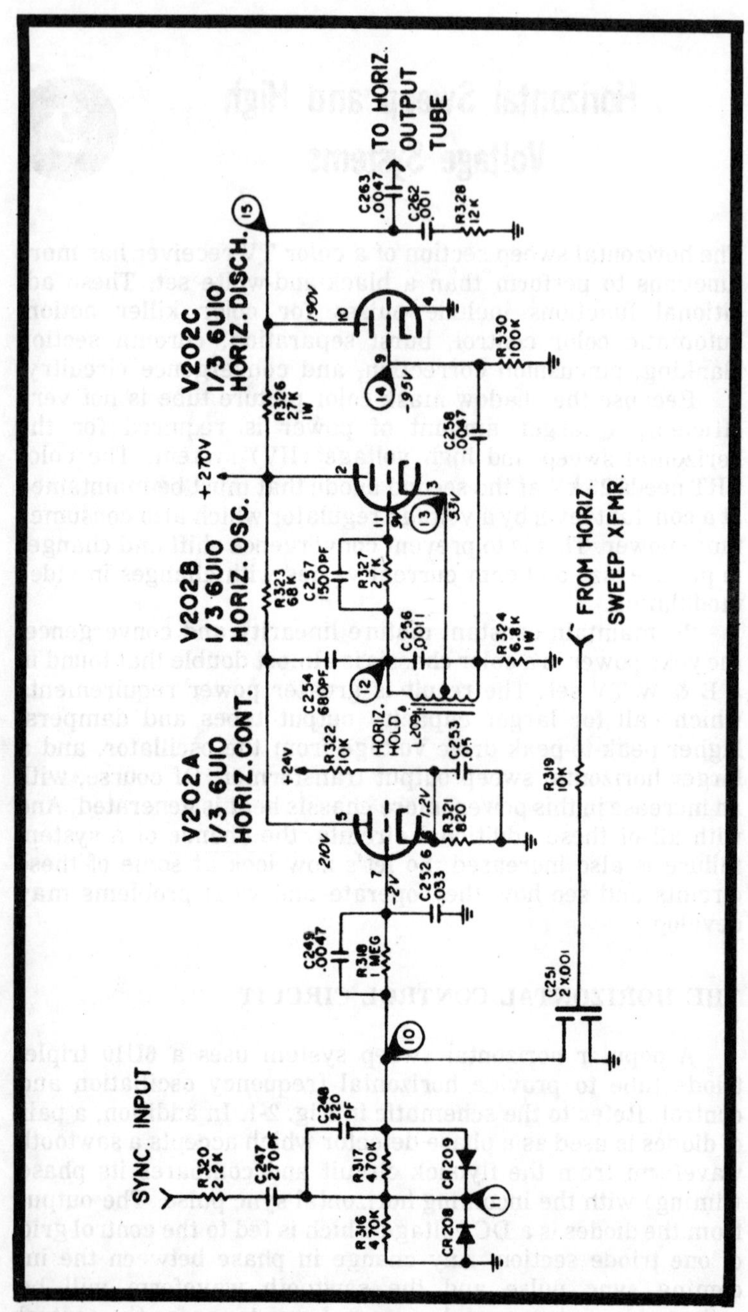

Fig. 2-1. Horizontal oscillator, AFC, and discharge circuit using 6U10 triple-triode tube.

triode is connected in series with a 680 pf capacitor (C254) across the horizontal oscillator tank coil. If the internal resistance of the triode increases, the value of C254 is, in effect, reduced and the oscillator frequency will increase. If the tube resistance is decreased, the capacitance of C254 is effectively increased and the oscillator frequency decreases. In other words, if the DC voltage output from the phase detector is negative, then the bias on the control tube will be increased, resulting in a higher frequency output from the horizontal oscillator. If the phase detector DC output is positive, the control tube bias will decrease and thus lower the frequency of the oscillator. If the phase detector output is zero, as is the case when the sync pulse and sweep sawtooth voltages are exactly in phase, then the control tube bias is not changed and the horizontal oscillator runs at its mean frequency of 15,750 Hz.

FUNCTION OF THE HORIZONTAL OSCILLATOR

The horizontal oscillator utilizes another one-third of the triple triode, 6U10, in a Hartley circuit. The remaining triode is used as the horizontal discharge tube stage.

This type of horizontal oscillator circuit has been basically unchanged through the years and can be considered conventional, being designed to produce a 15,750 Hz AC voltage across the tank circuit. Note all of the scope waveforms shown in Fig. 2-2 that will be found in this oscillator system. The oscillator plate current wave is distorted by the grid circuit constants R327 and C257 to develop a voltage waveform across unbypassed resistor R324 which contains a high peak of about

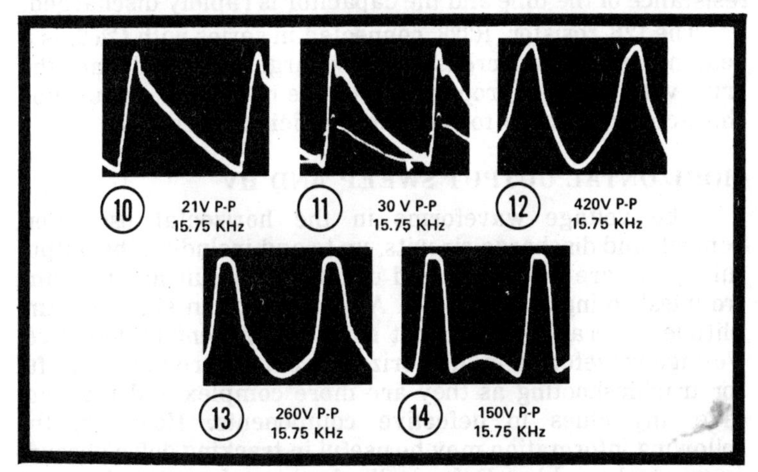

Fig. 2-2. Waveforms for circuit in Fig. 2-1. Circled numbers 10 through 14 are key to circled numbers on the schematic.

10 microseconds in duration. This voltage is applied to the grid of the discharge section which functions as the horizontal discharge tube.

HORIZONTAL DISCHARGE SECTION

The voltage waveform applied to the grid of the horizontal discharge tube is a positive pulse of around 170 volts peak amplitude and about 10 microseconds duration. When this pulse is applied to the grid, the tube is immediately driven into plate current saturation and the plate voltage is lowered to a value very close to ground potential. For the duration of the pulse the grid draws current which charges C258 (.0047 uf). After the pulse passes, C258 discharges through R330, a 100K resistor to ground, applying a high negative bias to the grid and causing the plate current to be cut off. The tube is held at plate current cutoff until the next positive pulse arrives at the grid. During tube cutoff time, the plate potential rises to B+ value. If only a load resistor was used in the plate circuit, the output, as viewed on the scope, would appear as a series of negative pulses that are approximately 10 microseconds in duration. The action of the discharge stage is similar to that of an electronic switch across the B+ supply, being alternately opened and closed 15,750 times a second.

The .001 uf capacitor, C262, is the sawtooth voltage generator. When the horizontal discharge tube is cut off, C262 charges in the direction of B+. The time constant selected is such that the portion of the charging curve used for deflection is essentially linear. When the positive pulse is applied to the control grid, C262 is, in effect, shorted by the low internal resistance of the tube and the capacitor is rapidly discharged.

The 12K resistor, R328, connected in series with C262, is a peaking resistor to increase the discharge speed and place the drive voltage at the proper point on the horizontal output tube characteristic curve for the most efficient operation.

HORIZONTAL OUTPUT SWEEP AND HV

The voltage waveforms in the horizontal oscillator, control, and discharge circuits, up to and including the output tube grid, are important and useful for circuit analysis and troubleshooting. See Fig. 2-2. Any deviation in shape or amplitude generally means that some component is defective. Voltage waveforms in the horizontal output are not as useful for troubleshooting as they are more complex and may not give any clues to defective components. However, the following information may be useful in tracking down obscure or unusual troubles. Refer to the horizontal output circuit in Fig. 2-3.

The sawtooth voltage fed to the output tube grid causes the plate current and yoke current to rise. If no previous sweep voltage has been applied to the output tube grid, the beam will be at about CRT screen center. The plate current of the output tube causes the beam to be deflected from the center to the right edge of the screen. When the beam reaches this point, the sawtooth is abruptly removed from the output tube grid, which cuts off tube plate current flow. In turn, the strong magnetic field around the yoke collapses and generates a voltage across the yoke windings. This magnetic field collapse starts a vigorous oscillation and current reversal in the yoke for one-half scan cycle. One function of this oscillation is to perform a rapid retrace so that the beam will be in the proper position at the left side of the screen to sweep out the next line of transmitted video.

During the one-half cycle of oscillation, the damper tube cathode will be positive with respect to its plate and will not conduct. By this time the half-cycle damper cathode will begin to swing negative. The tube will then conduct and place a low resistance load across the circuit. The oscillation is damped out and the current slowly decays, moving the beam from left to center of the CRT screen. Inductor L212 and capacitor C268 are connected in series with the damper tube plate. When the damper conducts, the rectified pulse charges C268 and since the damper tube is in series with B+, this charge is added to this B+ voltage and becomes the "boost" voltage (780V). The boost capacitor is C268. Diode CR213 is used to develop the "boosted-boost" voltage, 1200V.

At the same time the damper tube current approaches zero, and the beam is at screen center, the output tube is again coming out of cutoff and begins to supply current to the yoke. In order for the picture to be evenly spaced (linear) on the screen, the beam must be deflected at a constant rate from left to right. Since two deflection currents are involved in sweeping out the full line, there must be no change in deflection rate at the moment the output tube plate current takes over.

The voltage pulse developed across the yoke is stepped up by auto-transformer T205, then rectified by the HV rectifier and applied to the CRT anode as HV.

HIGH VOLTAGE REGULATION AND RECTIFICATION

High voltage is regulated in this system by the following method. A positive HV pulse is fed via capacitor C264 to the voltage dependent resistor (VDR), R333. This 1500V P-P pulse is obtained from a primary winding of the horizontal sweep transformer. This results in a negative potential ap-

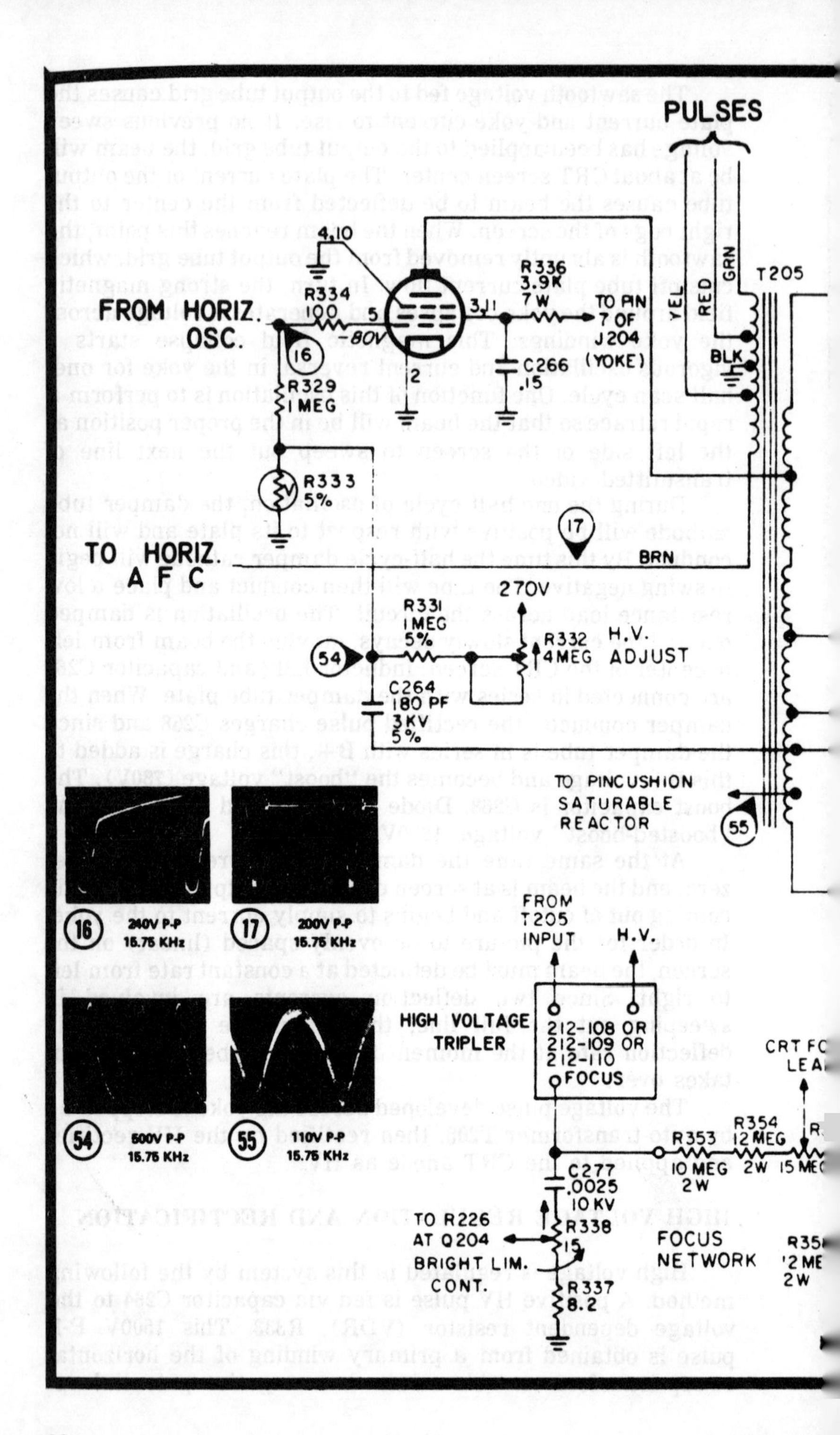
PULSES
4,10
R334
100
FROM HORIZ.
OSC.
5
-80V
16
R336
3.9K
7 W
3,11
TO PIN
7 OF
J204
(YOKE)
C266
.15
2
R329
1 MEG
YEL
RED
GRN
T205
BLK
R333
5%
17
TO HORIZ
A F C
BRN
270V
R331
1 MEG
5%
R332
4MEG
H.V.
ADJUST
54
C264
180 PF
3 KV
5%
TO PINCUSHION
SATURABLE
REACTOR
55
16
240V P-P
15.75 KHz
17
200V P-P
15.75 KHz
FROM
T205
INPUT
H.V.
HIGH VOLTAGE
TRIPLER
212-108 OR
212-109 OR
212-110
FOCUS
54
600V P-P
15.75 KHz
55
110V P-P
15.75 KHz
R354
12 MEG
2W
R353
10 MEG
2W
15 MEG
C277
.0025
10 KV
TO R226
AT Q204
BRIGHT LIM.
CONT.
R338
15
R337
8.2
FOCUS
NETWORK

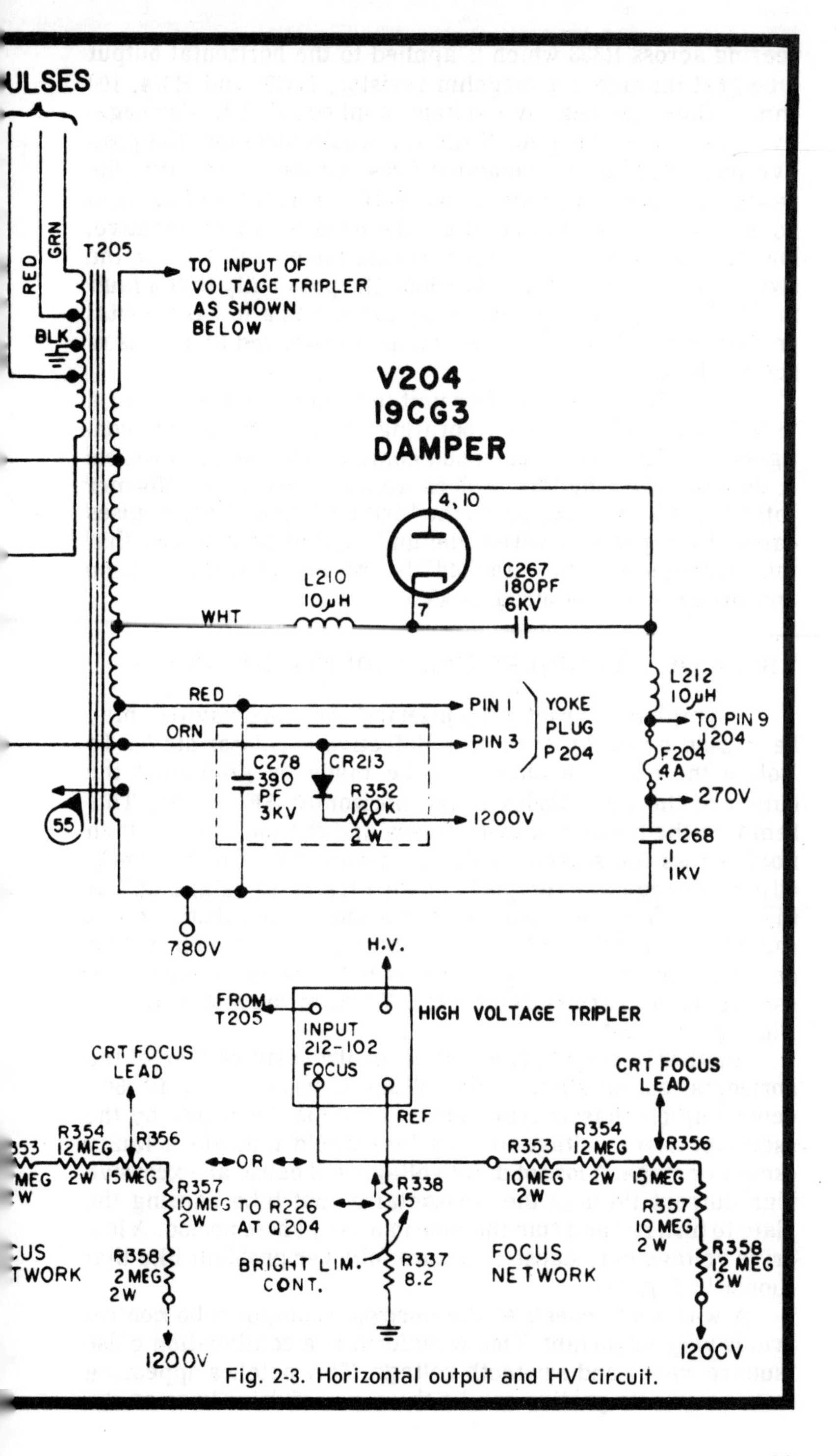

Fig. 2-3. Horizontal output and HV circuit.

pearing across R333 which is applied to the horizontal output tube grid through a 1 megohm resistor, R329, and R344, 100 ohms. Thus, the negative voltage applied adds to the negative potential at the grid. If the HV would increase, the positive pulse applied to capacitor C264 increases. In turn, the resulting negative voltage across R333 increases and the grid voltage of the horizontal output tube becomes more negative, lowering stage gain and thus lowering the output pulse to the sweep transformer. Do not overlook the possibility that a fault in this HV regulation system may appear as a low horizontal grid drive problem. The high voltage is adjusted by means of control R332.

The high voltage rectifier unit used in this system is of a "solid-state" design incorporating a tripler circuit configuration. The electrical circuit contained in the unit consists of diodes and capacitors. Pulses from the sweep transformer (about 8.3 kV) are coupled to the "tripler" input. These pulses cause the capacitors within the unit to charge in a way that their voltages become "stacked" to provide the tripling action and produce the required 25 kV.

TIPS AND TECHNIQUES FOR TROUBLE DIAGNOSIS

A symptom of "raster top jittering" or "flag waving" may be caused by faulty sync or a defective oscillator stage. To isolate the cause, a check can be made by grounding the output of the AFC diodes in the horizontal AFC stage. This removes the horizontal sync pulses and the picture will then float across the screen. If the horizontal hold can be slowly adjusted to cause a complete picture frame to "stand up" at the correct frequency point of operation without distortion, it can be assumed that all stages in the horizontal sweep system are in order. If not, the problem is most likely in the sweep oscillator, discharge circuits, etc. Refer to the schematics in Figs. 2-1 and 2-3.

Another important check is the control grid voltage of the horizontal output stage. This voltage is usually -50V to -80V depending on chassis type. This voltage is developed by the oscillator drive voltage and its loss would indicate a faulty oscillator stage. A loss of drive voltage will cause an excessive high current through the horizontal output tube causing the plate to turn red and ruin the tube in a very short period. A low drive voltage may cause a picture foldover problem like that shown in Fig. 2-4.

A waveform check at the horizontal output tube control grid is very important. This waveform is a combination pulse (square wave) and sawtooth voltage. Tiny notches appearing on the sawtooth portion can be the cause of drive lines on the screen. Look for faulty components between the plate of the horizontal discharge tube and the grid of the horizontal output tube.

Excessive or frequent failure of the horizontal sweep transformer or output tube is also an indication of excessive current being drawn by the sweep and output circuits. Check the current being drawn by the output tube. A meter, placed in series with the output tube cathode, will read the total current being drawn. This will vary, depending on chassis, from approximately 210 to 240 milliamperes.

The value of the RC network C257 and R327 in the grid circuit of the horizontal oscillator determines the width of the "pulse" part of the waveform at the grid of the horizontal output tube. A pulse that is too narrow can cause excessive sweep current. If C257 opens, the frequency of the oscillator will change. No HV will be produced and usual checks may not be effective. The voltage at the grid of the horizontal output transformer (H. O. T.) will be near normal, the waveform will have correct shape and amplitude, but it will be at the wrong frequency. This will be about the only clue.

If C254 (680 pf) develops a slight leakage, the oscillator also changes frequency. Adjustment of the horizontal hold control restores near normal operation. There will be some bending of vertical lines at top of the raster, and in some cases, shortened H. O. T. life.

Fig. 2-4. Horizontal picture foldover caused by low horizontal drive voltage.

Fig. 2-5. Drive line (middle of picture) caused by shorted winding in horizontal sweep transformer.

Many causes of intermittent oscillator operation can be traced to broken wires or cold solder joints on the horizontal hold coil, L209. Since this control can be physically moved by the set owner, excessive use or abuse can lead to a malfunction.

RINGING AND DRIVE LINES

Connecting the input probe of an oscilloscope to the insulation (using an alligator clip) of the unfiltered boost voltage lead will reveal a parabolic type waveform. A certain amount of low amplitude ringing will be present. The ringing should be smooth. Small "notches" in some of the scope waveforms is an indication of the presence of drive lines or a tendency for the set to produce drive lines. Note that only very slight drive lines are being produced and they may not even be noticeable in the picture but can be seen on this waveform. The drive line shown in Fig. 2-5 was caused by a faulty horizontal sweep transformer.

Resistance measurements of the horizontal sweep transformer are not of great value because the resistance of a particular winding will have negligible change if a shorted turn exists. But the shorted turn can greatly reduce the Q of the inductive winding, appreciably affecting performance. Pulse checks should be made around the windings of the sweep transformer as a most effective way to determine its condition. The scope probe can be held close to the insulated leads. **Use caution around these high peak pulse voltage circuits.**

SPOOKS, SNIVETS, "DIODEHAUSEN"

Another condition which can occur is called spooks or snivets. They appear as "blobs" of interference in the picture and can occur in any area of the raster. They are the result of undesirable oscillation (spray) being picked up by the video channel via inductive coupling. They originate in the horizontal output tube, damper tube, or HV rectifier tube. The chassis shown in Fig. 2-3 has a solid-state HV rectifier which eliminates the possibility of "diodehausen" which could be caused by a HV rectifier tube.

An oscillation within the horizontal output tube can contribute to a "snivet" condition. In some instances, changing the tube will cure this problem, but for others, the voltage around the tube should be accurately checked. An abnormal bias or voltage due to a component off-tolerance can cause the tube to create this undesirable oscillation.

Spook coils (L210 and L212) are inserted in the damper tube cathode and plate respectively to suppress any undesirable oscillation in this area.

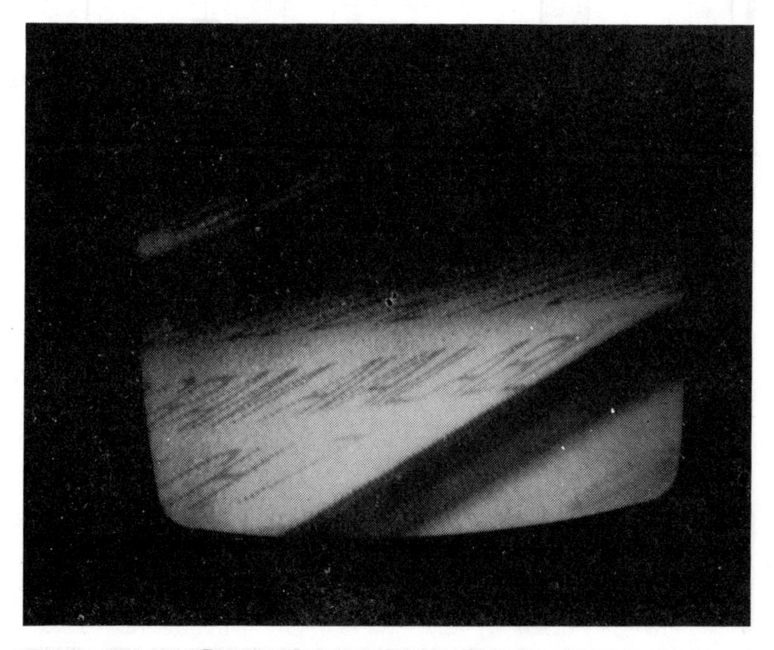

Fig. 2-6. Loss of horizontal lock-in.

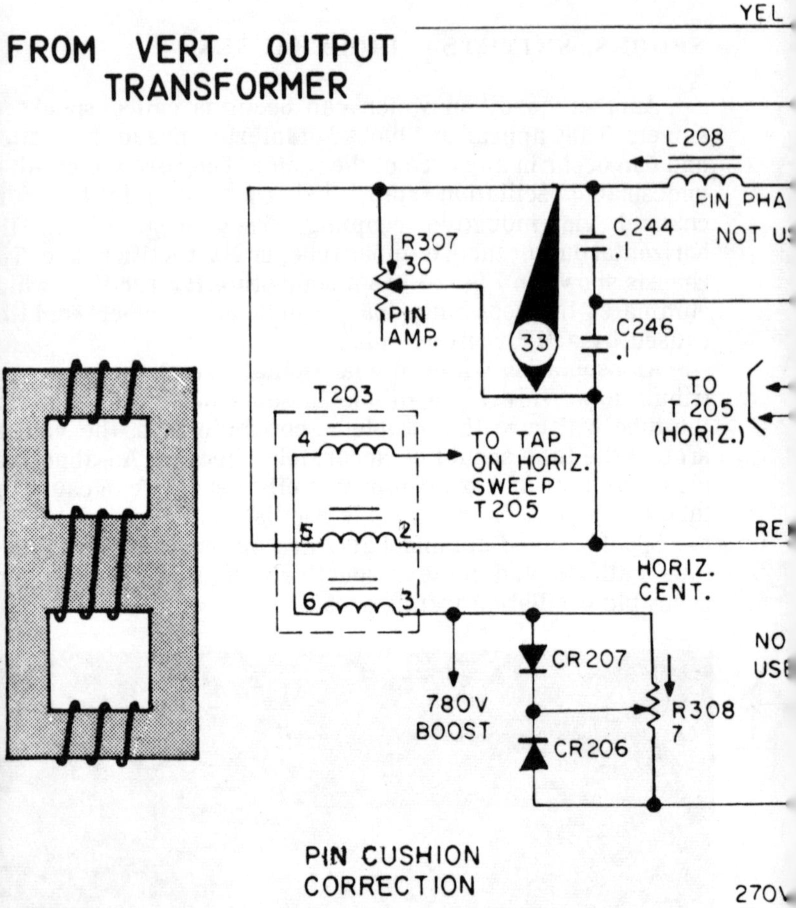

FROM VERT. OUTPUT TRANSFORMER

YEL

L208

PIN PHA

NOT U

R307
30

PIN
AMP.

C244
.1

C246
.1

33

TO
T205
(HORIZ.)

T203

4 1

TO TAP ON HORIZ. SWEEP T205

5 2

RE

HORIZ.
CENT.

6 3

NO
USE

CR207

780V
BOOST

R308
7

CR206

PIN CUSHION CORRECTION

270V

TO R336
OF 20LF6

33

80V P-P
60 Hz

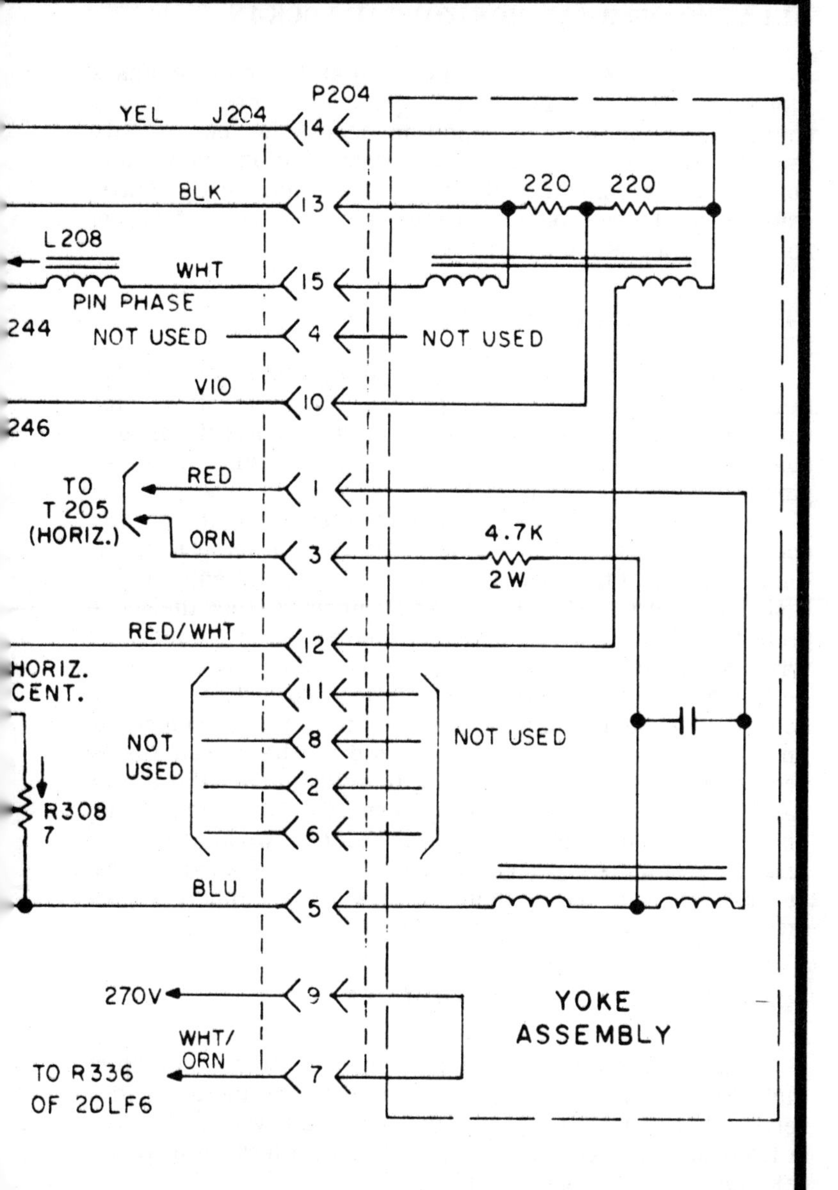

Fig. 2-7. Pincushion correction circuit showing yoke connections, highlighting possibility that what appears to be a horizontal problem may be caused by cross-coupling from vertical circuit through the yoke and associated circuitry, including wiring.

DELAYED OR WEAK HORIZONTAL LOCK-IN

In this instance it may appear that the picture has a "tendency" to lose horizontal sync; or, when switching channels, an out-of-sync condition occurs for one or two seconds. If the sync pulses are correct, the anti-hunt components in the control grid circuit of the horizontal control stage should be checked for correct values or vaults. A loss of horizontal lock is shown in Fig. 2-6.

PINCUSHION CIRCUITS

The pincushion circuitry is not too difficult to service due to the limited number of components. The saturable reactor with its adjustments are incorporated to correct for pincushion deficiencies which would otherwise exist. Note the pincushion circuit in Fig. 2-7.

A waveform that is very useful in determining the correct operation of this circuit is shown at test point 33 in Fig. 2-7. This is located at the phase and amplitude adjustments. In this case chassis ground is **not** used as a reference, thus the scope probes (both leads) are connected **above** chassis ground. The waveform appearing here shows the complete modulation for both sides and top-bottom correction. Adjusting the two controls (phase and amplitude) will affect this waveform. If an effect is noticed, and action noted on the screen (use a cross-hatch generator to make this adjustment), the circuit can be considered to be operating normally. Since the primary circuit consists only of the saturable reactor, two adjustments, and two capacitors, it could be said that just changing these components or checking them for lack of adjustment or poor adjustment should isolate the circuit fault.

CONVERGENCE CIRCUIT TROUBLE

This color set had developed an odd picture symptom (note Fig. 2-8) that was very misleading. The picture had a vertical bar down the center of the screen and was also pulled in from the left side. This appeared to be a horizontal phase detector or sync circuit problem.

However, this unusual picture trouble symptom was caused by a fault in the horizontal convergence circuit.

Now refer to Fig. 2-9A and let's see what happened. This trouble started when diode CR601D shorted out. When this diode shorted it also burned up coil L601, resistor R608 and control R609. We installed new components and then touched up the convergence controls (Fig. 2-9B).

DEFLECTION AND HV CIRCUIT CHECKS

Here's some more information for servicing horizontal sweep and HV circuitry in tube type color receivers. These are adjustments, key voltages, currents and trouble analysis to confirm proper operation. We will use an RCA CTC-38 chassis for these service checks.

Horizontal Circuit Adjustments (See Fig. 2-9)

Horizontal oscillator alignment consists of removing the incoming sync pulse, shorting the sine wave coil and adjusting the horizontal hold control for the correct free-running oscillator frequency. This assures that the oscillator is operating at the correct frequency and provides maximum noise immunity to the circuit.

Horizontal output and HV alignment involves the simultaneous adjustment and monitoring of horizontal output cathode current and HV. This assures maximum output stage efficiency and operation within the power dissipation rating of the output tube.

The efficiency coil, when properly adjusted, assures maximum deflection circuit efficiency with minimum power dissipation in the sweep output tube. This adjustment controls the decay of damper current and conduction of the output

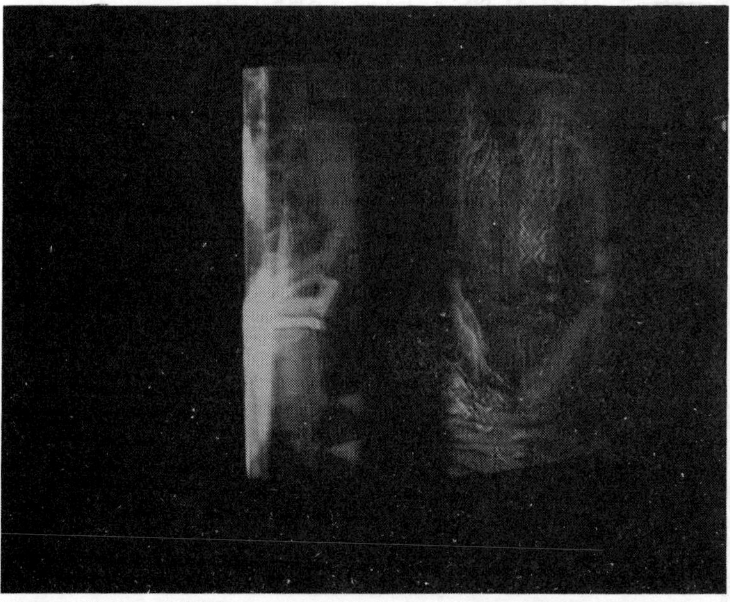

Fig. 2-8. Unusual CRT symptom. Horizontal phase detector or sync was suspect but trouble was found to be in horizontal convergence.

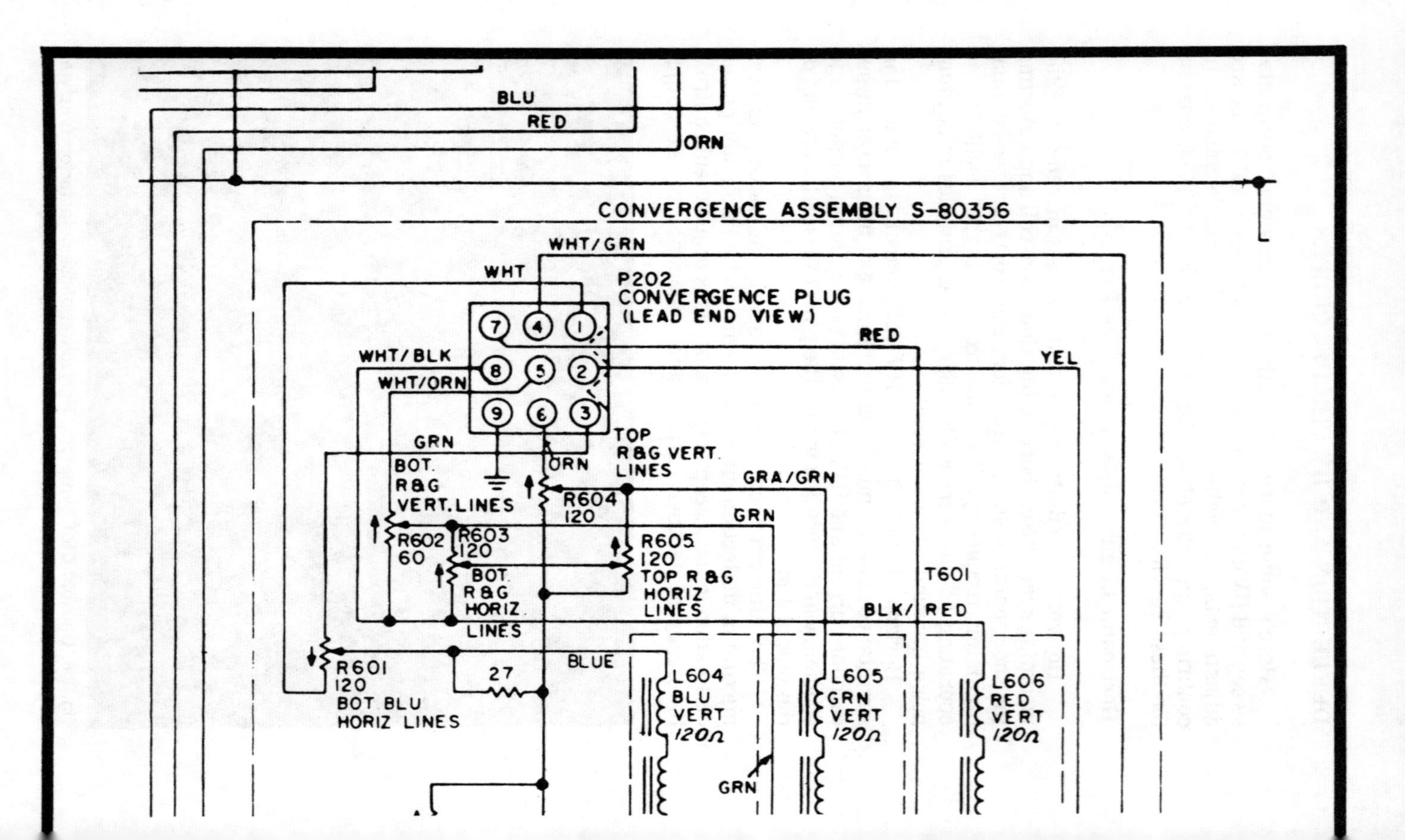

BLU
RED
ORN
CONVERGENCE ASSEMBLY S-80356
WHT/GRN
WHT
P202
CONVERGENCE PLUG
(LEAD END VIEW)
RED
YEL
WHT/BLK
WHT/ORN
7
4
1
8
5
2
9
6
3
GRN
TOP
R&G VERT.
LINES
BOT.
R&G
VERT. LINES
ORN
GRA/GRN
R604
120
GRN
R602
60
R603
120
R605
120
TOP R&G
HORIZ
LINES
BOT.
R&G
HORIZ.
LINES
T601
BLK/RED
R601
120
BOT. BLU
HORIZ LINES
BLUE
27
L604
BLU
VERT
120Ω
L605
GRN
VERT
120Ω
L606
RED
VERT
120Ω
GRN

A

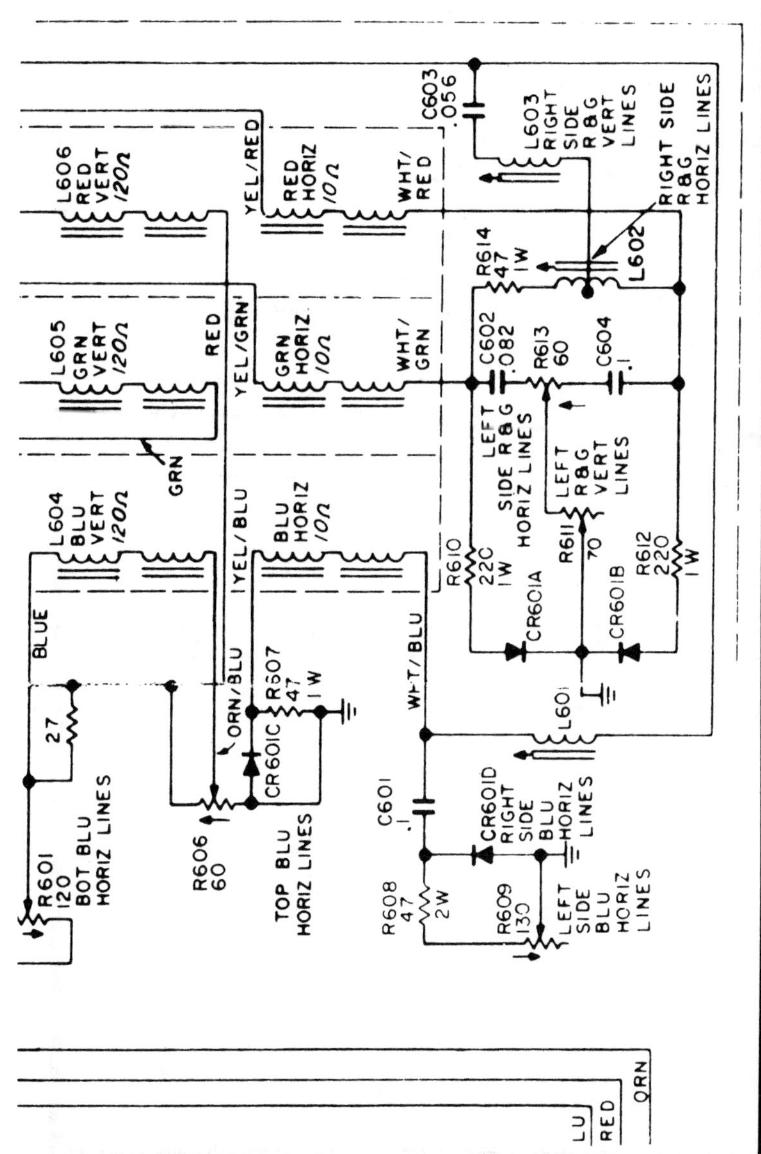

Fig. 2-9. Convergence circuit schematic (A) and block diagram for preliminary adjustments (B). (Part 1 of 2 parts.)

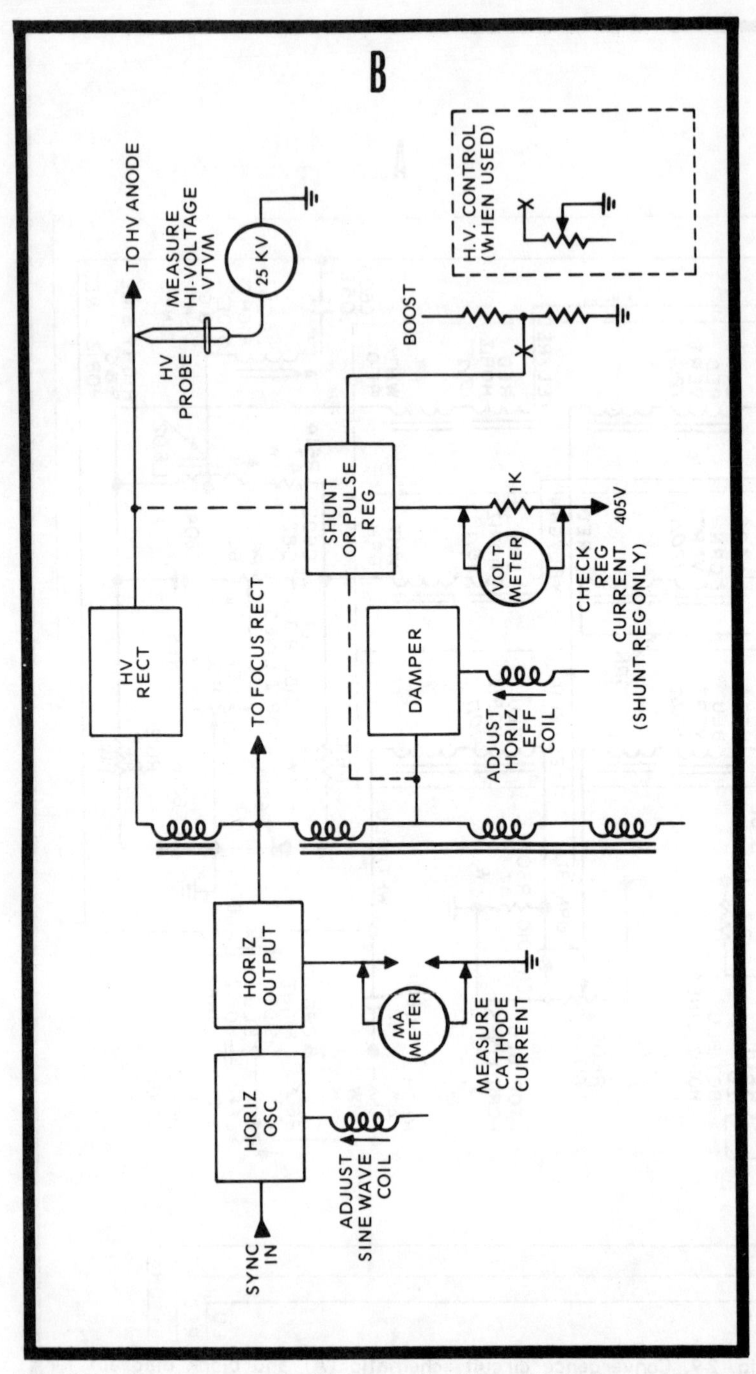

Fig. 2-9. Convergence circuit schematic (A) and block diagram for preliminary adjustments (B). (Part 2 of 2 parts.)

tube, to set the proper "crossover" transition between damper-tube current and output-tube current. A milliammeter is connected to measure the output tube cathode current, and the horizontal efficiency coil is adjusted for a dip (minimum current). Optimum horizontal linearity and efficiency usually occurs slightly offset from the "dip." Therefore, the coil should be adjusted to increase cathode current by 3 or 4 ma. However, the direction of adjustment must result in the following: Regulator current (voltage across 1K resistor) must **increase**, simultaneous with increasing output cathode current. Note: If regulator current decreases, the efficiency coil was adjusted in the wrong direction from the dip. This results in less efficiency, poor linearity, and more output stage power dissipation.

Sets equipped with a high-voltage control should be adjusted for the HV specified in the service notes for that chassis.

With the anode voltage at 25 kV and brightness at minimum, the regulator current must be at least 960 microamps (0.96V across 1K resistor). Typical current is approximately 1350 microamps (1.35V). As the picture tube current increases (brightness turned up), the regulator current decreases by the same amount to maintain a constant load on the HV system.

HORIZONTAL SWEEP CIRCUIT CHECKS AND DIAGNOSIS

Note: Before these steps are performed set the line input voltage to 120 V AC. Refer to Fig. 2-10 for these checks.

STEP 1 - CHECK DRIVE WAVEFORM

The grid of the horizontal output tube must receive a correct drive waveform for efficient operation. The grid-drive amplitude is normally 220 to 235 volts P-P. A correct waveform indicates normal operation of the horizontal oscillator stage.

Minimum P-P drive is approximately 200 volts. A lower reading indicates trouble in oscillator tube or associated components. Possible causes include:

1. Weak oscillator tube—does not discharge timing network properly, thereby decreasing amplitude of drive signal.
2. Defective component in timing network—most likely shorted C277 or off-value C278. Changed value R296, R297, and R173. RV301 and C127 can also affect drive voltage.

STEP 2 - CHECK GRID BIAS VOLTAGE

Nominal DC grid bias when measured with a TVM is around -54 volts, as determined by grid-drive signal and hold-down bias network. The hold-down circuit function is to limit HV to a safe value if the regulator tube fails.

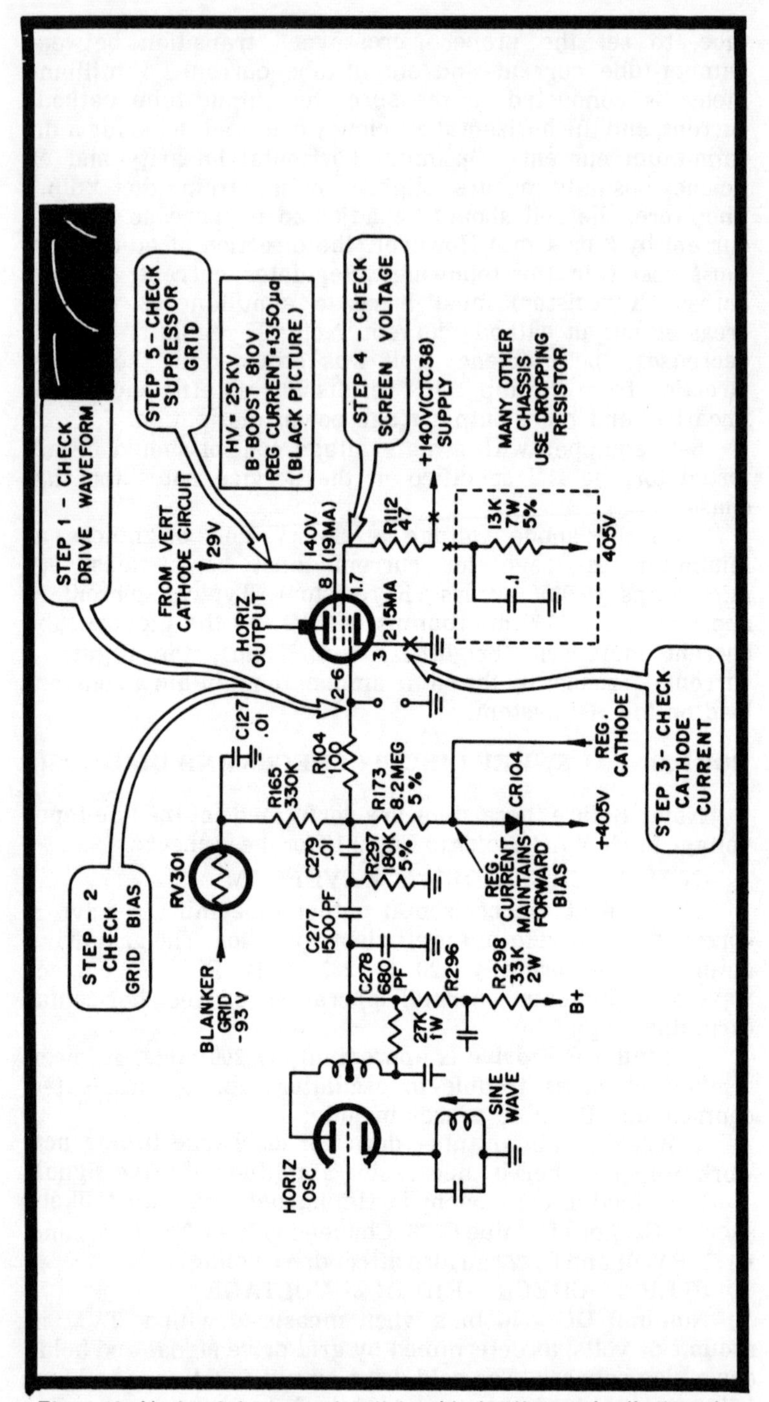

Fig. 2-10. Horizontal sweep circuit troubleshooting and adjustments.

Low grid bias (less negative), can be caused by:

1. Improper drive signal.
2. Fault in hold-down circuit.
3. Defective horizontal output tube.

High grid bias (more negative), can be caused by:

1. Faulty horizontal output tube.
2. Hold-down circuit action.
3. Excessive blanker grid voltage in chassis that use blanker grid to furnish output stage bias.

STEP 3 - CHECK CATHODE CURRENT

Nominal cathode current for the CTC-38 chassis is 215 ma (235 ma max.) with properly adjusted horizontal oscillator, efficiency coil, and high voltage. The maximum safe current of 235 ma will vary for a different chassis. Always check the service data. Total current depends on input drive, output load, efficiency adjustment, and screen grid operation. If current in near the maximum limit, perform monitor check as follows:

1. Operate set for 5 minutes, perform HV adjustments (if applicable), check cathode current.
2. Recheck cathode current "dip" and reset after 30 minutes of operation.
3. Let chassis cool down and repeat number 2 above.

Excessive current can be caused by: Defective horizontal output tube; low drive signal; screen grid load; misadjusted or defective efficiency coil; or overload from flyback system, boosted-B, or boosted-boost circuits.

STEP 4 - CHECK SCREEN VOLTAGE

Screen grid voltage is an important check point of output tube operation. Nominal voltage in the example circuit is 140 volts.

Nominal screen current under these conditions is 19 ma. Excessive screen current leading to short output tube life may be caused by:

1. Defective tube.
2. Decrease in load resistor value.
3. High B+ voltage.
4. Improper load on horizontal output by flyback system.

STEP 5 - CHECK SUPPRESSOR GRID

Voltage on the suppressor grid should be +20 to +40 volts, obtained from voltage divider in the cathode circuit of the vertical output tube. Positive voltage does improve the operating characteristics of the output tube. A wrong voltage here that could change the tube operation would be evident by an upset in the vertical sweep on the CRT.

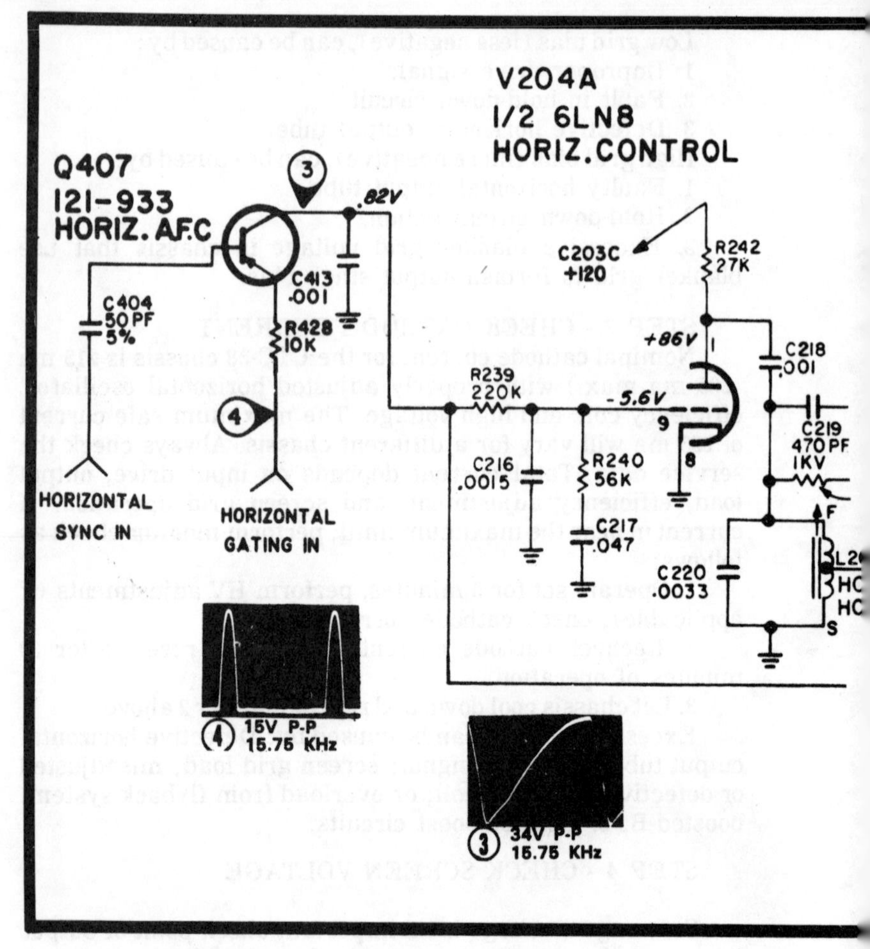

B & W HORIZONTAL SWEEP SERVICE TIPS

Troubleshooting the horizontal sweep system can be divided into four sections. These are the AFC, oscillator, sweep output and deflection yoke-damper. Since the oscillator is operated from the boost source, if there isn't boost voltage, the oscillator will stop functioning. Conversely, if the oscillator is not operating, boost voltage cannot be developed. In troubleshooting for "no raster," the oscillator must be isolated from the output circuit so that the defective component can be located.

With a symptom of "no raster" or "no high voltage," make the following checks. Refer to Fig. 2-11.

Measure the voltage at the grid of the output tube with VTVM. If it is near -60 volts, the oscillator is running and drive voltage is being developed. To verify this, check the grid drive

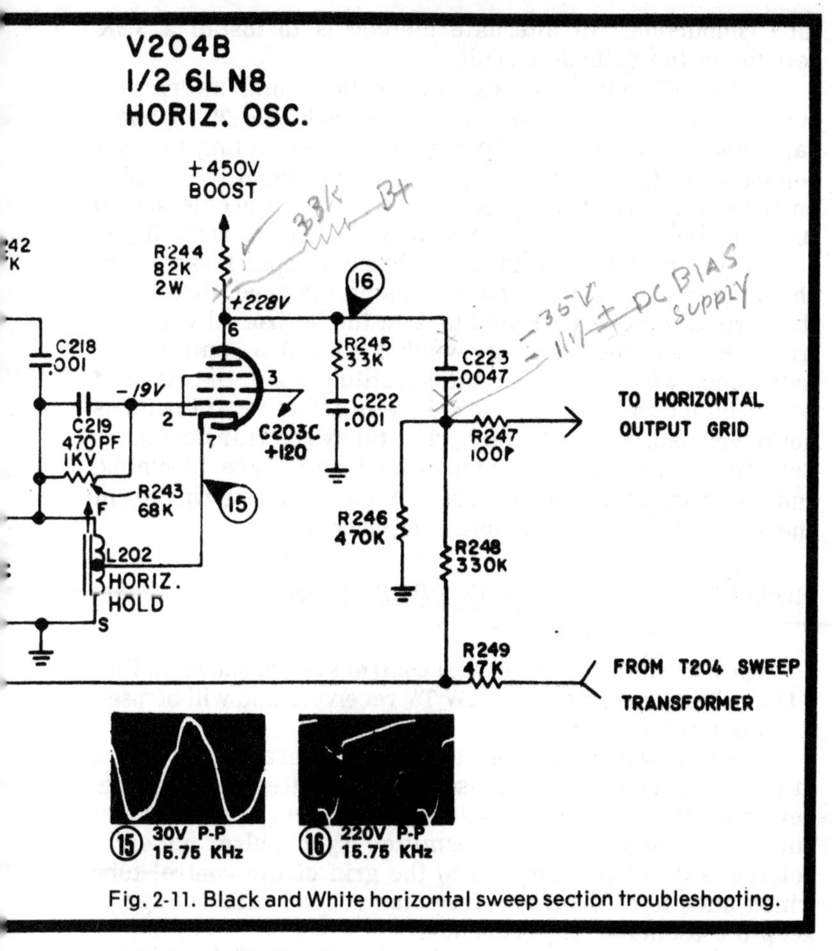

Fig. 2-11. Black and White horizontal sweep section troubleshooting.

waveform with an oscilloscope. It should be as shown in Fig. 2-12 and have an amplitude of approximately 260 volts peak-to-peak.

If the drive is missing, low in amplitude, or distorted, connect a 33K, 2 watt resistor in place of the 82K at the oscillator tube plate, pin 6. Connect the other end to the B+ line voltage supply. If a drive waveform can be obtained although lower in amplitude, the oscillator is okay and the problem is in the sweep output or damper circuit (which will cause low or no boost voltage to be developed). With the substitute B+ plate supply, drive will be about 150 V P-P and a -25 volt bias should be measured.

Note: While checking for oscillator operation it may be necessary to disconnect the coupling to grid of the output tube and apply -35 volts from a DC bias supply to limit excessive tube conduction. An alternate method is to install a 3.3K resistor in the cathode circuit.

Make DC voltage checks in the horizontal sweep and output stage as shown in Fig. 2-12. Check the .22 uf boost capacitor, C225. This can be done by disconnecting the end connected to the boost; with power applied read the voltage on the other end. If the capacitor is good the needle should jump to about +120 volts and then swing back to zero. If you read 5 volts or more, the capacitor has developed leakage and should be replaced. If the boost capacitor is good, disconnect the horizontal yoke leads and tack in the horizontal windings from a known good yoke. Any winding with 15 to 25 millihenry inductance will work. With the substitute yoke, the HV and boost voltage will return if the yoke was defective. If the HV is not developed, probably the horizontal sweep transformer is defective. For problems related to low-high voltage, blooming, and distortion of the raster, use voltage and scope waveform checks to locate the defective components.

HORIZONTAL OSCILLATOR OPERATION

The horizontal oscillator and control system shown in Fig. 2-11 has been used in many B&W TV receivers and will be used for this circuit analysis.

When horizontal sync pulses fed to the base of Q407 arrive in phase with the gating pulse on the emitter, a voltage is developed that keeps the oscillator on frequency. If the gating pulse leads or lags the transmitted sync pulse, an error voltage is developed and fed to the grid of the control tube which adds to or subtracts from the control tube voltage to keep the oscillator on frequency.

At the control tube, a network formed by R239, R240, C216, and C217 is used to filter the correction voltage before it is applied to the grid. The correction voltage adds to or subtracts from the negative bias on the control tube. As the DC current through the tube changes, the impedance changes. Since the control tube and C218 are in parallel with the tank circuit, the oscillator frequency will change as the impedance of the tube changes.

The pentode section of V204B, the 6LN8, is used as an oscillator and discharge tube. The oscillator is a Hartley type with L202 and C220 comprising the tank circuit. Voltage for the plate of the oscillator is fed from the +450 volt boost source through resistor R244. Resistor R245 and capacitor C222 comprise the sawtooth waveshaping network. The horizontal drive waveform from the waveshaping network is coupled to the grid of V205A, the horizontal output tube (Fig. 2-12).

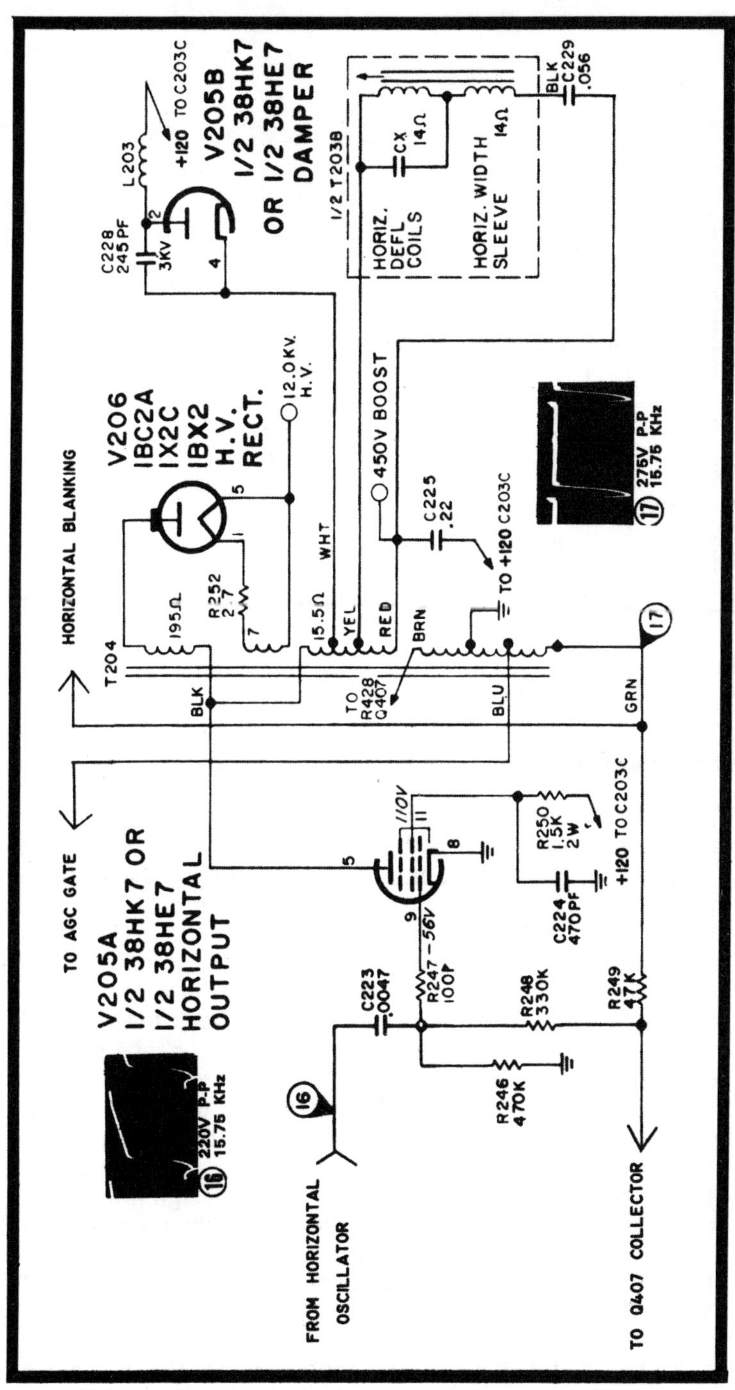

Fig. 2-12. Horizontal output, yoke, and damper, B & W set.

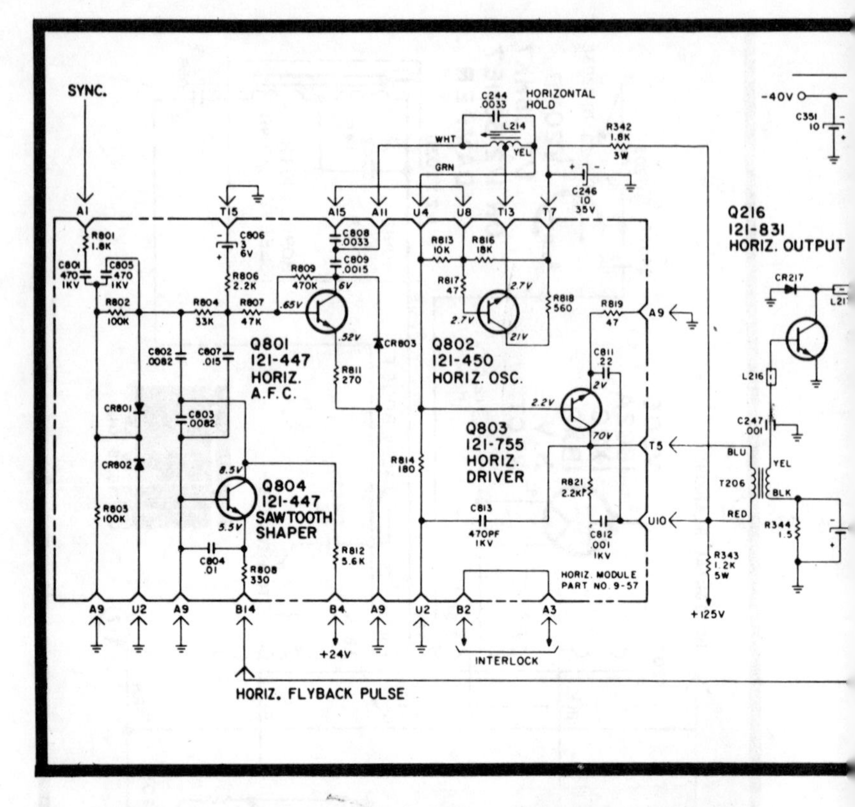

With drive voltage applied, V205 conducts to deflect the beam from the center of the screen to the right side of the CRT. At that point the tube is cut off, current ceases and the magnetic field in the transformer and yoke collapses. This generates an oscillation and current reversal in the yoke-transformer windings. This oscillation continues only for one-half of a cycle. During this one-half cycle, the beam is being retraced from the right edge of the screen back to the left. At the end of the retrace, the cathode of the damper tube becomes negative with respect to the plate, causing the tube to conduct. As the damper conducts, the oscillation is damped, and the beam is traced from the left to the center of the screen.

Conduction of the horizontal output tube starts before the damper current reaches zero. This circuit design allows for a linear sawtooth of current to be developed in the yoke.

SERVICING THE HORIZONTAL SOLID-STATE SWEEP SYSTEM

The solid-state horizontal module shown in Fig. 2-13 can be easily repaired by making some DC voltage checks.

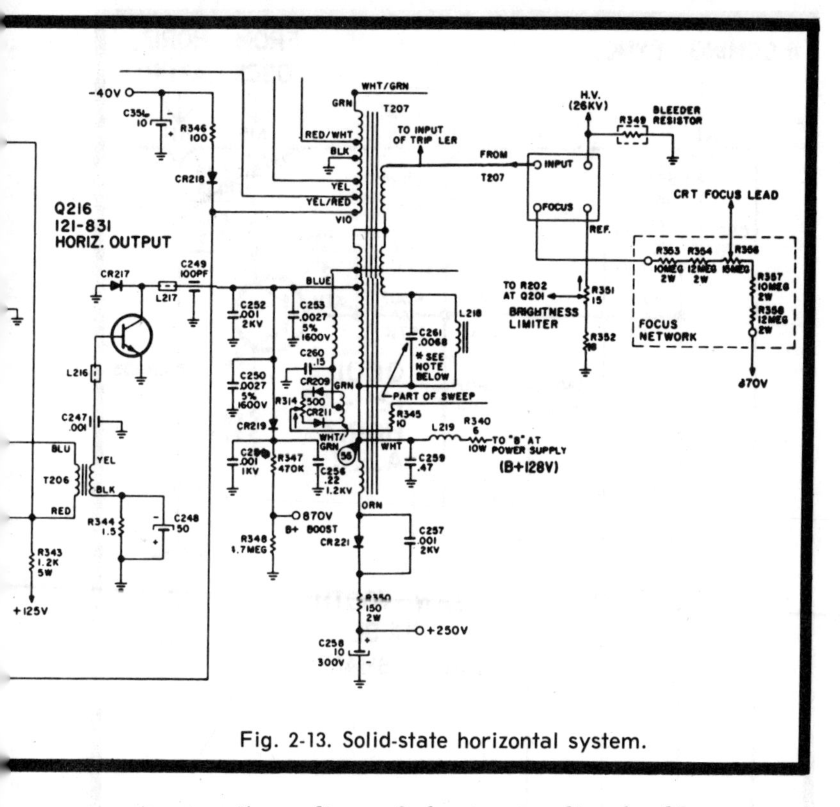

Fig. 2-13. Solid-state horizontal system.

A correction voltage of about ± 5 volts should appear at the base of the horizontal AFC transistor Q801 based on the comparison in phase of the incoming sync pulses and the feedback horizontal pulse which is shaped by sawtooth shaper transistor Q804. Both pulses occur at 15,750 Hz and are both required at the proper amplitudes for the comparator voltage. Either pulse alone will merely effect cancellation through the diode bridge (CR801 and CR802) resulting in no net DC level. See details in Fig. 2-14.

The sync input at the junction of the two diodes will develop a positive potential on the order of one volt from the random noise through the sync channel, even with the VHF tuner set "between channels." A scope check across the diodes will indicate waveforms and amplitudes. Normal performance of the overall system is indicated when a VTVM reading on the base of the AFC transistor (Q801) will swing through the range of about +3V to -7V, on a TV station signal, as the horizontal hold control is adjusted.

If horizontal locking is weak, unstable or nonexistent, a quick procedure to find the general area is to use a small variable bias supply. Clip the positive lead onto chassis and

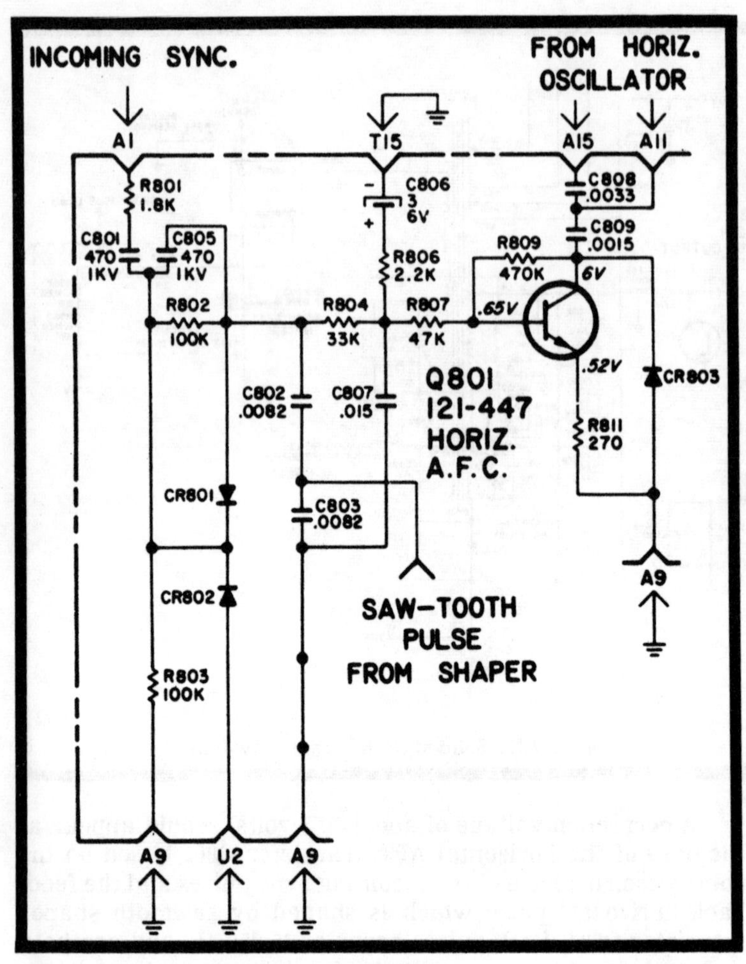

Fig. 2-14. Detail schematic of solid-state horizontal AFC stage and phase comparator diode circuitry.

the negative lead (set for 0 volt) to the base of the AFC transistor (Q801). Adjust the horizontal hold control (L214) for minimum raster shift, or for a setting giving a nearly locked-in raster. Next increase the bias to about -0.5V. Attempt to reset the horizontal hold control for a similar picture condition.

For several successive negative voltages, tracking of the horizontal hold control indicates the horizontal system following the Q801 input base is okay and the fault is ahead of the base. Check the diodes, a resistor or capacitor component, or the incoming sync or flyback pulse through the shaper.

If the horizontal hold control will not track, remove the bias network and jumper the base of horizontal AFC (Q801) to ground. Instability verifies the fault lies after the horizontal AFC input base and further tests will now be described.

A defective sawtooth shaper transistor (Q804) will probably have either an open or shorted junction. (See Fig. 2-15.) The base is at ground potential to terminal A9. Open junctions will cause non-conduction of the transistor so collector voltage will rise to the source potential.

The horizontal oscillator transistor (Q802) generates a sine wave at 15,750 Hz. (See Fig. 2-16.) The sine wave output, from the horizontal oscillator coil, is applied to the horizontal AFC transistor collector at terminal A11. Diode CR803, between the collector and ground, acts as a half-wave rectifier. Normal operation of the horizontal oscillator (indicated by measurement of +2.7V on the base), AFC stage and the diode, is indicated by the +6V value on the collector. When the horizontal oscillator is operating normally, but there is no AFC transistor collector voltage, the following are the cases:

The oscillator sine wave is not being applied to the AFC stage (capacitor C809, .0015 uf) is open.

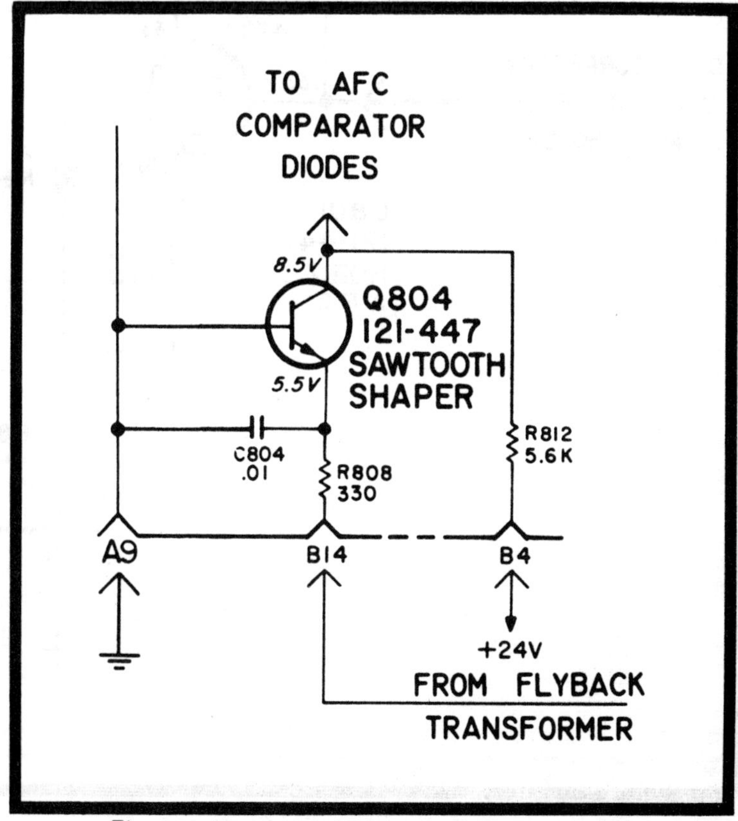

Fig. 2-15. Horizontal AFC sawtooth shaper circuit.

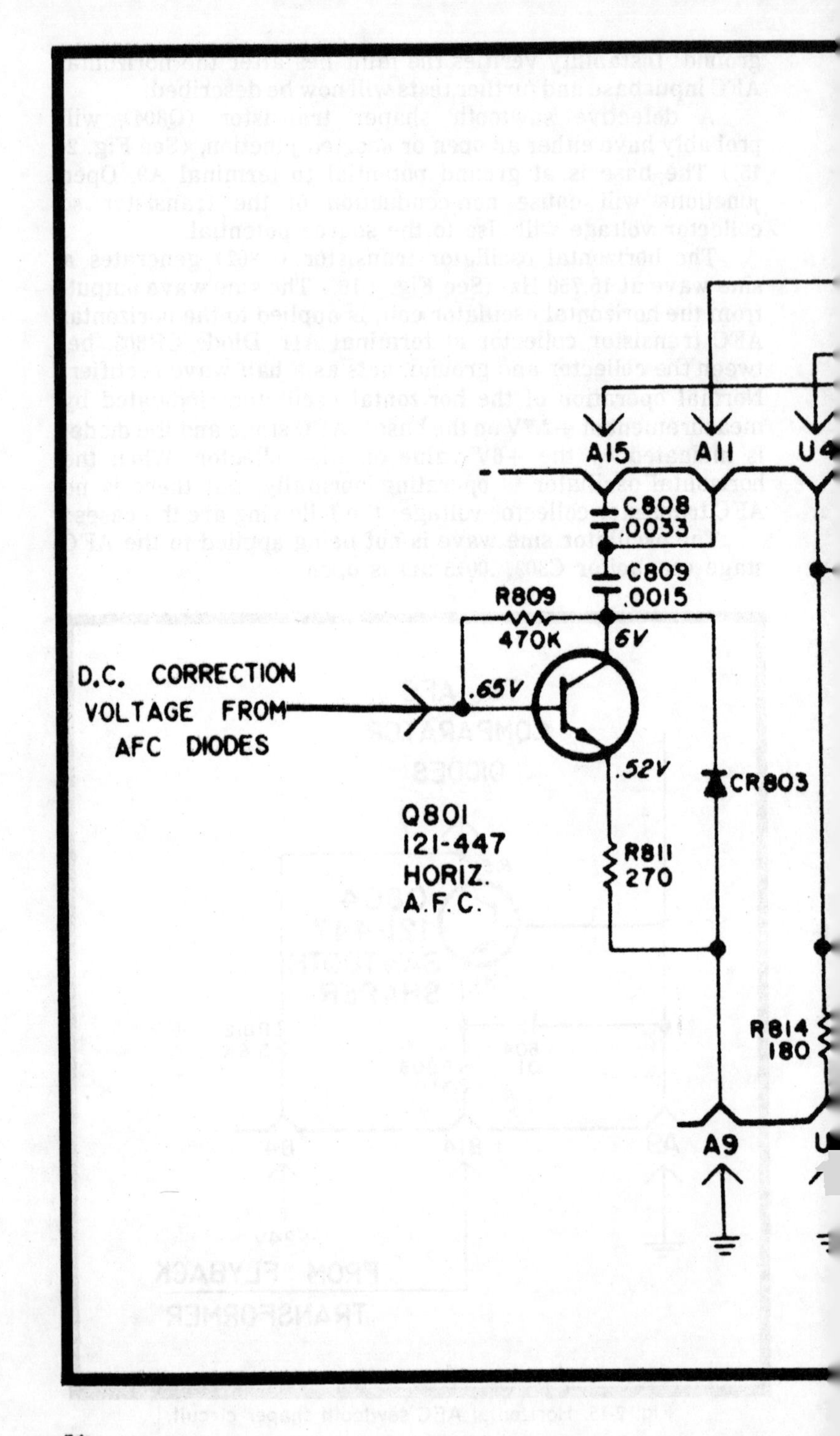
A15
A11
U4
C808
.0033
C809
.0015
R809
470K
6V
D.C. CORRECTION
VOLTAGE FROM
AFC DIODES
.65V
.52V
CR803
Q801
121-447
HORIZ.
A.F.C.
R811
270
R814
180
A9

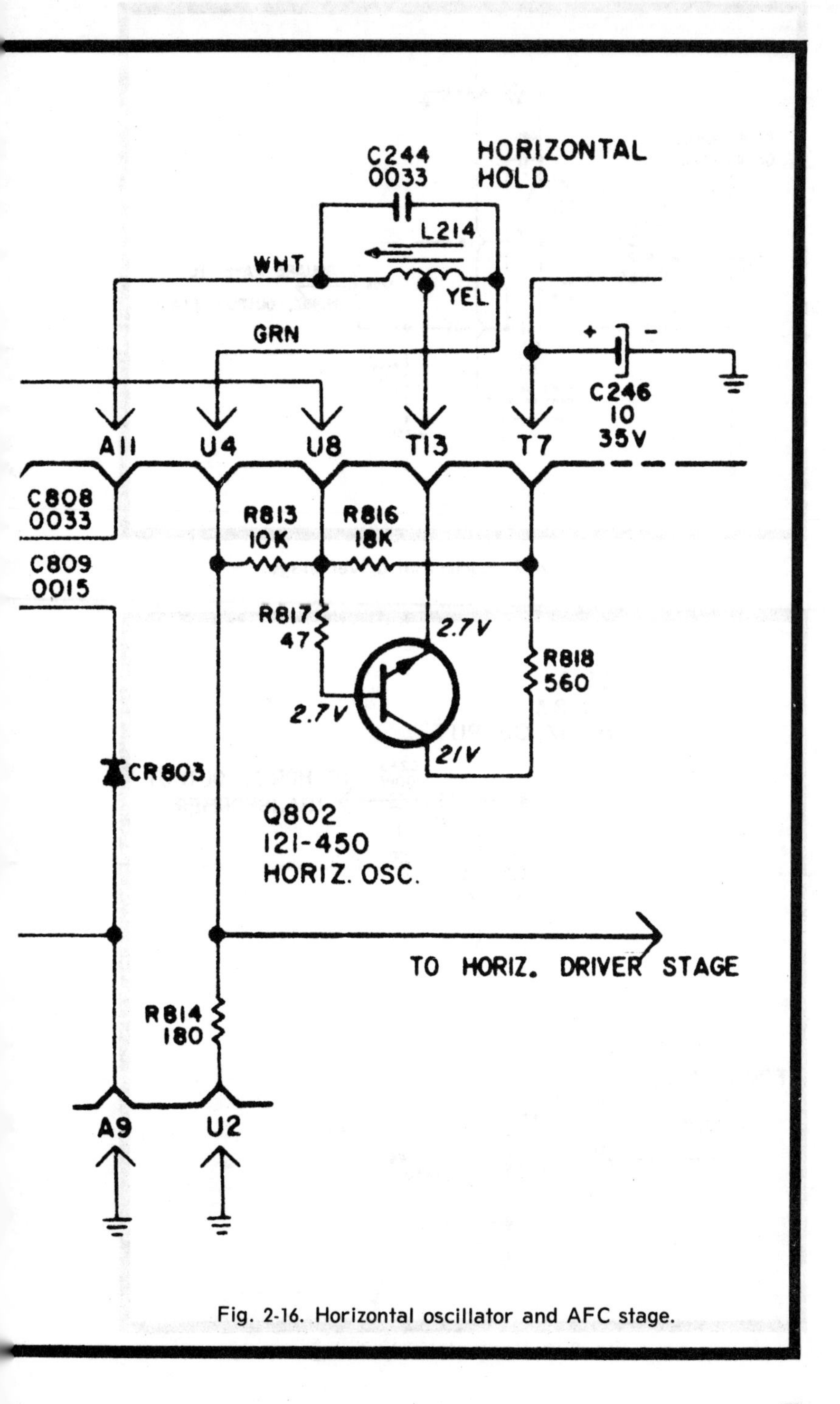

Fig. 2-16. Horizontal oscillator and AFC stage.

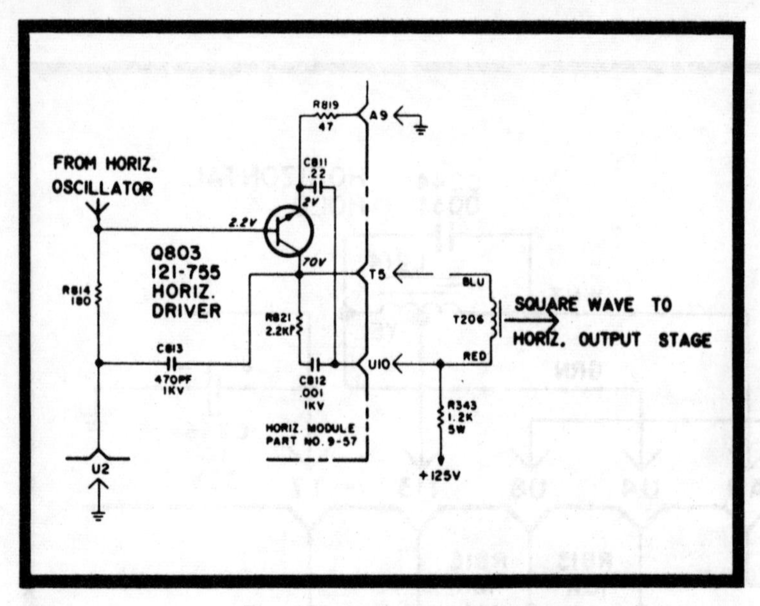

Fig. 2-17. Horizontal driver stage.

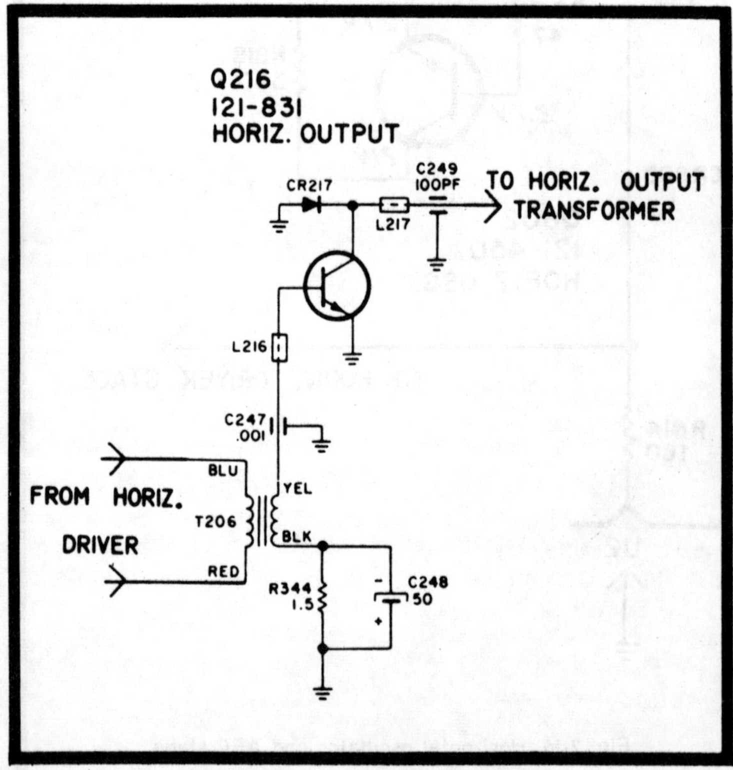

Fig. 2-18. Horizontal output stage.

The horizontal AFC transistor Q801 or the collector diode CR803 is faulty.

The horizontal oscillator sine wave output at terminal U4 is applied to the horizontal driver stage Q803, which is shown in Fig. 2-17. The abrupt swing from saturation to cutoff produces an essentially square wave output. If the oscillator stage is performing normally, driver performance can be checked by comparing voltage measurements shown on the schematic.

The horizontal output stage (Q216), in contrast to a tube-type counterpart, draws no current in absence of a drive signal. Note Fig. 2-18 and study this circuit.

The base circuit to ground is through the secondary of T209 and R344, a 1.5 ohm resistor. Forward bias to drive the transistor into conduction is developed by the horizontal drive square wave output through transformer T206. In the absence of drive, the output transistor loses forward bias. As a check, normal operation of the horizontal oscillator and drive stages will be indicated by measuring the +1.9V forward bias on the output stage base.

The collector of the horizontal output stage has a peak signal amplitude of around 1000V P-P and may be checked with a scope and proper probe.

Portions of the horizontal output and flyback transformer circuitry can be checked by use of the following steps:

If there is no raster, check B+ boost (870V) at high side of the G2 controls. Refer to the schematic in Fig. 2-19.

870 volts indicates the horizontal output system is performing normally, and the problem is in the tertiary area, the tertiary winding itself, or the tripler module.

125 volts indicates the flyback transformer is not functioning, the flyback sweep transformer is defective, or C219 is shorted, to name some possible faults.

Zero volts would indicate no B+ or a shorted component such as capacitor C254.

With a "direct short" (less than several ohms), from the white lead at junction of L219 and C259 to ground, check for a shorted capacitor (C252, C253, or C259), shorted feedthrough C249, shorted damper diode CR217, or shorted output transistor Q216. It may be necessary to check, in the power supply, electrolytic C243 at the emitter of regulator transistor Q215 as well as electrolytic C239A at B+ 125V power supply output.

In summary, here is a quick voltage checklist through the horizontal system. Refer to Fig. 2-13 again.

1. +2.7 volts on base of Q802: the horizontal oscillator is normal.

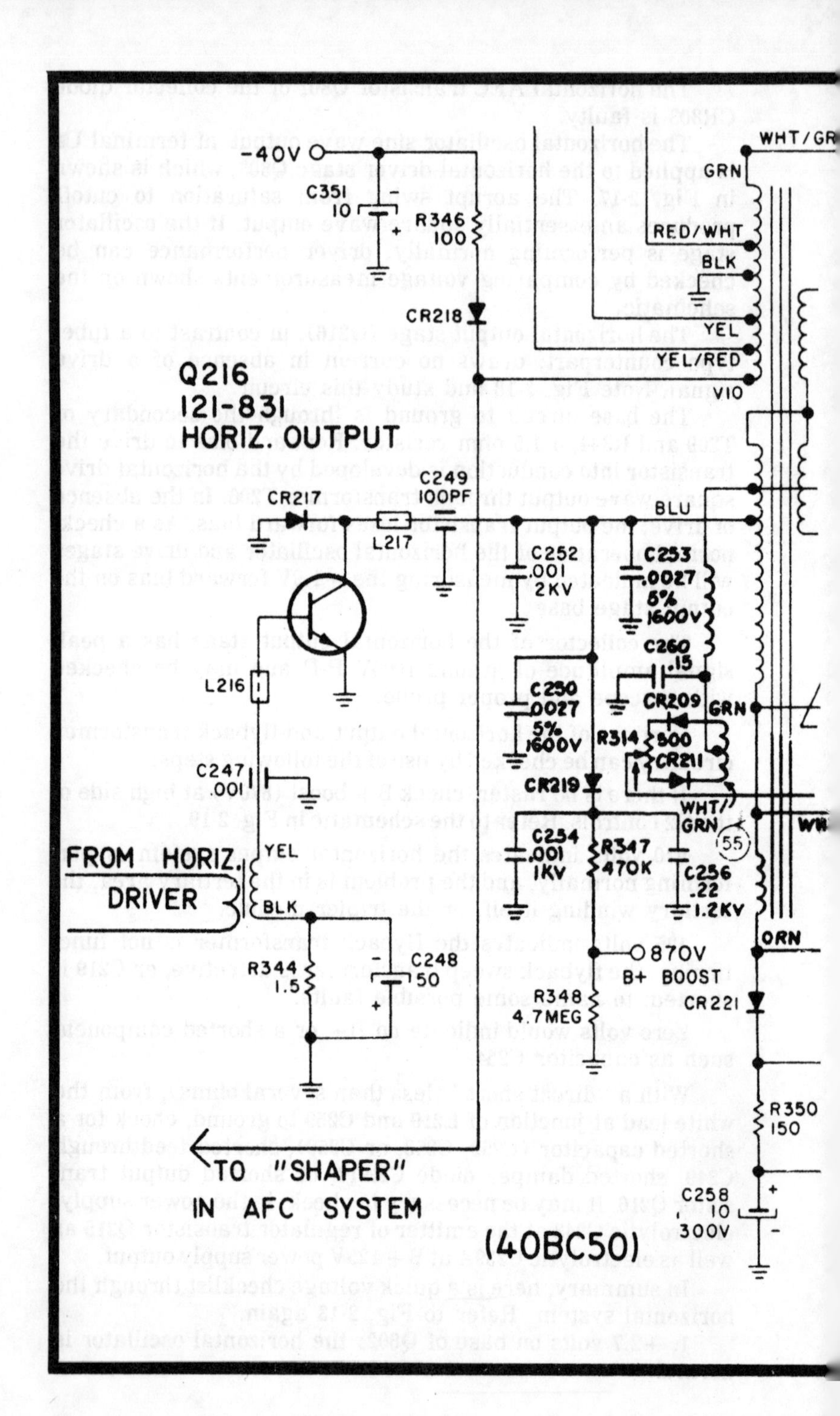

-40V
C351
10
R346
100
CR218
Q216
121-831
HORIZ. OUTPUT
C249
100PF
CR217
L217
BLU
C252
.001
2KV
C253
.0027
5%
1600V
C260
.15
L216
C250
.0027
5%
1600V
CR209
GRN
R314
500
CR211
CR219
C247
.001
WHT/
GRN
55
C254
.001
1KV
R347
470K
C256
.22
1.2KV
FROM HORIZ.
DRIVER
YEL
BLK
ORN
C248
50
R344
1.5
870V
B+ BOOST
R348
4.7MEG
CR221
R350
150
TO "SHAPER"
IN AFC SYSTEM
C258
10
300V
(40BC50)
WHT/GR
GRN
RED/WHT
BLK
YEL
YEL/RED
VIO

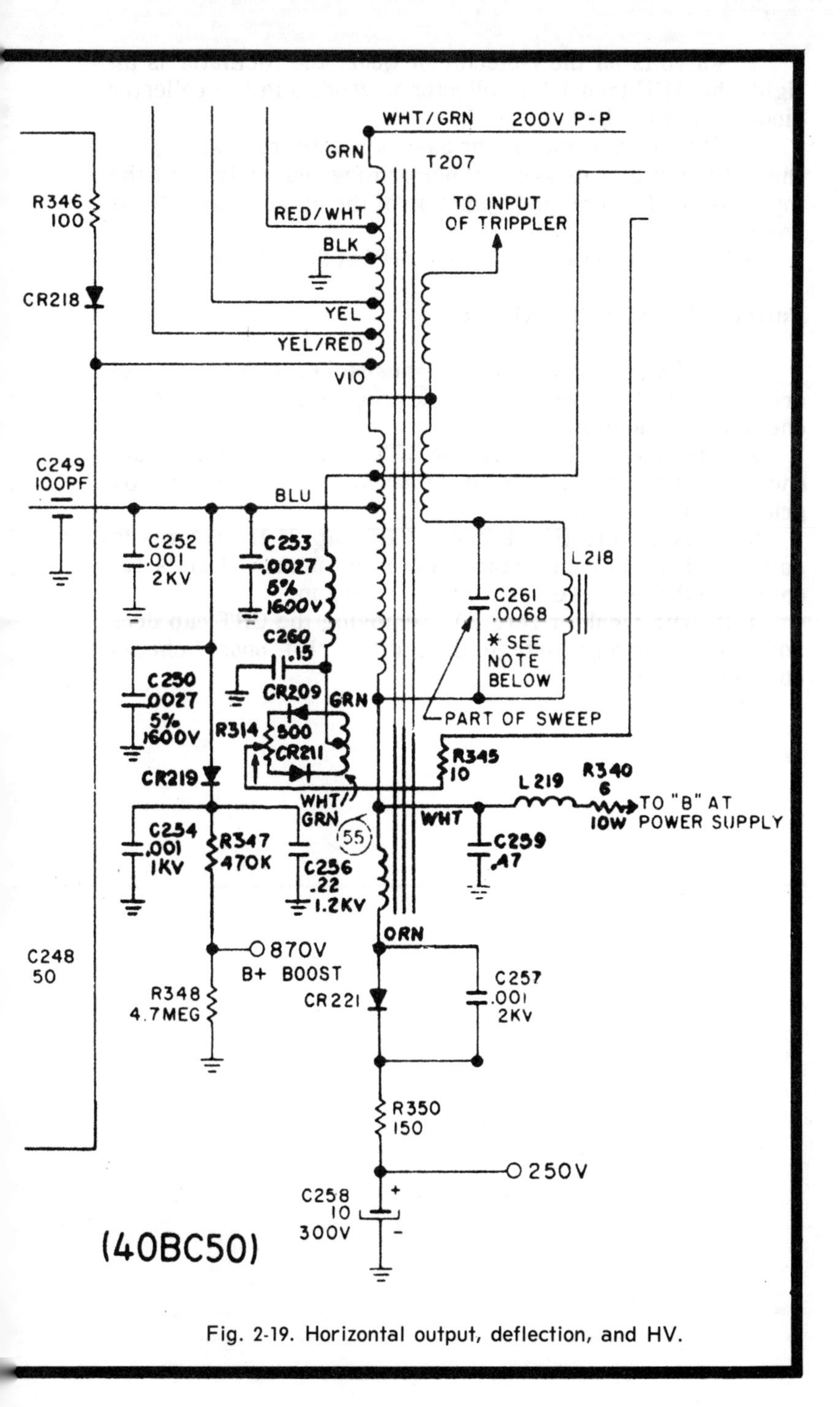

Fig. 2-19. Horizontal output, deflection, and HV.

2. +6 volts on the collector of Q801: the oscillator is all right, the AFC transistor collector is good, and the collector diode CR803 is okay.

3. +1.8 to +2.0 volts on the base of output transistor Q216: the oscillator and driver are performing normally, and the input (base-emitter) diode junction of the output transistor is normal.

4. If there is no raster, check for HV at the CRT anode.

Further checks in the HV area:

1. If Q216 (horizontal output transistor) mounting screws are loose, no HV or sweep will be developed. It's a good idea to check these mounting screws first.

2. Normal HV: Check that the CRT filament is lighting up check for normal voltages on the socket pins: cathodes, G1 grids, G2 grids, and focus.

3. Weak or zero HV: Remove CRT cap. If HV returns to normal, check out luminance and chrominance circuits for correct cathodes and G1 grid bias on each gun.

4. If, with weak or zero HV, removing the CRT cap does not recover the high voltage, then utilize "B+ boost" checks outlined above.

Vertical and Horizontal Sync

3

As you know, good sync action is required in order to lock the picture in vertically and horizontally. The sync signals start at the video detector and video amplifier, are separated and amplified in the sync circuits, then go on to the vertical and horizontal sweep oscillators. Of interest in the composite video signal for sync action are the vertical and horizontal equalizing pulses and blanking intervals. The mission of the sync system is to deliver sync pulses to the sweep oscillators, pulses which are correctly timed or synchronized with the picture transmitted from the TV station.

The following operations are performed by the sync system. Note the sync separator circuit in Fig. 3-1.

1. Removes all picture information from the composite signal and leaves only the sync pulses. Also removes all of the black level pedestals.

2. Amplifies the sync pulses to required level for solid picture locking action.

3. Clips the sync pulses to a uniform value and removes any pulses due to noise or interference. (These circuits are referred to as sync clippers or limiters.)

4. Shapes the horizontal sync pulses into sharp pips in order to operate the horizontal phase detector circuit and control horizontal oscillator frequency.

5. Develops the vertical pulses into relatively long sync pulses which will be used to trigger and lock-in the vertical sweep oscillator. RC differentiating and integrating filters are used for this purpose.

6. Produces the correct sync pulse polarity that is required to trigger the sweep oscillators or AFC control system.

SYNC CIRCUIT FAULTS

Let's now look at some probable sync circuit defects:

- *Soft lock or complete loss of vertical sync only.
- *Poor raster interlace.
- *Complete loss of vertical and horizontal sync.
- *Vertical or horizontal blanking bar across picture.
- *Pulling or bending.

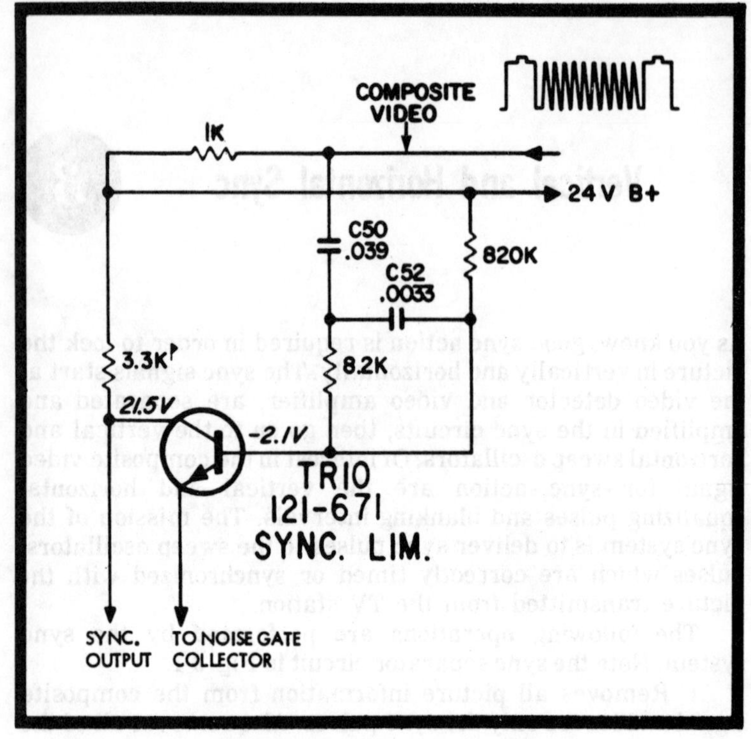

Fig. 3-1. Sync separator circuit.

* Flag waving or hook at top of picture.
* Vertical picture Yo-Yoing or bounce.
* "Pie-crusting" or "geartoothing."

Always keep in mind that the sync and AGC circuits are very closely related and a sync trouble may look like an AGC problem and vice versa. If adjustments of the contrast and AGC controls does not produce a normal picture, then what looks like sync trouble may actually be a fault in the signal section of the chassis, prior to sync take-off, or an AGC circuit malfunction.

Consider the following factors when troubleshooting the sync system:

1. If both the vertical and horizontal sync is lost, the trouble is probably in a common sync handling stage.

2. If the picture will not lock in horizontally but is vertically stable, it is then probable the fault is not in the common sync stage. The prime suspect would be the horizontal AFC circuits, horizontal oscillator, and any associated feedback circuits.

3. Should the picture be only vertically unstable, then the sync circuits past the sync branch-off stages should be

checked first, along with the vertical oscillator or output stage or any of its associated feedback networks.

SOME OBSERVATIONS ON SYNC CIRCUITS

As a starter, let's review some of the sync clipper and noise control circuit operation. Because the sync, AGC, and noise control circuits all must work together to keep a stable picture on the screen, you just about have to troubleshoot all three actions at the same time. A triple threat if I ever saw one. In some sets, all three circuits are found in one tube or on one module.

On the newer Zenith solid-state chassis, the sync clipping, AGC, and noise immunity circuits are all located on one module. Several components are used in both systems and the noise immunity system is used by both AGC and clipping circuits. For this reason it becomes very practical to service and observe (with oscilloscope) the sync clipping action first.

Sync clipping is the action of removing from the composite video those parts not required for the synchronization process. The unwanted portions are as follows:

All video picture information.

All blanking information.

The top and bottom sections of the actual horizontal and vertical sync pulses.

The top section of the pulses is undesirable because noise pulses would be present at this point, while the bottom part would be too close to the blanking and video areas. The only "wanted" part is a tiny portion of the sync pulse as illustrated in Fig. 3-2.

A WORKING SOLID-STATE SYNC CIRCUIT

In many TV receivers the sync clipping action is coupled to the AGC operation. The reason for this can be seen by referring again to Fig. 3-2. If the DC level of the signal changes, or the amplitude varies, the predetermined clipping action is defeated. In one instance it will include the noise pulses on top of the sync and at another time it will bite into the blanking or even the video portion. These actions should be taken into consideration when troubleshooting both sync and AGC in the circuit module shown in Fig. 3-3.

AGC troubles can be detected by observing the sync information at the output of the sync clipper. The only active component in the sync clipping stage is NPN transistor Q401.

Input signals are fed through a dual time constant filter to the base; with proper biasing parameters, clipping will occur. The clipped sync will be seen at terminal B12, the collector of

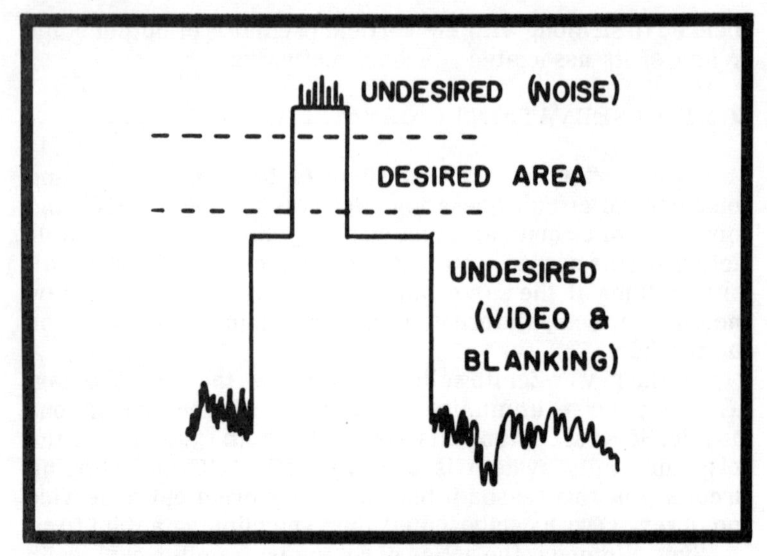

Fig. 3-2. Why sync clipping is necessary.

Q401. (See Fig. 3-4.) There is always a good possibliity that, due to an AGC fault, the clipped sync section is not operating properly.

To add more probable faults to these systems, many modern color sets now use noise immunity circuits. In Fig. 3-3, this circuit has been added around the sync clipper Q401, being noise driver Q403 and noise gate Q404. These transistors and associated circuitry have the specific function of making the horizontal and vertical oscillators less susceptible to noise pulses. However, since these circuits are connected to the clipper, they can, if defective, jeopardize the sync clipper action. (More details about this circuit operation later.)

Sync separation and amplification is performed by the sync limiter transistor Q401. For this sync operation refer to the circuits in Figs. 3-3 and 3-5. The emitter of Q401 is "grounded" (collector of noise gate transistor). The collector is connected to the 24 volt supply through CR404. Composite video (positive-going) appearing at test point C3, is fed to the base of the sync limiter through capacitor C401 and a parallel RC combination consisting of R403, C402 and diode CR404. This dual time constant filter provides the sync limiter with a high degree of immunity to aircraft flutter. A small amount of forward bias-emitter bias is applied to Q401 through R402. However, if the stage were to conduct at all times, video as well as sync information would be coupled to the horizontal and vertical sweep circuits. This is overcome by having a small amount of reverse bias (negative voltage on the base) that is proportional, but not equal to, the amplitude of the

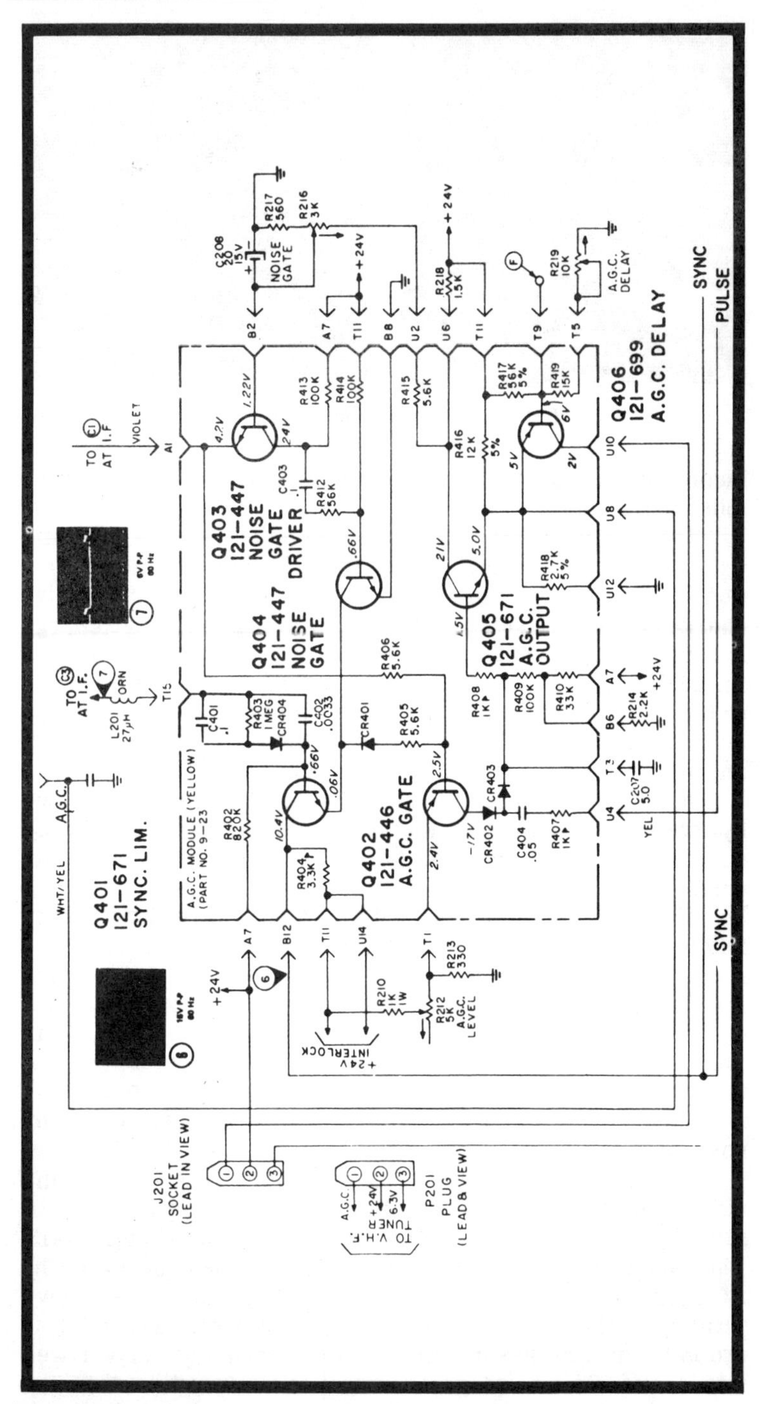

Fig. 3-3. Solid-state AGC and sync module.

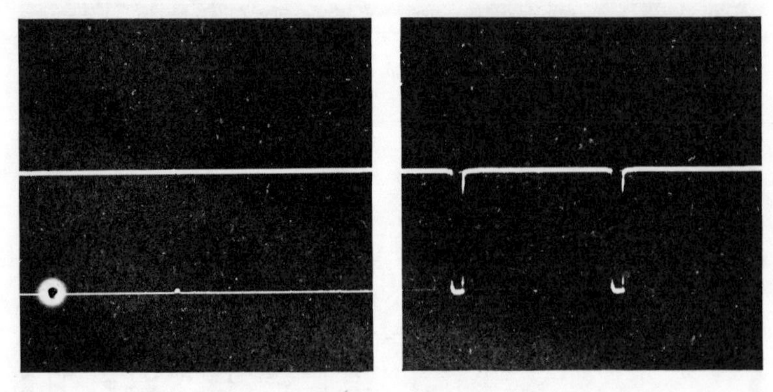

Fig. 3-4. Clipped sync waveforms.

incoming sync pulses. This negative voltage is developed as follows:

The positive composite video is coupled through C401 and C402 and appears at the base of Q401. The base-emitter junction acts as a rectifier to the sync pulses, setting a net negative charge on the base side of capacitor C401. The negative charge is reduced somewhat by the positive voltage dropped across the 820K resistor. The negative voltage reduction is designed so that just the uppermost positive excursion of the sync tips overcomes the reverse bias, causing the transistor to conduct and amplify only the sync information.

Noise immunity for the sync and AGC stages is provided by two transistors and their associated circuitry. These transistors are termed the noise gate, and the noise gate driver.

The bias conditions for the noise gate transistor are as follows. The emitter is returned directly to ground. The base receives a positive voltage coupled through R414, thus keeping the stage in saturation. The collector is coupled to the emitter of the sync limiter, and the base circuit of the AGC gate. With the transistor in saturation, the collector voltage is very low (0.6 volt), thus providing a low resistance to ground for normal operation of the sync and AGC stages in absence of any noise pulses.

The emitter of the noise gate driver receives DC bias and video (negative sync) information from test point C1 of the IF module. The collector is returned to +24 volts through R413. The base bias for this stage is developed at the collector of the AGC output transistor. The source voltage is divided down through resistor R415, the noise gate control, and R217 to ground. The voltages now set have the noise gate driver well into cutoff. The noise gate control is adjusted so that the conduction point of the transistor is just beyond sync tip level.

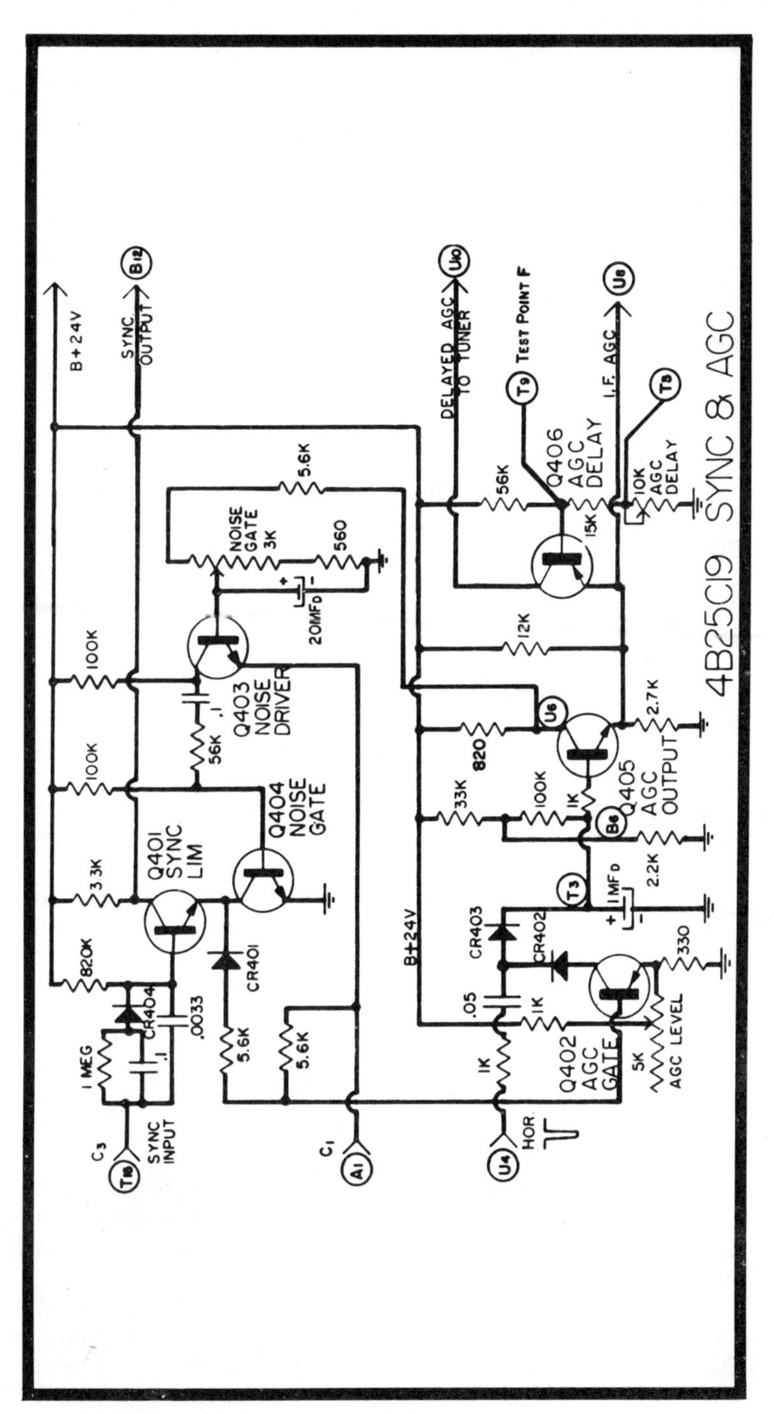

Fig. 3-5. Sync - AGC circuits of Fig. 3-3 redrawn and simplified.

When noise pulses just beyond sync tip level appear at the emitter, the transistor is driven into conduction. The noise pulses are amplified and coupled to the noise gate transistor through capacitor C403. The negative pulses momentarily bring the noise gate out of conduction. With the noise gate out of conduction, the collector voltage increases, thus cutting off the sync and AGC functions during the noise pulse duration. (Remember, the collector is the ground return for the AGC and sync stages.) Shutting down the sync limiter in this way keeps the noise pulses from tripping the vertical and horizontal oscillators.

SYNC AND NOISE SYSTEM SERVICE GUIDE

The symptom of complete loss of vertical and horizontal sync usually points to a fault in the sync or AGC circuits. Note Fig. 3-3. For this chassis, however, there is an exception as the first video stage (which could be defective) is located in the IF module. A voltage check at test point C3 (with lead disconnected at the IF module) will help isolate the difficulty. The voltage at C3 should measure +12.5 volts. If this voltage is off by more than 2 volts, trouble is indicated in the first video stage. Faults in this area may interrupt the video to the sync limiter. With the problem narrowed down to the sync circuits, then make the following checks.

1. Ground the collector of the noise gate transistor. If the sync has been restored, the trouble is in the noise gate system. If the difficulty is still present, go on to the next check.

2. Measure the base voltage on the sync limiter Q401, with no signal applied to the receiver. It should measure +0.66 volt. If this voltage is high (more than 1 volt), the transistor is probably open. A shorted transistor will generally be indicated by a low collector voltage reading.

3. Marginal sync problems such as horizontal pulling may be caused by either the sync limiter or the noise gate system. Again, grounding the collector of the noise gate transistor will narrow down the trouble spot. Use the scope to look for distorted, missing, or below-par sync pulses. In some instances when this type of marginal symptom is isolated to the sync stage, a check of individual components around the circuit is desirable since voltage checks may not indicate any fault.

The effects of an open noise gate transistor, Q404, is a complete loss of vertical and horizontal sync (note symptom in Fig. 3-6), plus what appears to be AGC overload. The same effect will appear if the transistor is not receiving forward bias through R414.

A less obvious problem which may develop within this noise gate system is its inability to respond to noise pulses. This will be evident in weak and/or noisy signal areas as unstable sync. To determine if the system is working, turn the noise gate control fully counterclockwise. Next, adjust the AGC level control so that the receiver is about to overload. Then slowly turn the noise gate control clockwise. If the video breaks up, the system is working. If turning the noise gate controls has no effect, proceed to the following.

Adjust the AGC level control back to "normal," and turn the noise gate control fully counterclockwise. Measure the collector voltage of the noise gate driver. It should measure 24 volts. A low voltage reading would indicate a shorted transistor. If the voltage is 24 volts, adjust the AGC level just under overload and turn the noise gate control fully clockwise. This will cause the transistor to conduct on the sync information, and if the transistor is not open the collector voltage will come down to about 12 volts. If the voltage does come down as just described, check for a shorted noise gate transistor.

SYNC LIMITER AND NOISE BIAS CIRCUIT

Let's look at the operation of the sync limiter and noise bias circuit shown in Fig. 3-7. Sync is extracted from the incoming composite video by Q404. This stage is designed so that

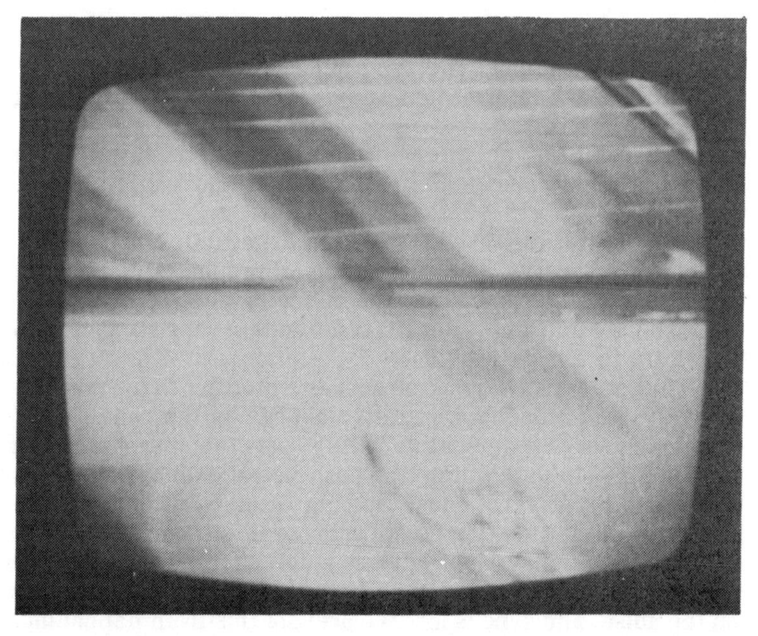

Fig. 3-6. No sync condition caused by open noise gate transistor.

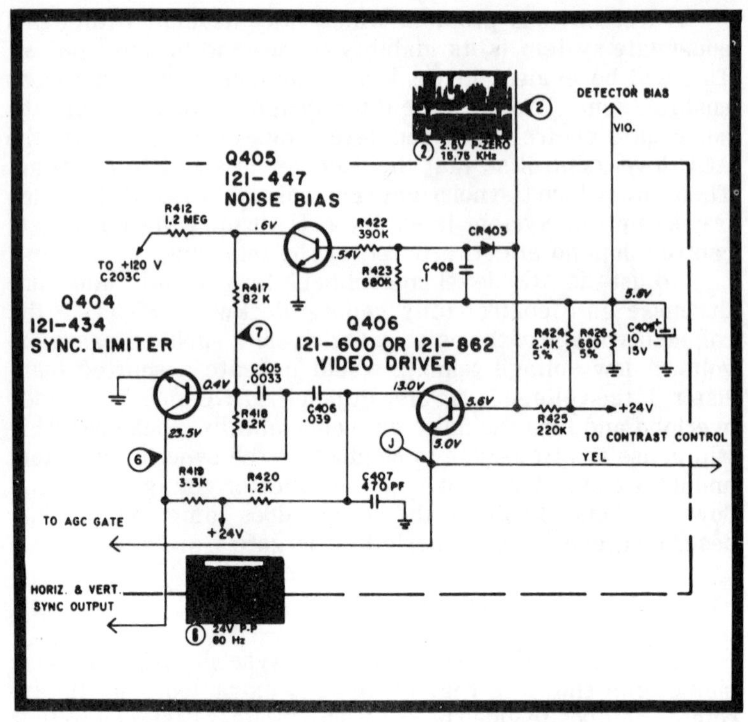

Fig. 3-7. Sync limiter and noise bias circuit.

the uppermost positive portions of the sync tips cause the transistor to conduct. The sync pulses are clipped from the composite video information and coupled to the vertical and horizontal circuits.

Action of the sync limiter stage will be covered first, since the noise bias transistor is essentially a short circuit on a strong signal.

The emitter of Q404 is connected to ground. The collector is connected to the +24 volt supply through R419. Positive going composite video from the collector of Q406 is coupled through C406 and a parallel RC combination consisting of C405 and R418 to the base of Q404.

The base-emitter junction acts as a rectifier to the positive sync pulses, producing a negative charge on the base side of capacitor C406. This negative voltage sets the operating point for the limiter, so that only the positive parts of the sync tips overcome the reverse bias, causing conduction of the transistor and amplification of only the sync information.

On weak signals, noise pulses riding on the sync can shift the operating point of the sync limiter so that clipping occurs on the noise, and sync is lost. To prevent this from happening, a noise bias transistor Q405 is used.

Now let's follow the action of the noise bias transistor. The voltage divider consisting of resistors R424 and R426 provides about +5.5 volts, which appears on the anode of CR403. The cathode is fed the +5.6 volts from the +24-volt supply via R425. This positive voltage on the base keeps transistor Q405 in saturation, effectively shorting the collector of Q405 to ground. Noise pulses that appear on the sync cause the cathode of CR403 to become negative with respect to the anode. When this happens, CR403 conducts and a bias voltage opposed to the positive bias is fed to the base of Q405 to bring it out of saturation. As conduction decreases, the voltage on the collector increases. This voltage change is fed to the base of the sync limiter to oppose the increase in negative bias caused by the noise on the sync. This bias change keeps the limiter at the proper operating point, so that clipping occurs on the sync and not on the noise pulse.

TROUBLESHOOTING THE SYNC LIMITER AND NOISE STAGE

With a symptom of complete loss for the horizontal and vertical sync make the following checks:

Measure the base and collector voltages of sync limiter Q404. With a strong signal, base voltage should be around —2 volts. Collector voltage should be around +23 volts. If voltage on the base is zero or slightly positive, composite video is not reaching the base circuit. Check capacitors C407, C406, and the collector voltage at Q406. If C407 is leaky or R420 has increased in value, the video driver stage may be clipping on the sync pulses.

For a symptom of poor horizontal sync accompanied by picture pulling on a strong signal, short the collector of noise bias transistor Q405 to ground. If the problem clears, Q405 may be defective or some other defect exists in the noise bias stage.

With a symptom of poor horizontal and vertical sync on weak signals, check Q405, CR403 and associated components.

TUBE-TYPE SYNC SYSTEM

The tube-type sync system that is used in many Zenith TV receivers is shown in Fig. 3-8. A composite video signal is fed from the detector, which has negative-going pulses of about 1 volt P-P and is coupled to pin 7 of V203A, the sync amplifier stage. In this stage the signal is amplified from 1 volt P-P to 44 volts P-P for sound takeoff through transformer L211 and to feed the sync and AGC stage of V204. The correct scope trace that you should find at test point 6 is shown in Fig. 3-9. The signal at test point 8 will look about the same but should measure 37 volt P-P for this chassis. Always refer to

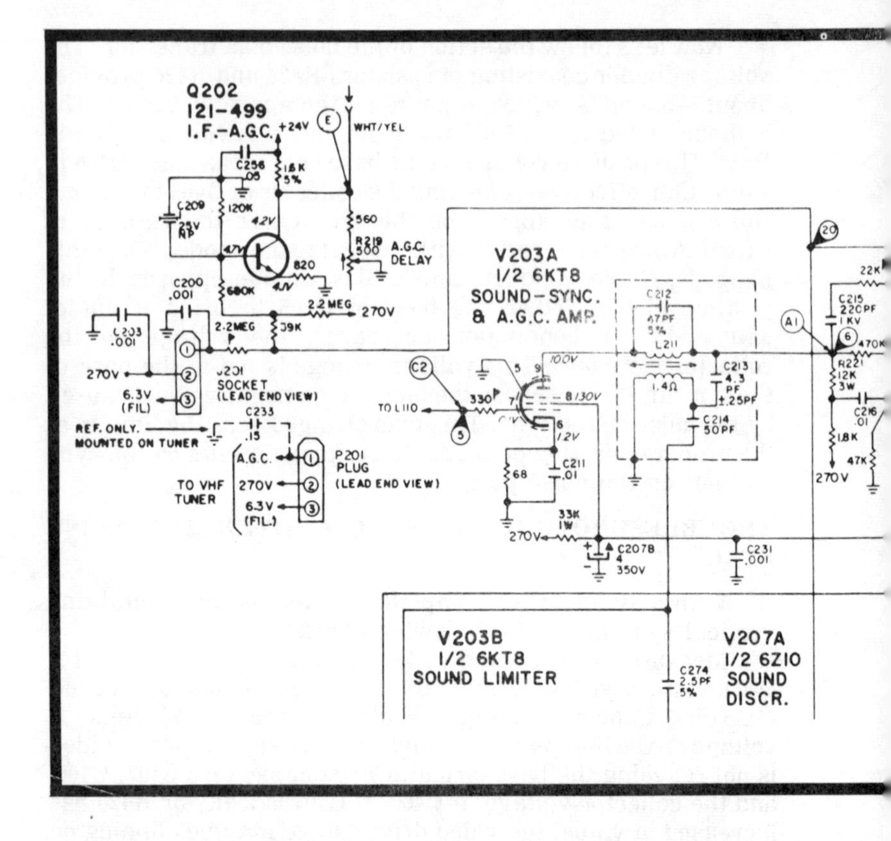

the service information for correct scope pulse and peak-to-peak voltage measurements.

The sync clipper and AGC tube V204 has a single cathode, control and screen grids, but separate suppressor grids and plates. The point of conduction is determined by the AGC level control (R22) in the cathode circuit. The DC-biased control and screen grids have no applied AC signal and are used as accelerating electrodes. The AC signals on the suppressor grids determine the current flow in both the sync separator and the AGC plate circuits. The sync signal (note Fig. 3-10) out of the separator is AC-coupled by C222 and through intergrator A201 to the cathode of the vertical oscillator. The vertical integrator filters out the horizontal sync pulses while shaping the equalizing pulses that will be used to lock-in the vertical oscillator. Note this integrator action in the Fig. 3-11 scope trace patterns. The bottom trace is the signals going into the integrator network and the top trace is the resultant output vertical sync pulse. Another capacitor, not shown, is used to feed the horizontal pulse from test point 9 onto the horizontal AFC stage.

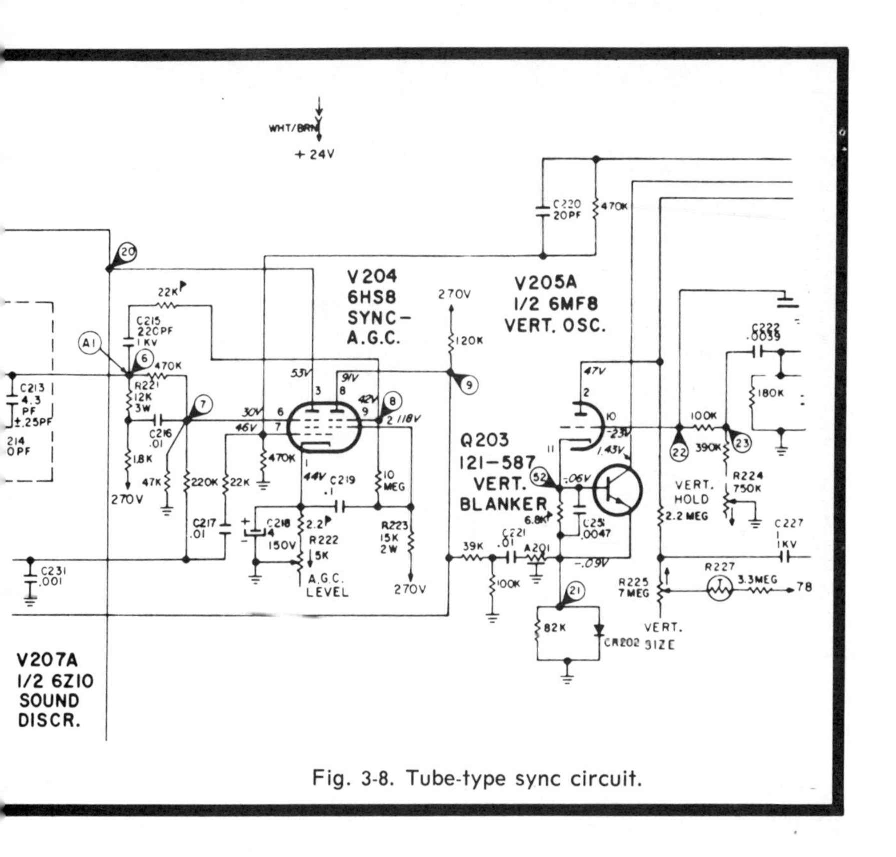

Fig. 3-8. Tube-type sync circuit.

If both the vertical and horizontal locking are affected, the fault is either in the sync amplifier or clipper stage. If the picture will sync in momentarily with hold control adjustments, this indicates the oscillators are not at fault. Complete loss of vertical and horizontal sync would call for using the scope to trace the sync signal throughout the entire sync system. However, in most cases, if only one oscillator is out of sync you should concentrate on the input circuits of that oscillator stage. Check the DC potentials of all sync circuits and see if they are normal. Be on the alert for resistors that have changed value and capacitors which have developed some leakage.

Vertical jitter is usually a fault encountered in the vertical output stage. However, if this checks out good, suspect slight capacitor leakage or a defective integrator circuit. Critical vertical sync can be caused by a shift in component values in the vertical oscillator input circuitry or some capacitor leakage. Sometimes the only way to locate these faults is to "sub-out" the suspected components (clip one lead and temporarily tack in a known good part).

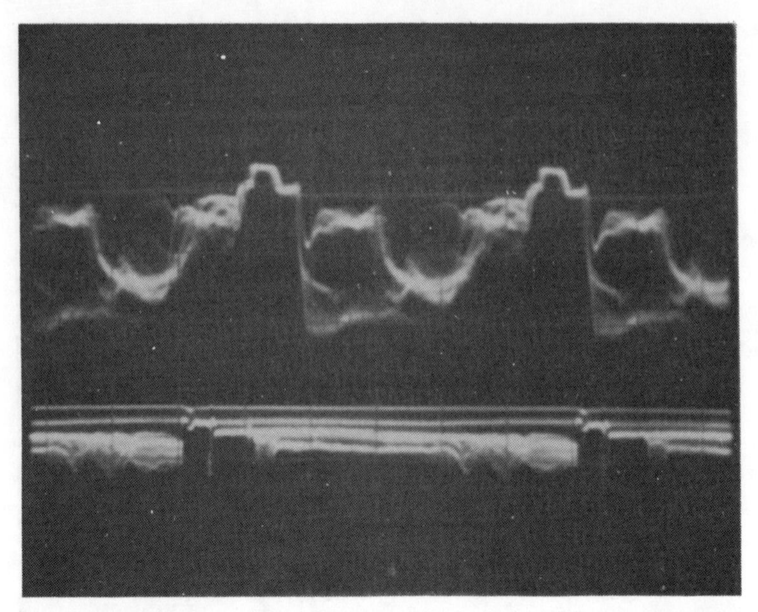

Fig. 3-9. Sync pulses at TP6 of Fig. 3-8. Top trace, 44 V P-P at horizontal rate. Bottom trace, 44 V P-P vertical rate. These are the correct and normal waveforms.

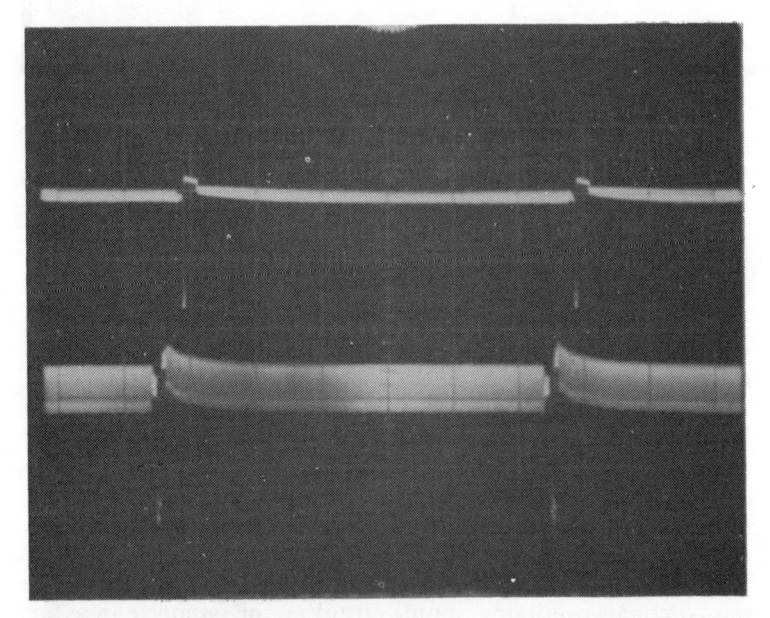

Fig. 3-10. Normal pulses at TP9 of Fig. 3-8. Both traces 70 V P-P. Top trace at horizontal rate, bottom trace at vertical rate. The horizontal and vertical sync pulses branch from this point to their respective oscillators. This is a key point for trouble diagnosis.

Remember this tip: When tracing pulses with a scope throughout the sync circuits, it is advisable to disable the vertical and horizontal sweep oscillator stages. This eliminates any of the sweep pulses from being fed back into the sync circuits and causing some confusion. On most sets it is easy to stop the vertical sweep action by simply flipping the "setup" switch to the service position.

HORIZONTAL AFC CIRCUITS

A differentiating circuit to separate the horizontal pulses from the vertical pulses was not feasible because of poor noise immunity and no phase control. That is why automatic frequency control (AFC) circuits are now used in all modern color sets to control the horizontal oscillator. Composite sync is fed directly into the AFC circuit that is shown in Fig. 3-12. This circuit compares the incoming sync pulse with the horizontal sweep oscillator feedback pulse of the receiver. If a phase or frequency difference occurs, a DC "correction voltage" is developed and fed back to the control grid of the control stage, which will then adjust the oscillator frequency to the incoming sync pulse. This control voltage may be either positive or negative depending upon the type of oscillator circuit.

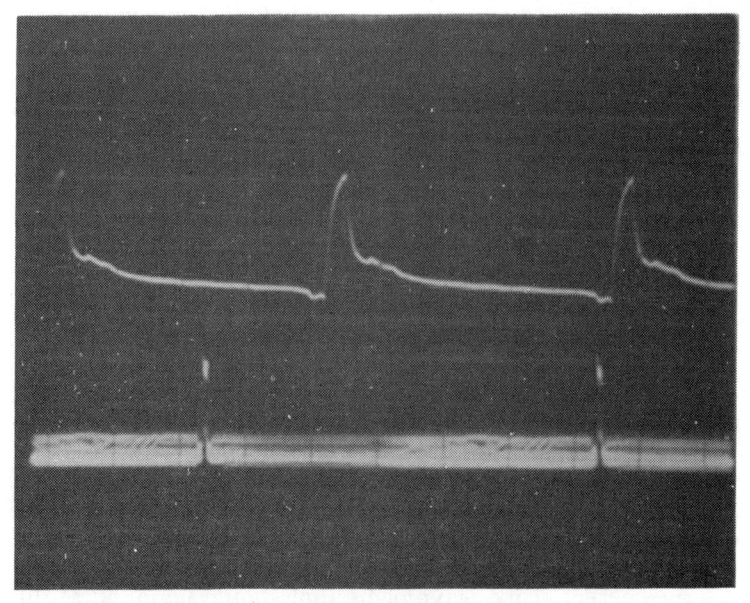

Fig. 3-11. Vertical sync pulses (scope set at vertical rate). Top trace, sync pulse after passing through integrator A201 at TP21 of Fig. 3-8. Bottom trace, sync pulse before being processed through integrator. 10 V P-P.

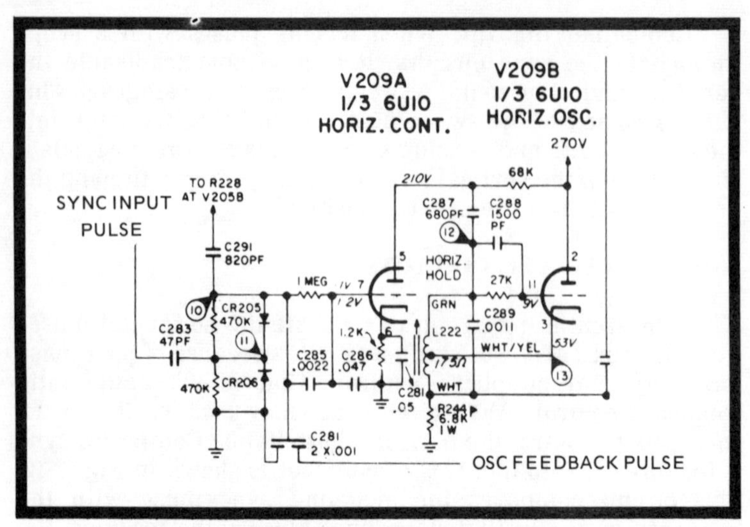

Fig. 3-12. Horizontal AFC and oscillator control circuit.

Some sets use a push-pull input type AFC system while others are of the single-ended type. The AFC diodes are matched with parallel resistors in order to maintain an equal, balanced system. Most phase detector circuits are the common cathode type shown in Fig. 3-12. These solid-state dual diodes are simple components, containing two closely matched diodes to assure electrical balance, but will cause critical horizontal sync troubles should they become faulty. They differ from conventional diodes by being highly insensitive to temperature variations. These diodes should be checked first for any type of AFC fault. They should also be checked by substitution. Forward and reverse resistance readings are not conclusive unless the diodes are open or shorted.

Some symptoms of defective phase detector diodes are as follows:

1. Complete loss of horizontal sync.
2. Bending or horizontal picture movement.
3. Tearing or jitter.

Refer back to the AFC circuit in Fig. 3-12, and note that a negative sync pulse is needed to drive the diodes into conduction. The pulse from the sweep transformer is fed to diode CR206 by capacitor C281 and appears as a sawtooth form. The shifting of the input sync and horizontal sweep pulses to these diodes creates a changing DC voltage that speeds up or slows down the oscillator via the reactance control stage. In Fig. 3-13 are the correct scope waveforms that should be found at the AFC diodes. The bottom trace was taken at test point 11 while the top trace was scoped at test point 10.

A quick and simple check for proper AFC action is to rock the horizontal control back and forth while monitoring the output control voltage with a VTVM to determine if the AFC voltage will vary. If the DC voltage on the output of the AFC stage remains the same or varies nonlinearly, the AFC circuit is not working properly.

HORIZONTAL SYNC CIRCUIT TIPS

Pie-crusting is a symptom that is easily recognizable by the shape found in the vertical lines of the picture information. Usually, all vertical-line objects take on a "wiggle-look" and both sides of the picture edges look like a pie crust.

Pie-crust symptoms are caused by the oscillator "hunting" between a range of frequencies on both sides of the exact frequency. The fault will be found in the anti-hunt network of the AFC circuitry. Found in the horizontal AFC circuit in Fig. 3-14, the anti-hunt components are C268, R284, and C269 at grid, pin 2, of V204, 6FQ7.

Almost any component in the horizontal sync or AFC section can cause a critical sync problem. Many times a careful component check is the easy way out. Prime suspects are always the phase detector diodes. If you do replace them make sure they are not put in backwards. The picture may lock in, but the horizontal hold will be critical if the diodes have been reversed.

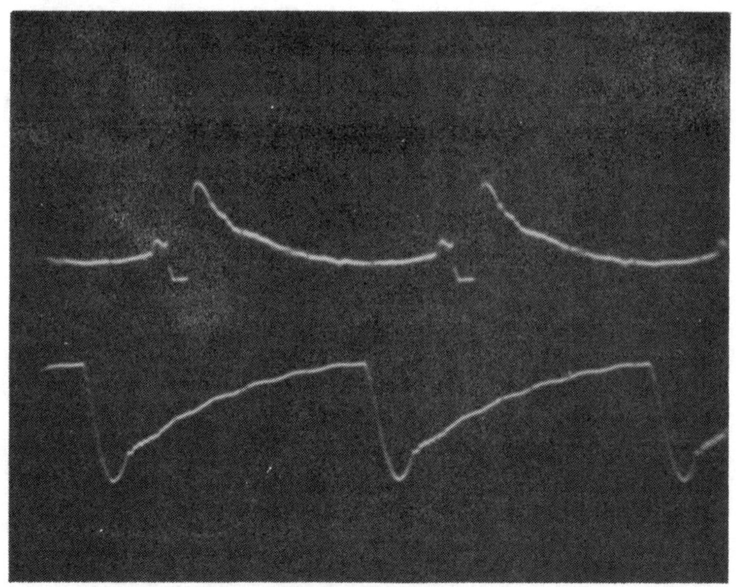

Fig. 3-13. Correct waveforms at AFC diodes. Top trace, TP10 of Fig. 3-12, 21 V P-P. Bottom trace, TP 11, 30 V P-P. Both traces at horizontal rate.

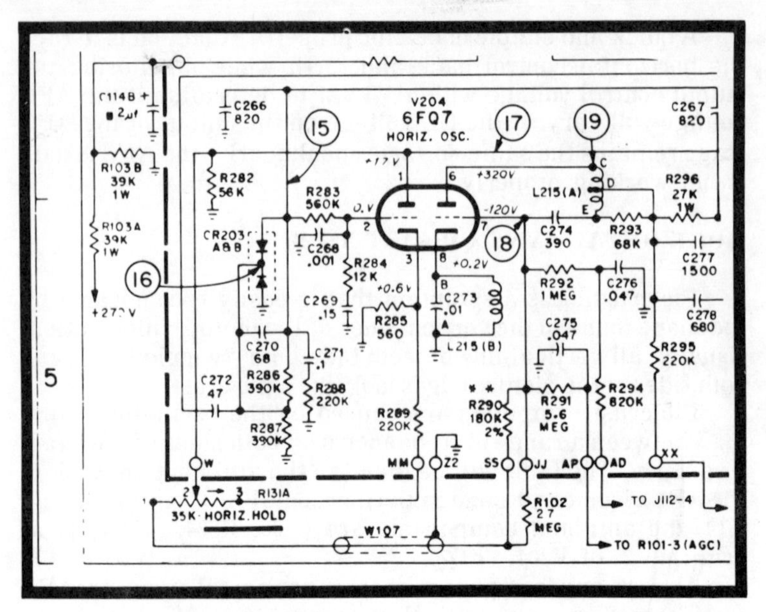

Fig. 3-14. Horizontal AFC circuit anti-hunt components.

A Christmas tree effect is the result of the horizontal oscillator operating far off its normal frequency. The AFC circuitry is a prime suspect with this type symptom. In some cases, however, loss of sync combined with improper oscillator adjustments will cause the Christmas tree effect. In some chassis, an open filter capacitor in the power supply will cause the same symptoms.

VHF Tuner Characteristics

4

The VHF tuner (also referred to as the front end or head end) of a TV receiver functions as a station channel selector, RF amplifier, and mixer or converter which produces the IF frequency. The schematic in Fig. 4-1 shows the usual stages found in the VHF tuner system. They are as follows:

Stage number one is the RF amplifier. This is tuned to the desired TV station frequency and amplifies the incoming signal.

Stage number two contains the local oscillator. This circuit generates a CW signal whose frequency is different for each channel.

Stage number three is the mixer or converter system. This stage receives the amplified RF channel signal and beats it with the signal from the local CW oscillator. Or stated another way, it converts the TV station signal down to the correct IF frequency.

The new signal frequencies obtained are a result of a heterodyne process that includes the harmonic combinations of the CW local oscillator and the incoming TV channel signal. The plate coil circuit of the mixer is tuned to the difference, or intermediate frequency, and feeds this signal to the IF amplifiers. This is the same principle on which the superheterodyne radio receiver operates. The VHF tuners are designed to receive 12 channels and if UHF is built in, one position is used to couple in the UHF tuner. (Before the FCC made all-channel UHF / VHF mandatory, older sets only had 12 channels, from 2 through 13.)

The two most used VHF tuners are the turret and bandswitch type. Some of the very early TV sets used the continuous tuning arrangements. A few of the newer color chassis are using the varactor or varicap all electronic diode bandswitching tuners. The turret or bandswitch tuner may use either tubes or transistors. The solid-state ones may use either bipolar and MOSFET transistors or utilize both types. An exploded view of a bandswitch tuner is shown in Fig. 4-2 while the turret type is illustrated in Fig. 4-3.

SOLID-STATE VHF TUNER OPERATION

The antenna input impedance for the solid-state tuner shown in Fig. 4-4 is 300 ohms. The step down to 75 ohms is

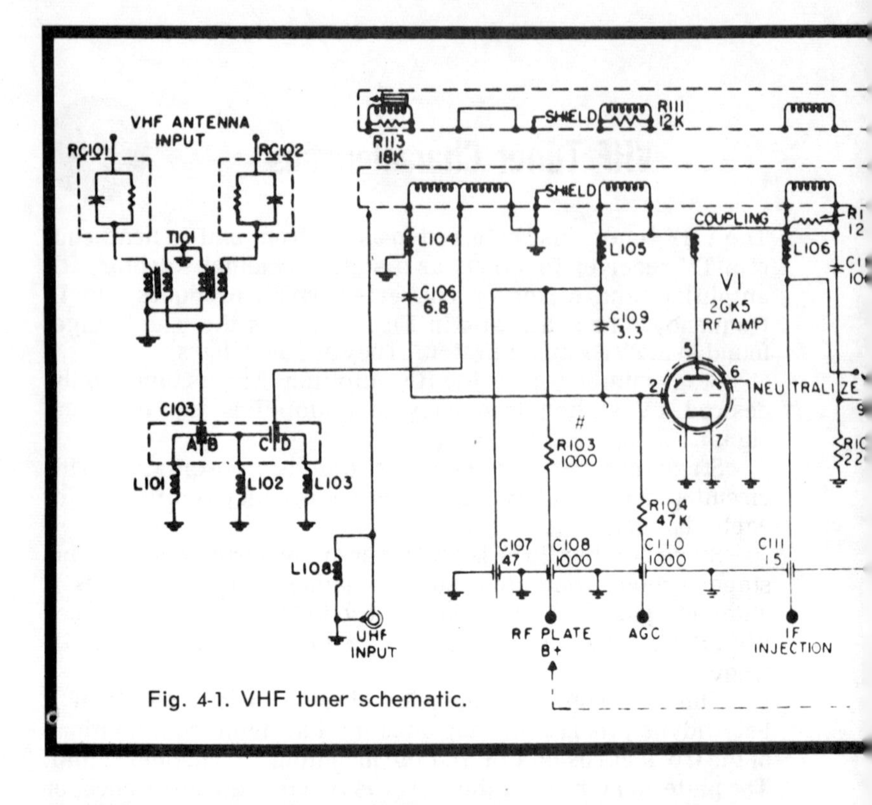

Fig. 4-1. VHF tuner schematic.

provided by the antenna balun, T1. The balun output is coupled to a series of LC networks that provide FM and IF signal trapping. These traps are fix-tuned and generally do not require service unless damaged.

The signal is coupled from the antenna filter assembly to the antenna preselector circuit through capacitor C3. When operating on channels 2 through 6, the antenna preselector consists of each individual channel antenna coil paralleled by capacitor C1. Inductor L provides RF impedance matching on the low channels. When operating on the high (7 through 13) channels, each antenna coil is placed in parallel with L4. Capacitor C1 then becomes a coupling capacitor in series with C2 to couple the signal to the RF amplifier transistor. In UHF the antenna input is shorted to ground, and the UHF IF signal is coupled through C2 to the RF amplifier transistor.

The RF stage is operated as a common base amplifier and utilizes forward AGC. The bias voltage is set for maximum RF gain by resistors R2 and R3. This divider supplies about 1.9 volts to the base. When strong stations are received, the chassis supplies a more positive AGC voltage (up to +7 volts) for gain reduction.

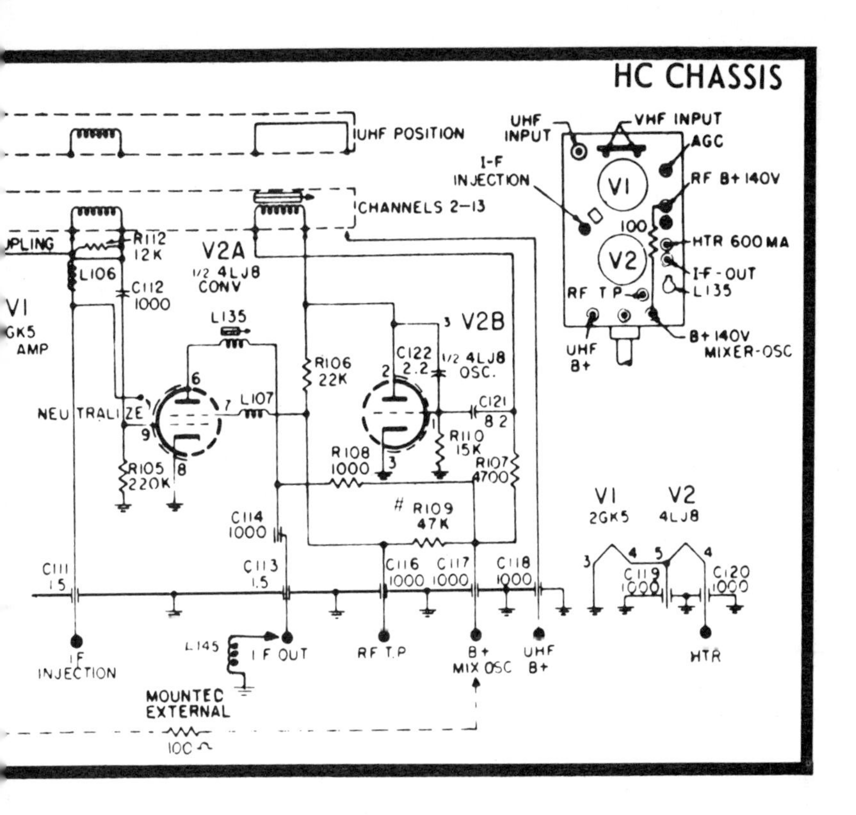

RF collector tuning is provided by the RF coil in each channel strip. Similarly, tuning of the mixer input is accomplished by the converter coil in each channel strip. RF to mixer coupling takes place due to the proximity of these coils. Oscillator injection to the mixer is coupled through C17. The output of the mixer is coupled to the IF module through coil L3.

The oscillator stage is a Colpitts configuration. The tank, or frequency-determining components, consists of the individual channel oscillator coils and the capacitor divider network comprised of C18 and C19. Automatic frequency control (AFC) of the oscillator is performed by varactor diode X1. As an example, if the oscillator drifts higher in frequency, the AFC unit in the chassis would sense this and alter the nominal (or "no correction" voltage of +3 volts) AFC voltage. In this example, the voltage would drop below 3 volts, thus causing the capacity of the varactor diode to increase. The increased capacity is reflected through capacitors C20 and C24 to the base of the oscillator transistor, thereby lowering the oscillator frequency back to "normal." Similarly, if the oscillator drifted lower in frequency, the AFC voltage would increase, thus increasing the oscillator frequency.

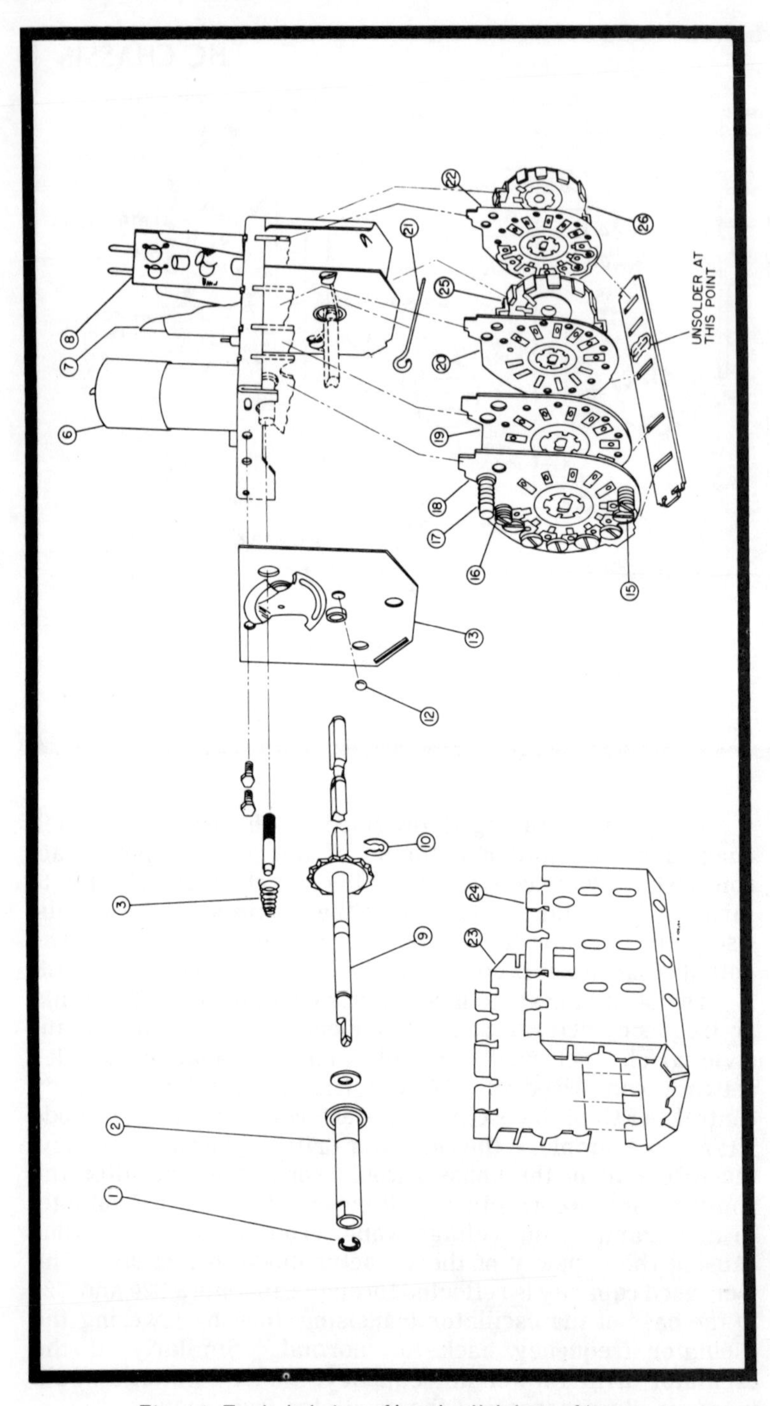

Fig. 4-2. Exploded view of bandswitch type of tuner.

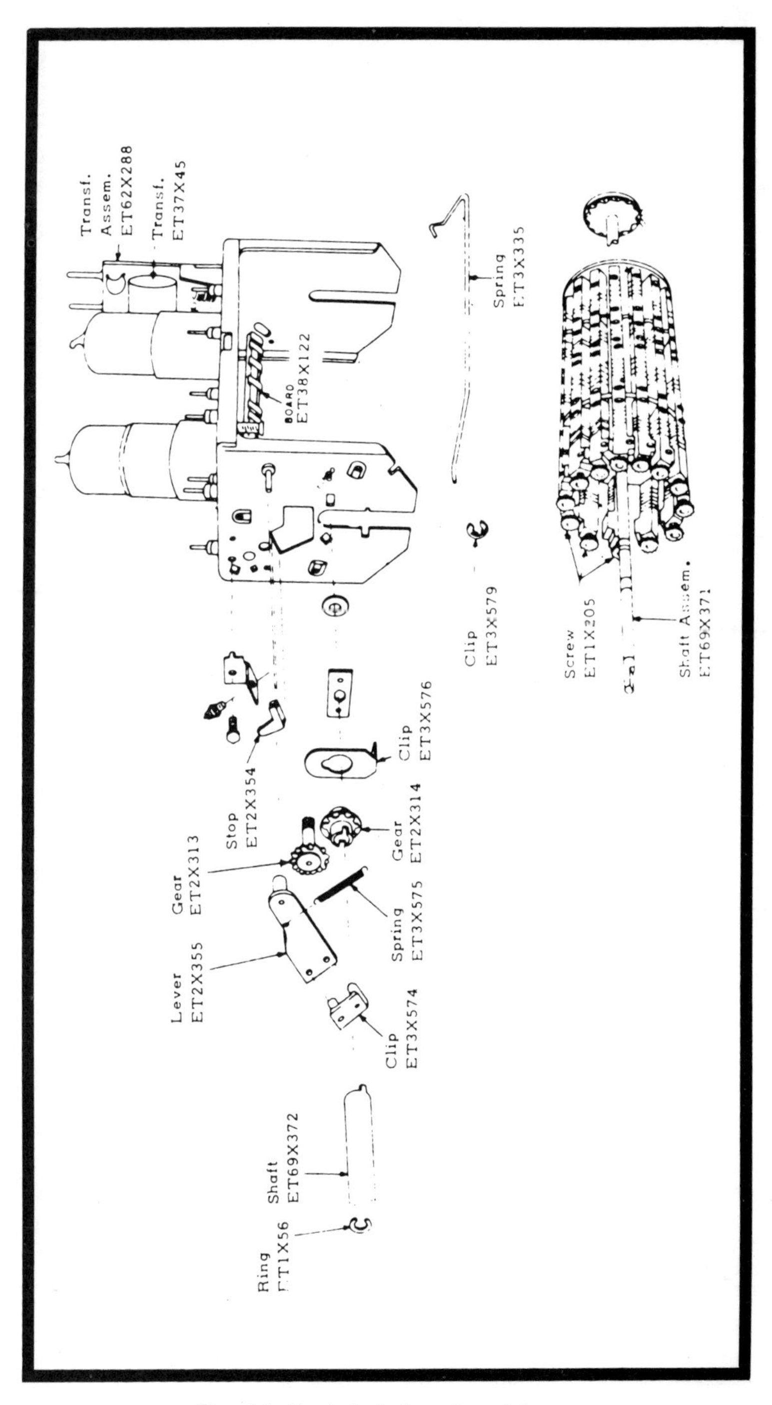

Fig. 4-3. Exploded view, turret tuner.

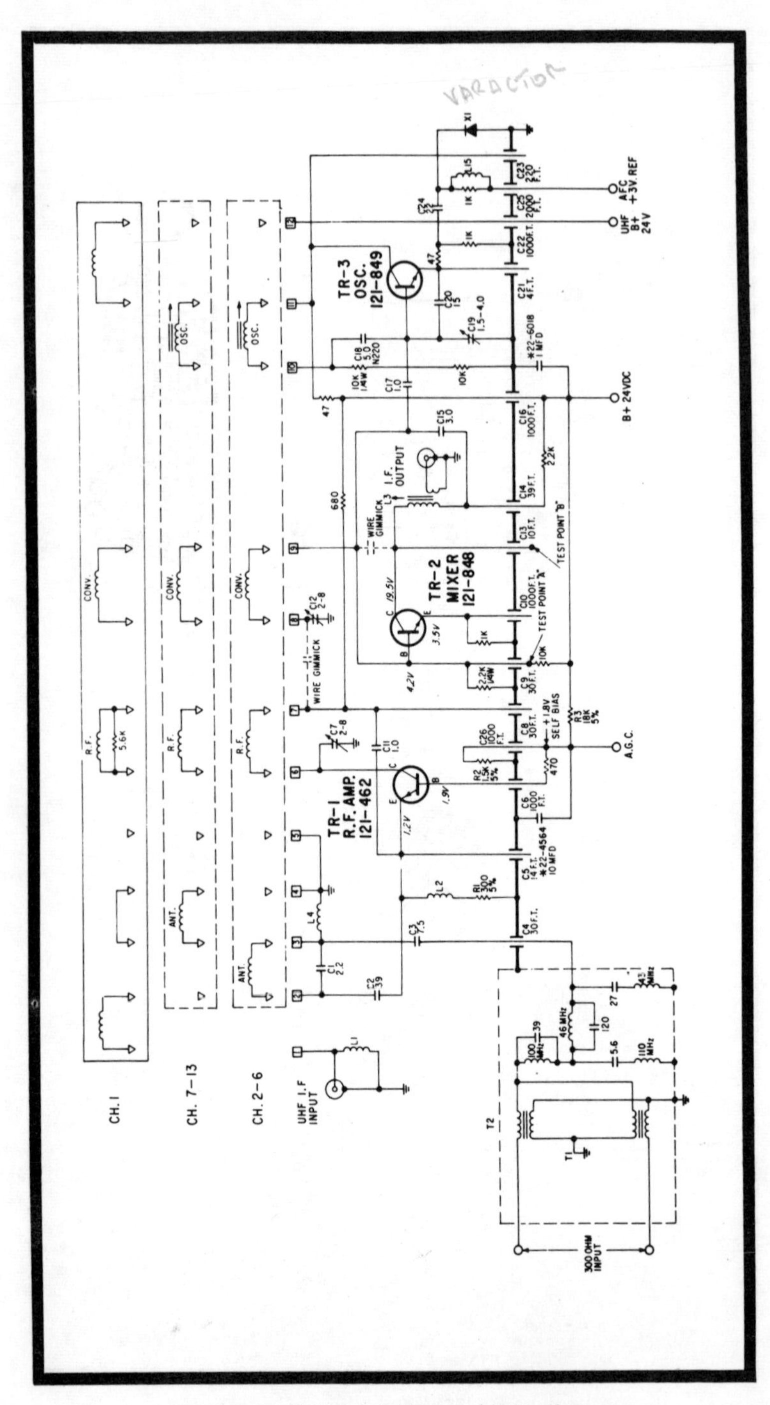

Fig. 4-4. Solid-state tuner schematic.

TUNER TROUBLESHOOTING HINTS

A snowy picture is normally associated with a fault in the RF stage. Refer to the tuner schematic in Fig. 4-4. With the AGC lead connected, check the base bias voltage of the RF amplifier transistor TR-1. This voltage is the maximum RF gain bias developed by resistors R2 and R3. The base voltage should be +1.9 volts. If the base voltage is normal, check the emitter and collector voltages. They should read +1.2 volts and +22 volts respectively. If the emitter voltage is very low or reads zero, the transistor may be open. A shorted transistor is indicated when the base and emitter voltages are several volts higher than normal.

A blank, milky raster (no video or snow) usually indicates trouble with the mixer or oscillator stages. Voltage checks for the mixer transistor may be performed like those described for the RF stage. That is, look for the normal forward bias that indicates the stage is conducting. Voltages around the oscillator are somewhat different. The collector should read 24 volts, base 11.4 volts, and emitter 11.3 volts. Rather than the typical 0.7 volt forward bias, there is a 0.1 volt difference between the base and emitter. This is an indication that the stage is indeed oscillating. If the stage stops oscillating because of a defective capacitor in the tank circuit, the forward bias will increase towards 0.7 volt. A defective transistor will yield voltage discrepancies similar to the RF and mixer stages.

THE ELECTRONIC TUNER

Now for a brief look at the all electronic VHF varactor tuner. See Fig. 4-5. In this system, channel selection is achieved by electronic, rather than mechanical means, that is, there are no moving parts. A DC control voltage is applied to varactor diodes to accomplish the tuning. The advantages are apparent as there are no switch contacts or gear trains, no torque is required, channel changing is almost instantaneous, and channel selection is completely silent.

The VHF tuner used for this system is a four circuit, four transistor device employing a dual insulated gate field effect transistor (DIGFET) in the RF stage. This tuner employs five switching diodes that change the circuit characteristics so that the tuner will cover both the low and high VHF bands with approximately the same varactor tuning voltage.

AUTOMATIC FREQUENCY CONTROL (AFC)

The AFC circuitry (note Fig. 4-6) in this tuner has as its main element the variable capacity diode BA142 whose

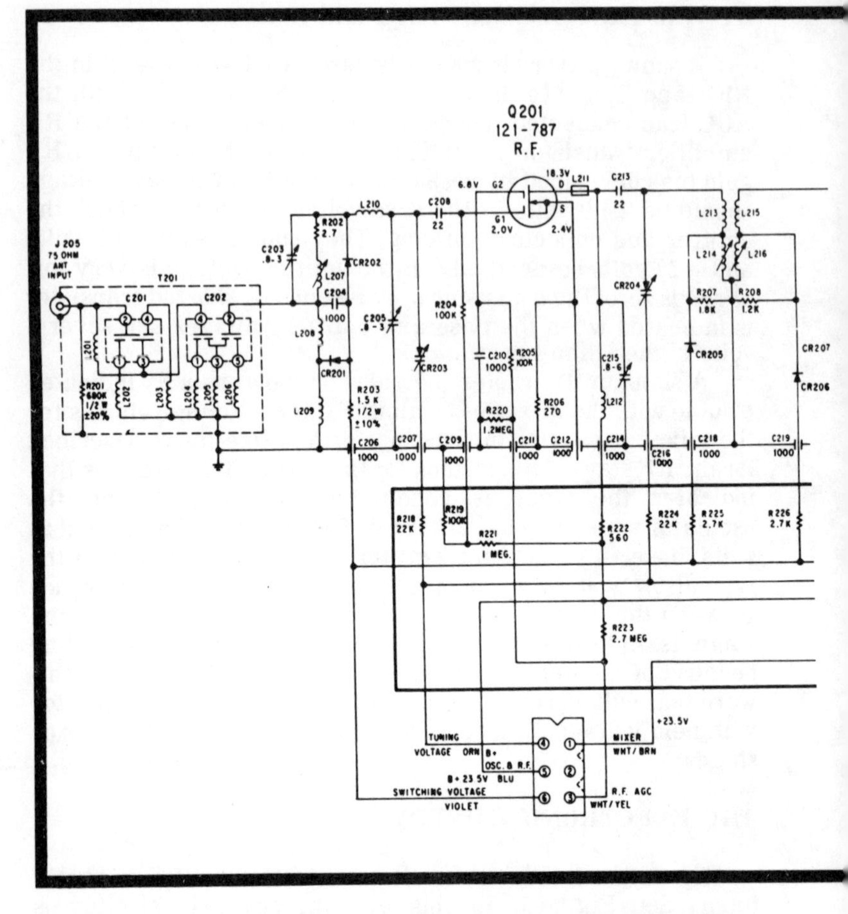

capacitance varies with an applied voltage change. When reverse biased, the diode acts like a capacity which is inversely proportional to the applied bias. If the voltage is high, the capacity is low and vice versa.

The diode is connected to the plate of the oscillator, V2B, through a 10 pf capacitor so that the maximum capacity effect of the diode is less than 10 pf. An AFC control voltage is applied to the anode of the diode. Any change in the control voltage will change the diode capacity and therefore the frequency of the oscillator. A fixed back-bias level is applied to the cathode of the diode. The bias voltage is derived from the bleeder network R118, R119, connected between +140 volts DC and ground. Resistor R116 provides isolation against oscillator radiation.

The AFC control voltage is also applied to a varicap diode in the UHF tuner. The two diodes are then fed in parallel. When the VHF tuner is switched to the UHF mode, R117 is

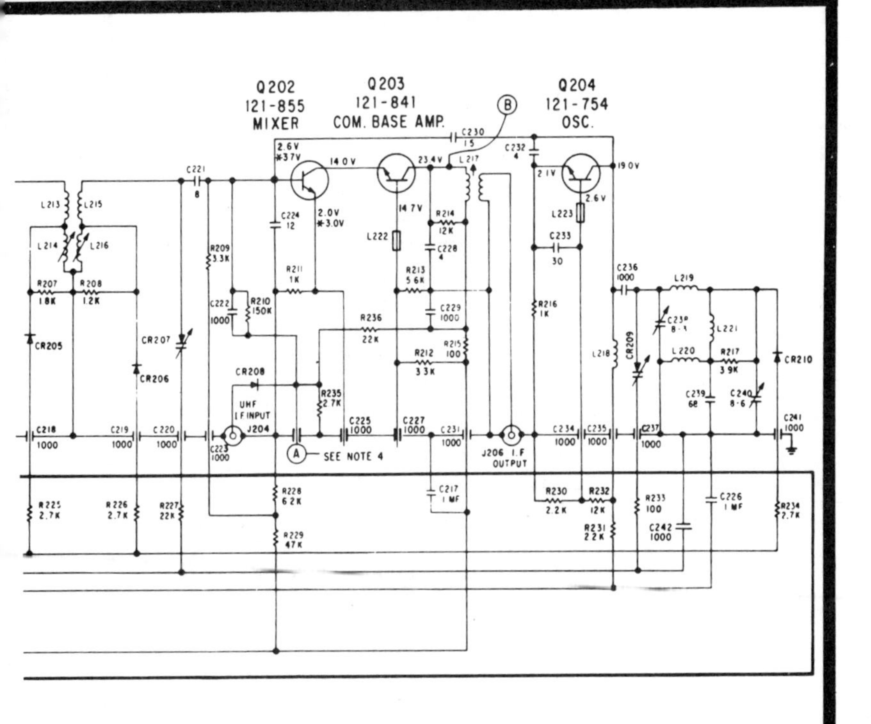

Fig. 4-5. Electronic (varactor) tuner.

switched into the back-bias circuit in parallel with R119. This increases the back-bias on BA142 so that this diode is ineffective during UHF operation. The control voltage comes from the AFC module located on the main chassis.

The AFC control circuit (Fig. 4-7) is on a module which contains integrated circuit IC303, tuning coils L354, L356, and other components. IC301 contains an IF amplifier, AGC amplifier, detector, DC amplifier, and a voltage regulator. A 45.75 MHz IF signal fed to terminal 12 of the IC is amplified by the IF amplifier section, and appears at terminal 3. The signal is transferred to the detector stage by means of "tapped" capacitance in the tuned circuits, L354 - C360 - C362, and L356 - C358 - C364. Coil L356 is the tuning adjustment and L354 is the crossover adjustment. The detector outputs are DC voltages whose amplitudes are proportional to the amount by which the input signal has shifted from 45.75 MHz. These DC outputs are amplified by the DC amplifier and appear at terminals 5 and 8

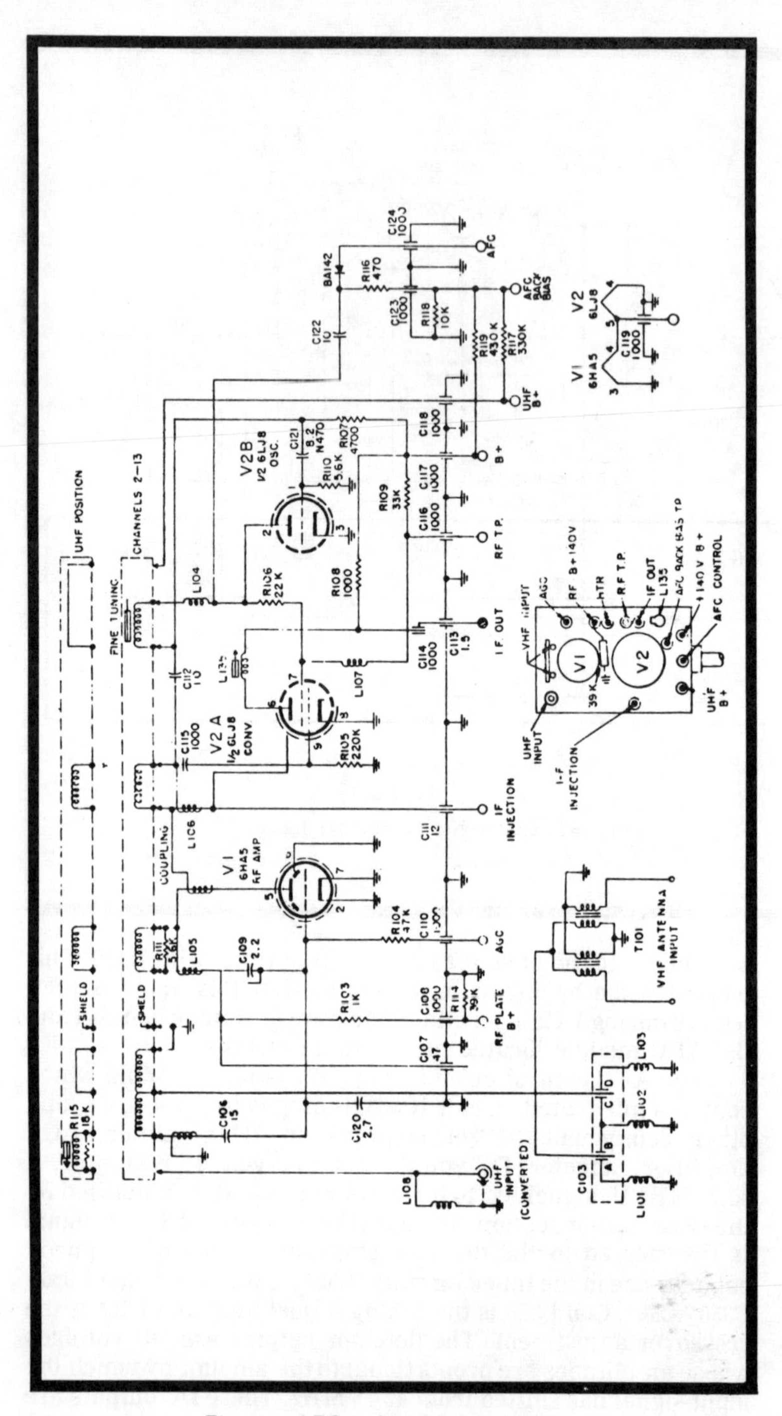

Fig. 4-6. AFC in bandswitch tuner.

of the IC. These terminal voltages "center" around 6 volts. That is, if the incoming signal is exactly 45.75 MHz, both voltages will be about 6 volts. If the frequency of the incoming signal shifts from 45.75 MHz, one of the voltages will become more positive than 6 volts, and the other less positive than 6 volts. These voltages are coupled to the anode and cathode of a varicap diode in the tuner, and cause the diode to change capacitance in a direction which will correct for frequency shift. Because one of the voltages is coupled to the cathode of the varicap diode, and the other is coupled to the anode, the varicap diode "sees" double the voltage change that it would if the correction voltage were "single ended". Hence, the tuning accuracy is greatly improved. Other systems use one control voltage or a "single end" type circuit.

TV TUNER SUBSTITUTION CHECK

Many times the technician is not certain if the tuner is at fault or if the problem is in the IF section and possibly the AGC system. A fast way to find out for sure is to use a TV tuner "subber." The one we have used with great success, is built by **Castle TV Tuner Service, Inc.** and a photo of this unit is shown in Fig. 4-8.

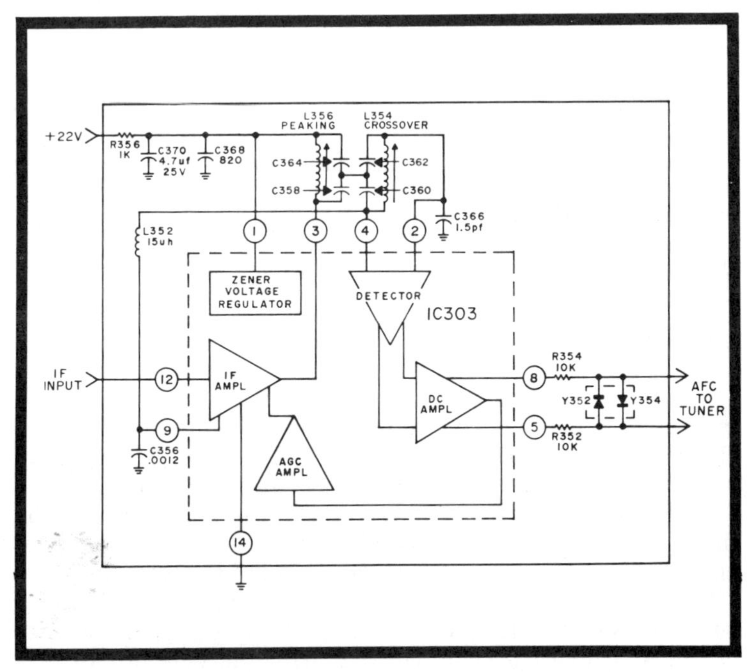

Fig. 4-7. AFC control module; integrated circuit.

Fig. 4-8. Tuner "subber."

Using The TV Tuner "Subber"

If the receiver tuner output terminates in a phono jack, unplug the IF cable plug and insert it into the IF output jack of the "subber." If the receiver tuner output terminates in a feed-through terminal, disconnect from tuner and use the extension cable/adapter with alligator clips to connect the receiver IF cable to the "subber." Connect antenna lead to "subber" antenna terminals. Switch on TV receiver and "subber," select channel on test tuner and adjust fine tuning and RF gain control.

If receiver performance is now normal the receiver tuner is probably defective. Check operation of the AGC control and if it is normal remove the tuner for repairs.

Checking the IF and AGC Systems

With the "subber" still connected in place, clip the leads from a variable bias box between the AGC line and ground. Negative goes to the AGC line of tube sets, but observe polarity specified by receiver manufacturer of solid-state sets. Vary the bias box voltage and note if reception is restored by thus overriding the AGC supply. If it is, analyze the AGC circuits for any faults. If performance is not restored, use a VTVM to measure the DC voltage developed by the video detector. Should this detector voltage be absent at all settings of the AGC override voltage, and with the bias box disconnected, then the IF amplifier system is probably defective and should be checked out. If detector voltage is present with the AGC bias box disconnected and varies with adjustment of receiver AGC control or changing channels on "subber," then the IF and AGC systems are functioning and the fault is elsewhere. Check amplitude of detected video signal against normal signal specified in the service notes.

Quality Of Reception

When using the "subber" the quality of reception may not be as good as with the TV tuner, due to the change in bandpass of the tuner IF coupling. In the receiver, the self-capacitance of the IF cable is usually used as part of the common, or low side, "C" component of the bandpass circuit. This circuit is comprised of the tuner mixer coil, first IF coil and the low "C" capacitance as illustrated in Fig. 4-9.

When original connections are restored after service of the tuner (if this was the case), the bandpass will also be restored to normal. However, while servicing the tuner, if the mixer coil was adjusted, or if the tuner has been exchanged or replaced by another (even exact part number), then it will be necessary to adjust the mixer coil, for best bandpass. **Do not adjust the first IF coil.**

Of course, this test tuner is only intended to prove whether the original tuner works or not. It may not always indicate the quality of performance for a tuner that is working but producing a poor quality signal. It would be best to replace or repair any tuner suspected of such performance.

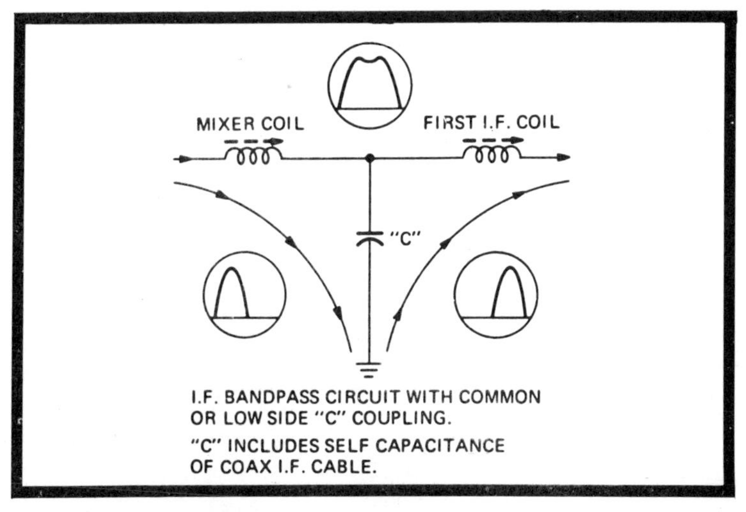

Fig. 4-9. Why "subber" quality of reception may not be as good as the tuner itself.

This "subber" is used to check tuners in receivers with sound and video IF carriers of 41.25 MHz and 45.75 MHz respectively. It should not be used with older receivers having an IF in the 20-MHz range, nor with "off shore" sets having an IF in the 20-30 MHz range.

Testing The UHF Tuner

Because most modern VHF tuners are equipped with a 40 MHz amplifier function (channel 1, or UHF position), the 40 MHz UHF input jack of the "subber" tuner is brought out to a jack on the back of the case. By plugging the output of the receiver UHF tuner into this jack, and switching the "subber" to the UHF position, it is possible to test operation of the UHF tuners.

In some cases the VHF tuner in the receiver works normally when receiving VHF channels, but the UHF tuner does not work. If the UHF tuner will work when the "subber" is used instead of the TV VHF tuner, the 40 MHz amplifier function of the VHF tuner is defective and requires service. For this test, the TV receiver VHF tuner must be switched to the UHF position because the B+ supply for the UHF tuner is switched on only in this position.

This tuner "subber" is of a solid-state, battery operated design.

Video Amplifier Circuitry

5

Video amplifiers in black and white or color TV receivers are used to increase the amplitude of the composite video signal which appears at the video detector. This amplified video signal, with correct polarity, is fed to the grid or cathodes of the CRT to intensity-modulate the electron scanning beam. Due to the low efficiency of the shadow mask color CRT, a larger signal amplitude is required for a color receiver than in a B & W set. Also, the chroma circuits introduce some time delay and the color chassis must have a delay line for the video amplifier to match the timing of the video signal to the chroma signal. With correct deflection circuit sync, the desired color picture will be seen.

The video amplifiers that are used in television receivers are RC or direct coupled. Some color sets use only tubes to increase video gain while others use tubes and transistors (hybrid) in these stages. The newest color chassis now coming off the production lines are designed with solid-state in mind for the video amplification stages.

SOLID-STATE VIDEO AMPLIFIER SYSTEM

Let's now look at one of these newer all solid-state video amplifier systems (Fig. 5-1). For a Zenith 4B25C19 chassis the negative-going video information is obtained from the emitter (test point C1) of the first video amplifier transistor in the IF module. The signal is coupled by C213 and the contrast control through R232 to the base of Q205 second video transistor. R231 determines the amount of DC coupling from the video detector through the video channel. C214 provides video peaking and R232 gives some isolation for Q205.

Video information is coupled from the collector of Q205 through the delay line, L202 and L204 to the base of Q206, third video amplifier. Because of the DC coupling in the video circuits, any change in bias on the second video transistor causes a change in CRT bias and thus varies picture brightness. The contrast control, besides varying the amount of video applied to the second video transistor, also varies base bias of this transistor. As the contrast is increased, the

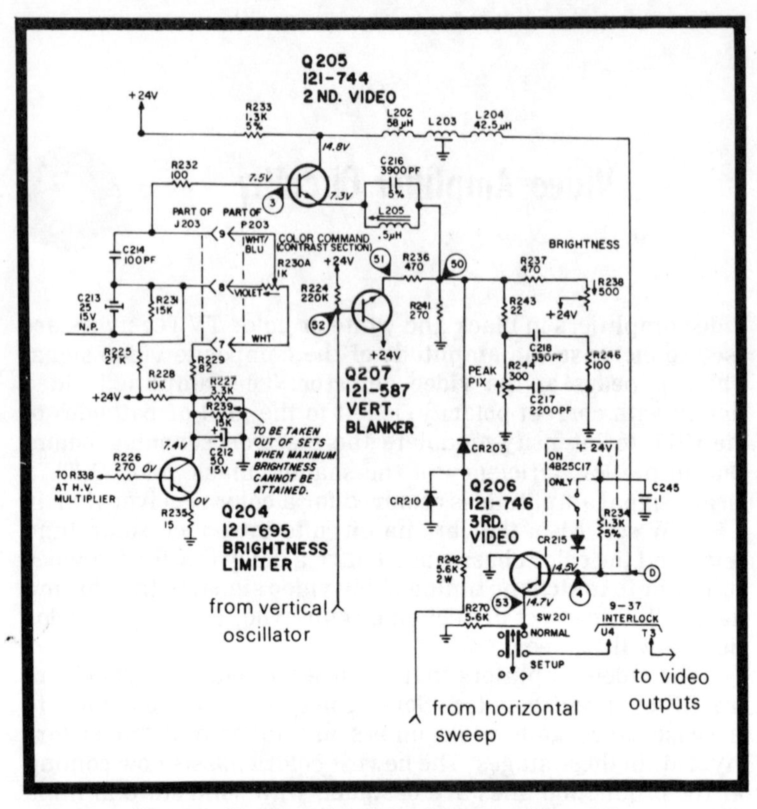

Fig. 5-1. Zenith solid-state video amplifier strip.

brightness is also increased, keeping a proper ratio between brightness and contrast. In the same manner, brightness range is controlled by varying the emitter voltage on Q205. This is accomplished by R237,and R238, the brightness control.

Brightness limiting is accomplished by Q204 in the base of the second video stage. Changes in CRT brightness are reflected to the brightness limiter transistor as a series of pulses from the high voltage tripler circuitry. An instantaneous increase in CRT brightness is seen at the brightness limiter transistor base as an increase in the size of the pulse applied to it. This larger pulse causes greater conduction of the brightness limiter transistor, thus lowering its collector voltage. The reduced collector voltage, coupled through the contrast control to the base of the second video transistor, serves to vary the bias on this transistor. This will lower the brightness of the picture because of the DC coupling from this point to the CRT.

A brightness limit control at the high voltage tripler serves to vary the amplitude of the pulses fed to the brightness limiter circuitry. This allows adjustment for the amount of brightness limiting.

A zener diode at the high voltage tripler protects the transistors should an arc occur within the picture tube gun.

The brightness limiting function also prevents excessive current from being drawn from the high voltage tripler. Essentially, the maximum current limit is 1.5 milliamperes.

Vertical and horizontal blanking signals are applied to the emitter circuit of second video transistor Q205. A positive pulse from the sweep transformer is coupled to diodes CR203 and CR210. These diodes admit, shape, and clamp the pulse. During this time, the emitter is driven positive enough to cut Q205 off. This causes the collector voltage to increase. This increase in voltage is coupled to the CRT through the remaining video stages and the picture will be driven to cutoff during horizontal retrace time. During the vertical sync pulse interval, a positive voltage from the vertical oscillator is applied to the base of vertical blanker transistor Q207. During this time, Q207 conducts heavily and reflects a positive voltage at the emitter of Q205. This causes Q205 to be cut off and the positive increase in its collector voltage functions in the same manner as horizontal blanking. This causes the CRT to be cut off during vertical retrace time.

As stated before, positive-going video with blanking is coupled to the base of third video transistor Q206. This PNP transistor couples the video information to the common emitters of the three color video output stages, Q209, Q210, and Q211. Transistor Q209 also functions as the emitter return circuit of these three transistors.

VIDEO OUTPUT AMPLIFIERS

The video output amplifiers consist of three transistors, Q209, Q210, and Q211. Refer to Fig. 5-2. These three stages amplify both the monochrome video (Y) signal and color signals, as well as matrixing.

The three demodulator output signals consisting of -(R-Y), -(B-Y) and -(G-Y) are coupled respectively to the bases of Q209, Q210, and Q211. Since the monochrome signal is also fed to all three amplifiers, matrixing is performed in these stages, producing red, green, and blue signal voltages at the amplifier outputs. All three amplifiers operate in an identical manner.

The DC operating potentials for Q209, Q210, and Q211 are taken from the 270-volt B+ supply. Collector voltages are established from the voltage drop across the 18K and 27K resistors and gain controls in each collector circuit. Base biases are obtained from voltage drops across the 27K and 56K resistors to the base and from the base through the 1K and

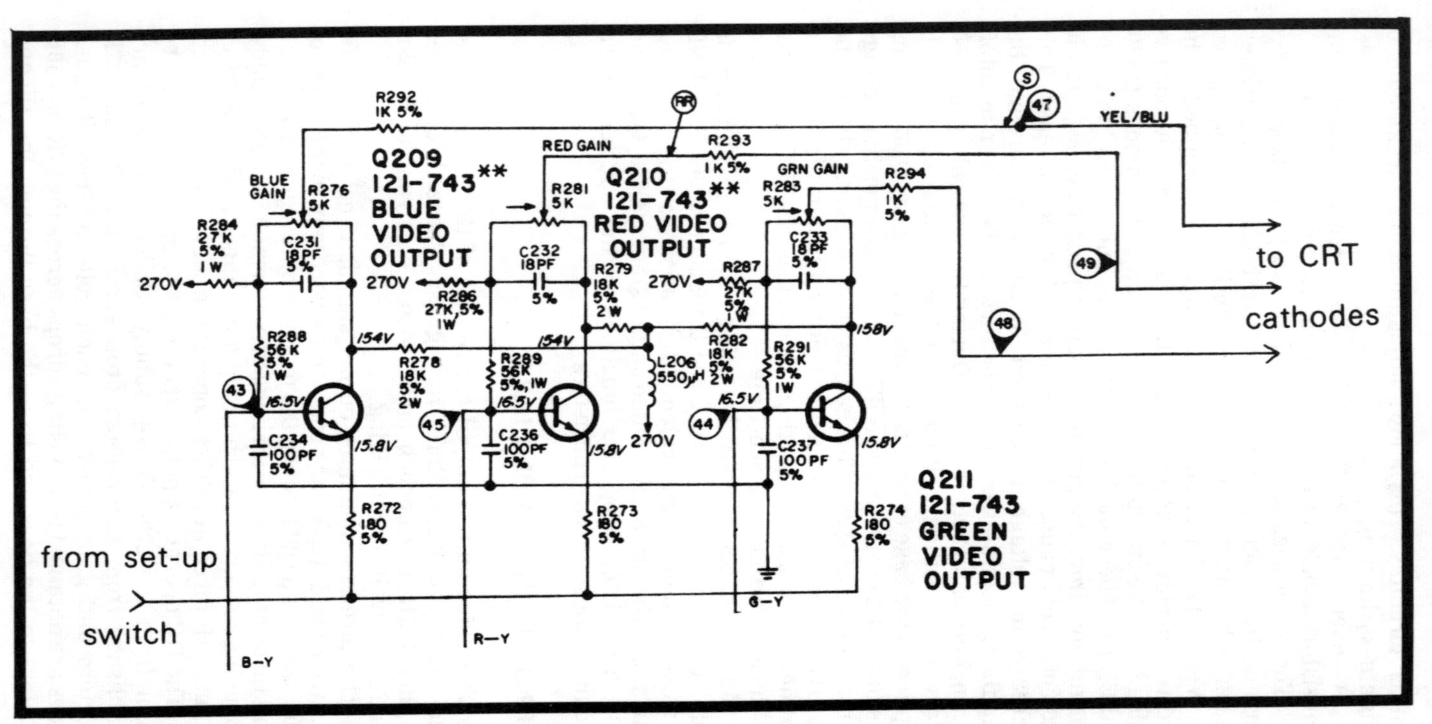

Fig. 5-2. Video output amplifiers.

2.2K resistors to ground (on chroma module). Emitter voltages are developed by the 180 ohm resistors in each emitter circuit and the Q206 transistor which is common to each emitter.

Each gain control in the color amplifier output circuit is frequency compensated by an 18 pf capacitor shunting each control.

The outputs of the color amplifiers represent R, G, and B signal voltages and they are coupled to the cathodes of the CRT as shown in Fig. 5-3.

TROUBLESHOOTING THE VIDEO OUTPUT STAGE

Should some but not all colors be missing, the video output transistor circuitry should be checked out first. One color may predominate or be lacking in the red, blue, or green fields.

The video output transistor collector voltages are in effect the cathode bias voltage for the corresponding guns of the CRT. If the collector voltage for one of these transistors is too high, the gun of the CRT connected to it will tend to be biased off and that color will be missing from the screen. If the collector voltage of a transistor is too low, the corresponding

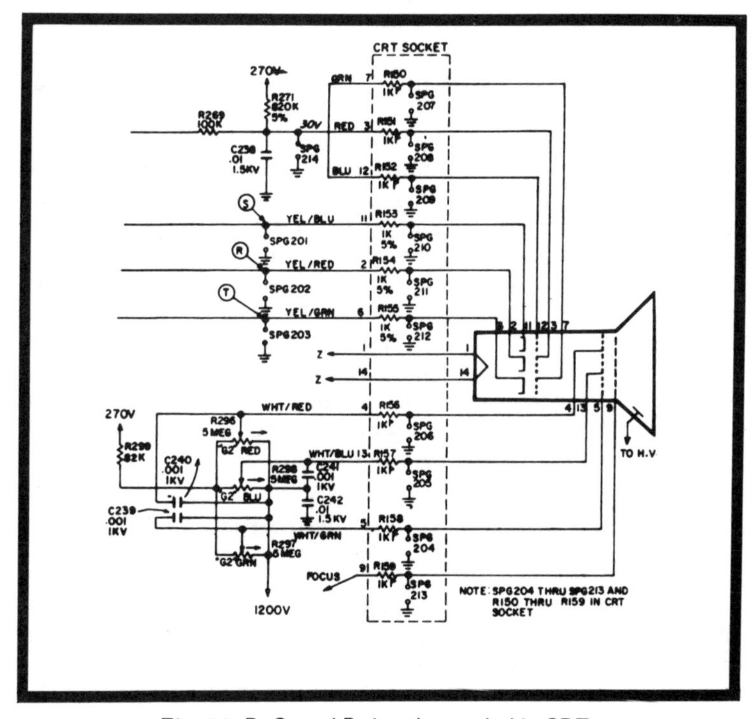

Fig. 5-3. R, G, and B signals coupled to CRT.

gun of the CRT will conduct heavily. This makes the raster predominantly the color of the heavily conducting gun.

If resistors R292, R293, or R294 in series with the CRT cathode opens (either 1 or 2.2K), the symptom is that of a defective picture tube. With the normal / service switch set for "service" position, the gun of the CRT with proper resistance in the cathode circuit will now produce little or no line across the screen as the G2 control for that gun is advanced. With the switch in the "normal" position, the color corresponding to the gun in question will be missing from the raster.

Preliminary voltage checks may fail to reveal the cause of the problem. A good test is to note which gain control does not vary the cathode voltage with the resistor open. When the resistor is normal, some change in cathode voltage will be noted as the appropriate gain control is adjusted.

In this circuit there are two sources of collector voltage for the video output transistors. One part is through L206 and R278, R279, and R282. The other part is through the individual resistors R284, R286, and R287.

Check all waveforms and DC voltages (another good use for the AC-DC scope) associated with the inputs and outputs to the chroma demodulator IC902. See Fig. 5-4. Check for both R-Y and B-Y signals at demodulator IC terminals 6 and 7. Check the two 470 pf capacitors C918 and C919. Oscilloscope amplitude may appear normal on both sides of these capacitors, but color reproduction may still be incorrect if either capacitor is open.

Use a 560 pf capacitor to check the 3.58 MHz signal coupling. Jump this capacitor,using a clip lead,from IC 221-46 pin 3 to edge connector B6 (Fig. 5-4). Red should appear on the screen. Connecting the capacitor from pin 3 to edge connector A7 should produce a blue field. This proves that the demodulator system is operating and the 3.58 MHz signals are present. If the 3.58 MHz signal is missing, check the subcarrier regenerator module phase shift components L1001, L1003, C1014, and C1016. An "open" circuit in this area produces a picture missing one or more primary colors. Also, do not overlook the possibility of a defective color demodulator IC causing a video picture symptom fault.

ACTUAL VIDEO AMPLIFIER TROUBLES

Here are some specific tips, and examples of video amplifier troubles that may not be found by normal troubleshooting techniques. These symptoms include poor video reproduction, weak color, and intermittent color. Refer to Figs. 5-2 and 5-4.

1. When the "chroma" control is advanced, color appears first at the left hand edge of the screen, gradually spreading to the right. This is probably caused by an open B+ filter capacitor (C220) that is used for filtering the +24 volt line into the chroma demodulator duramodule 9-37. With a scope connected to this B+ line, a spike will be visible.

2. Set has a pale, negative picture and no color. Check for an open choke coil L-218, in series with the B+ 24 volt line to the 9-37 duramodule.

3. Intermittent color sync. May be caused by an "open" ground connection at the color level control. Check or resolder this connection if needed.

4. Brightness level varies. An intermittent short in one CRT gun (grid G1) can cause this condition as all three G1 grids are common. When one gun becomes faulty, all guns are affected simultaneously while holding good grey scale tracking but producing a brightness change.

5. Picture is lacking red content. This can be caused by a defective spark gap at the red CRT grid. The same fault will happen to the blue and green CRT guns.

6. Raster is bright blue and is out of focus. Shorted spark gap at CRT socket (blue gun) cathode pin number 11. The blue video output transistor collector voltage measures +40 volts instead of 150 volts.

7. No blue in the raster for either black and white or color reception. Open 1K series resistor in cathode circuit of the CRT.

8. Screen will develop a slight grey scale shift. Probably caused by a faulty video output transistor. Use this service tip to determine if this is the cause: Switch to the "set-up" position and only one color will appear in the set up line. Within about one minute, the other two lines will become superimposed. Switch back to the "normal" position, the field will appear whitish, but gradually two colors fade slightly.

9. Clean pure fields cannot be obtained with individual G2 controls. Caused by defective CRT (internal short). To check, extinguish all three fields. Bring one G2 control up to view a dim field, then turn back slightly until just extinguished. Now turn up another G2 control. If a field begins to appear while the second G2 control is still nearly fully counterclockwise, there is an internal gun structure short.

10. The convergence panel can influence color shading of individual fields, boost voltage, and overall picture sharpness. As a quick check, pull off the convergence plug and note any improvement in picture quality.

11. There are several coupling bypass capacitors in the G2 control area. A short or leakage within a capacitor can affect G2 control operation, purity of individual fields, or boost

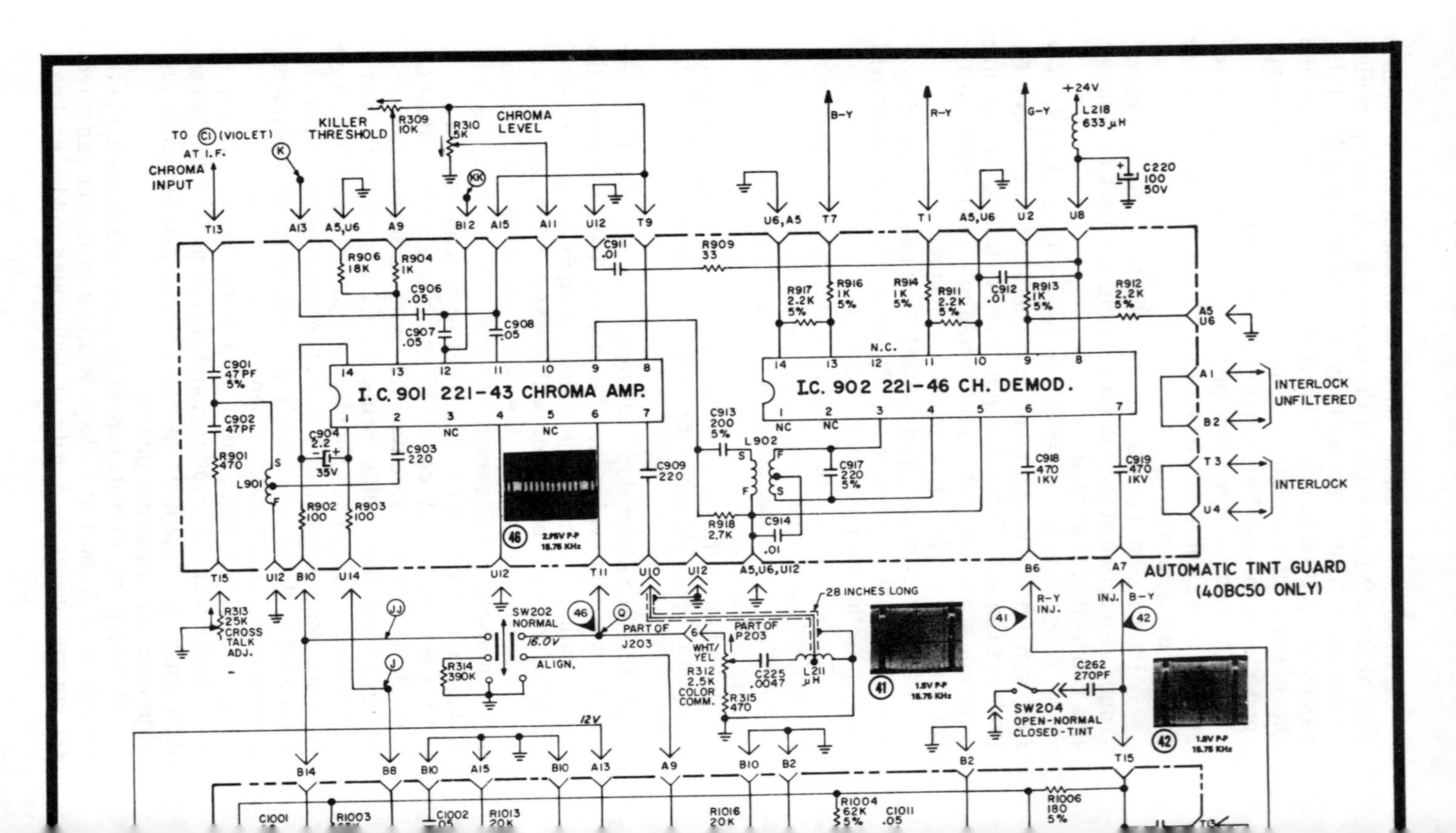

TO (C1) (VIOLET) AT I.F.
CHROMA INPUT
KILLER THRESHOLD
R309 10K
R310 5K
CHROMA LEVEL
B-Y
R-Y
G-Y
+24V
L218 633 μH
C220 100 50V
T13
A13
A5,U6
A9
B12
A15
A11
U12
T9
U6,A5
T7
T1
U2
U8
R906 18K
R904 1K
C906 .05
C907 .05
C908 .05
C911 .01
R909 33
R917 2.2K 5%
R916 1K 5%
R914 1K 5%
R911 2.2K 5%
C912 .01
R913 1K 5%
R912 2.2K 5%
N.C.
I.C. 901 221-43 CHROMA AMP.
I.C. 902 221-46 CH. DEMOD.
NC
C901 47PF 5%
C902 47PF
R901 470
L901
C904 2.2 35V
C903 220
R902 100
R903 100
2.75V P-P 15.75 KHz
C909 220
C913 200 5%
L902
C917 220 5%
R918 2.7K
C914 .01
C918 470 1KV
C919 470 1KV
INTERLOCK UNFILTERED
INTERLOCK
A1
B2
T3
U4
T15
U12
B10
U14
T11
U10
A5,U6,U12
B6
A7
AUTOMATIC TINT GUARD (40BC50 ONLY)
R313 25K CROSS TALK ADJ.
SW202 NORMAL
16.0V
ALIGN.
R314 390K
PART OF J203
PART OF P203
WHT/YEL
28 INCHES LONG
R312 2.5K COLOR COMM.
C225 .0047
L211 μH
R315 470
R-Y INJ.
INJ. B-Y
1.5V P-P 15.75 KHz
C262 270PF
SW204 OPEN-NORMAL CLOSED-TINT
12V
B14
B8
B10
A15
A13
A9
B2
C1001
R1003
C1002 .05
R1013 20K
R1016 20K
R1004 62K 5%
C1011 .05
R1006 180 5%

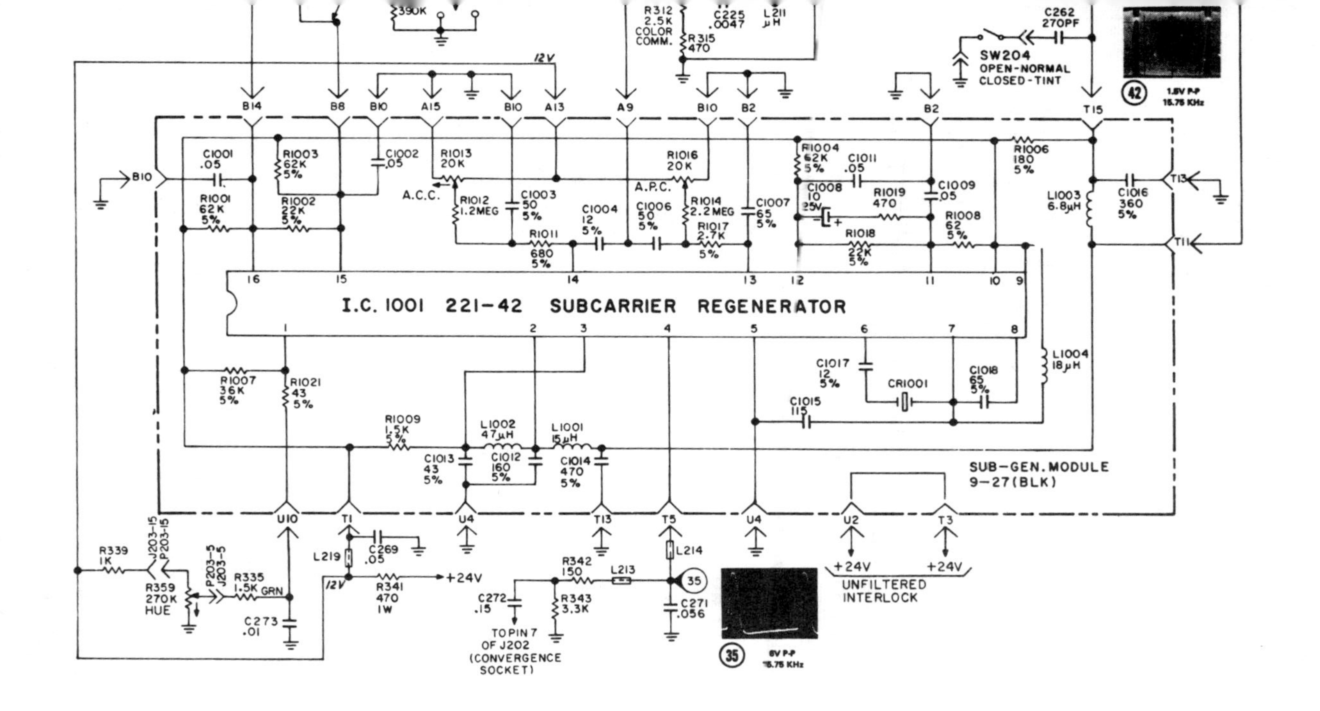

Fig. 5-4. Color signal processing circuits (subcarrier regenerator, chroma amplifiers, and chroma demod).

voltage. For an example, leakage in C242 will greatly reduce "boost" voltage causing that field to be extinguished. Note the circuit in Fig. 5-3.

12. As known for the color type TV sets, an open delay line will cause a loss of the video information. In the solid-state DC coupled system, an open delay line will upset the CRT bias voltage and cause picture blackout. Don't overlook a possible open delay line L203, shown in Fig. 5-1, with this type symptom.

VIDEO AMPLIFIER TROUBLESHOOTING

Let's go into greater detail for checking the DC coupled solid-state R.G.B video amplifier system. We will be using the Zenith 25CC55 chassis for these checks. See Fig. 5-5. However, these tests can be applied to any solid-state R.G.B system, because the video circuitry is similar and can be regarded the same way. As has been seen, a large degree of DC coupling is used and troubleshooting can become more difficult.

Referring to Fig. 5-5, note the location of the brightness and/or brightness range control. It's obvious that the bias on the second video transistor is changed to obtain a variation in brightness of the picture tube. The brightness control (and its voltage divider) affect the bias at the emitter of the second video amplifier.

Let's now look at some of the actions or reactions that take place in these video circuits.

1. Any malfunction from the detector to the CRT will show up as a brightness change (either too bright or too dark).

2. A "voltage check" method is not always best as transistors are basically current-operated devices. Voltage checks can become meaningless unless the proper setting of all controls is observed.

3. Not only are video and DC voltages used, but blanking signals, both horizontal and vertical, are also injected.

4. These blanking signals are also DC-coupled into the video system and have an affect on the brightness component.

Thus, all of these combined factors may result in some "trouble spots" for the service technician.

Use the following techniques to simplify the complexity and help give a positive indication of where the defective component is located. When 8 transistors and several diodes are combined in a circuit like this, tips to facilitate servicing are most helpful.

The "set-up" switch found on many color sets is an excellent troubleshooting tool. Its basic function is to collapse the vertical sweep and remove the Y signal from the picture tube to perform the brightness tracking procedure. With the introduction of R.G.B, and DC coupling, this "set up" switch has

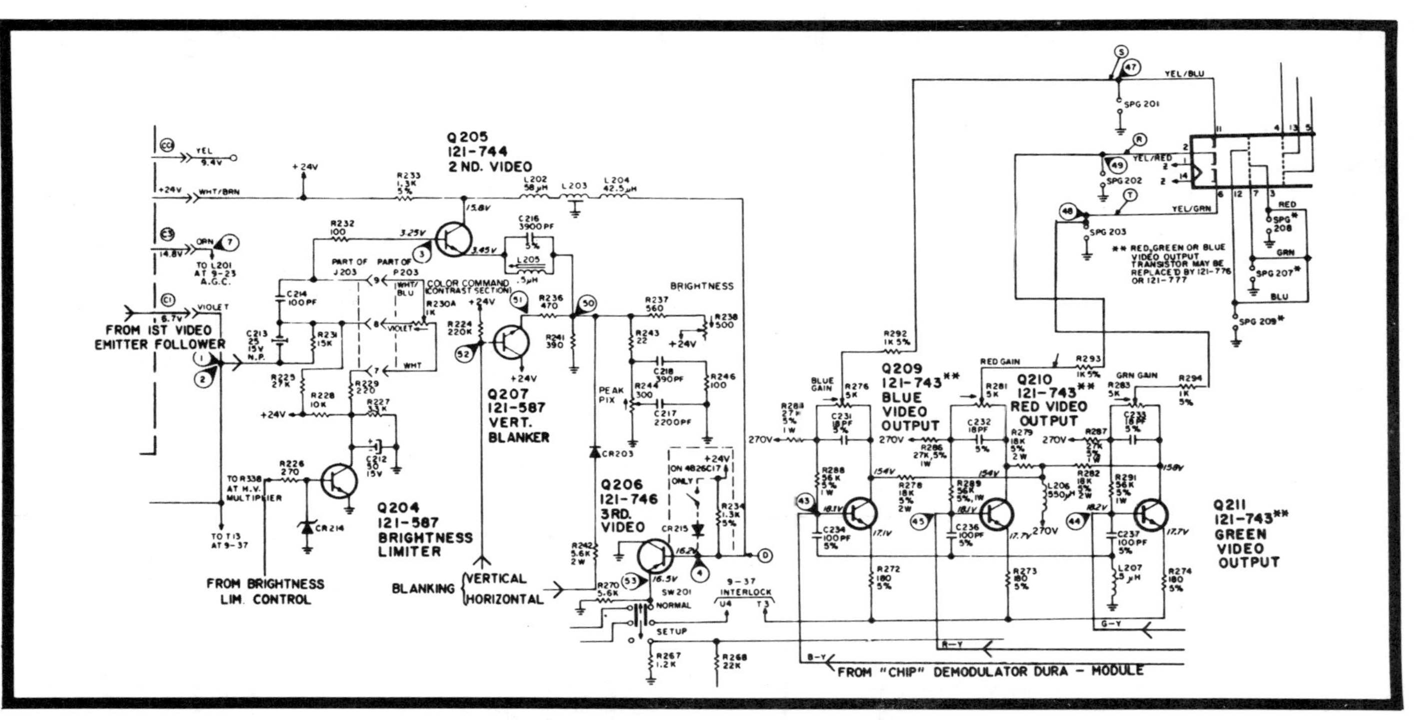

Fig. 5-5. Complete solid-state video amplifier system.

become more useful. Note this switch SW201 in Fig. 5-5.

When a brightness problem occurs use the following check list:

Place the "set up" switch in "set up" position and notice if a horizontal line can be produced by adjusting the G2 controls. If a line of sufficient brightness can be made, this would indicate the following conditions:

1. High voltage is present and the horizontal sweep system is operative.
2. The voltages or bias on the CRT are close to normal.
3. The R.G.B output transistors with switch in "set up" position are working normally.

If no line can be made, the R.G.B output transistors should be checked as well as the operating voltages for the CRT circuitry. Make routine voltage readings at this point.

If a line can be made in the "set-up" position, yet in "normal" position there is still excessive blooming or no brightness at all, then the trouble is probably in the video amplifier.

As mentioned before, voltage readings can be made, but are meaningless unless the controls are set at predetermined levels. The best method for troubleshooting these video amplifiers is to trace through the circuits with an oscilloscope.

Oscilloscope tracing of the video signals is not very complicated. There are only a few points where signals of any importance are to be found. These are the emitter, base, and collector of the second and third video amplifiers. However, these signals may be confusing if one is not familiar with the circuit operation.

The first step is to find the most accessible test point in the video chain. This is the delay line, which is located between the second and third video amplifiers. Now connect the scope (set at the vertical rate) to either the input or output of the delay line and view the waveform shown in Fig. 5-6. This display can be somewhat misleading. At the vertical rate, the actual video signal is small, but the apparent sync pulses are large and out of proportion.

At the horizontal rate setting of the scope controls, the display in Fig. 5-7 should be evident. Here the video signal appears large in amplitude and no apparent sync pulses are present. The difficulty in observing these waveforms is that what appears to be sync pulses are not sync pulses at all.

Remember that in the emitter of the second video amplifier these following signals are added:

Horizontal blanking
Vertical blanking
DC voltage for brightness.

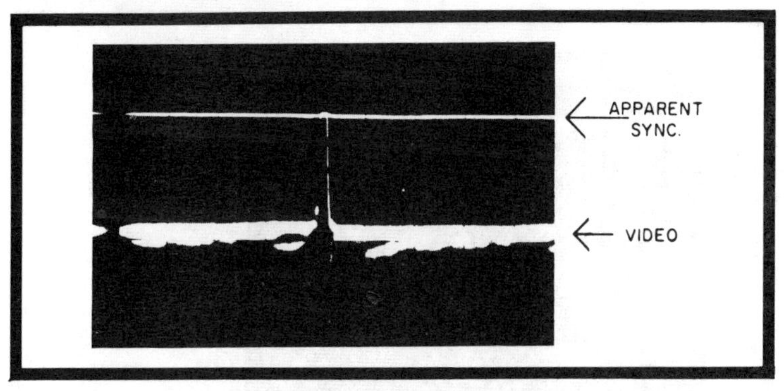

Fig. 5-6. Waveform at delay line. (Scope set at vertical rate.)

What appears as sync pulses in Fig. 5-6 are actually horizontal and vertical blanking pulses injected into the emitter of the second video transistor. It is possible to see this at the vertical rate only, by turning off the vertical sweep with the "set-up" switch. Now the video signal and the sync pulses, and the top line, are formed by the horizontal blanking pulses, as shown in Fig. 5-8.

There is misleading information here as the horizontal and vertical blanking signals obscure (blank out) the sync pulses. Another factor is the influence of the DC level from the video as it is detected. This point is not accessible since the first video amplifier is contained in the IF module unit. This detected carrier and resulting voltage variation will influence the brightness.

To service this part of the circuitry without the influence of the video detector, as well as the blanking signals, a known value test signal may be injected into the base of the second video amplifier. This will allow for accurate gain measurements.

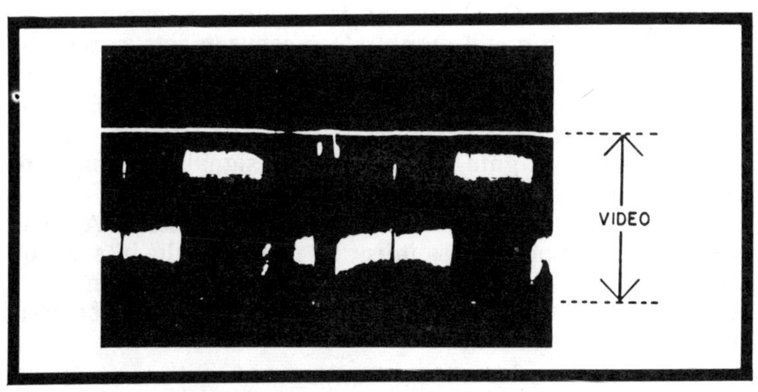

Fig. 5-7. Waveform at delay line. (Scope set at horizontal rate.)

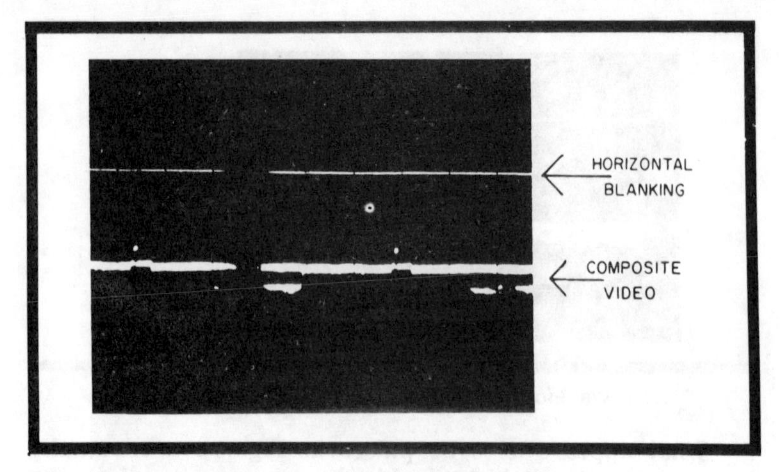

Fig. 5-8. Composite video and sync pulses referenced to horizontal blanking. (Vertical sweep turned off, scope at vertical rate.)

The test signal should be low enough in amplitude so that the transistor will not saturate, yet high enough to be viewed on the scope. The best way to do this is use a pulse- or square-wave generator if you have one on the bench. The input test pulse should be calibrated with the scope at 1.4 V P-P as will be shown in a later example.

A low AC voltage can be used, such as 6.3 volts AC from the filament supply. However, this signal is too high in amplitude. Also, the amplitude may vary due to line voltage changes. Rather than 6.3 V, let's limit this AC sine wave by means of two diodes and the resulting waveshape is similar to a square wave fixed at 1.4 V P-P amplitude. The back-to-back connected diodes are shown in Fig. 5-9. Due to the symmetry of the signal, the zero voltage line is centered in the middle. This is a symmetrical waveform as illustrated in Fig. 5-10.

Locate the C1 plug on the IF module and disconnect the purple lead from C1. With the test signal injected into C1 (purple lead), gain and amplitude measurements can easily be made. In Fig. 5-11, the combination of the fixed reference input signal and the added horizontal and vertical blanking pulses are shown.

DC gain, AC gain, and blanking signals can now be tabulated as in Fig. 5-12.

This is a tabulation of DC gain in the system. A 1-volt variation at the brightness injection point, results in a 90-volt variation at the cathodes of the CRT, or a DC gain of 90. Addition of the 1.4V P-P square wave test signal is shown in the Fig. 5-11 scope trace.

Increasing or decreasing the contrast (with the contrast control) will cause only the AC square wave change in am-

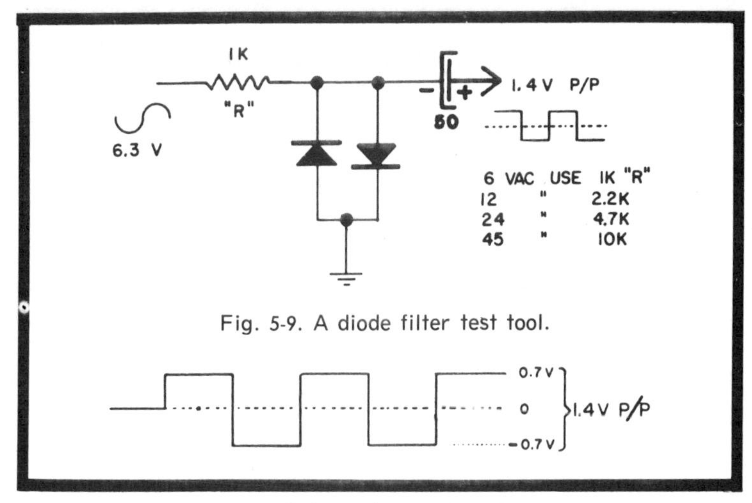

Fig. 5-9. A diode filter test tool.

Fig. 5-10. Output of circuit in Fig. 5-9.

plitude. The blanking level will remain constant. When the brightness is changed, the level of the blanking pulse is varied, while the AC waveform square wave remains constant as shown in Fig. 5-13.

Observations of the square wave and blanking pulses should make troubleshooting simpler. The waveforms will remain the same in the subsequent test points. Use the table in Fig. 5-14 to tabulate the amplitude changes. This is a tabulation of AC gains in the video system. The blanking amplitude gains are tabulated in Fig. 5-15. Note: These tabulations were made on the Zenith 25CC55 color chassis. Other chassis will have different readings.

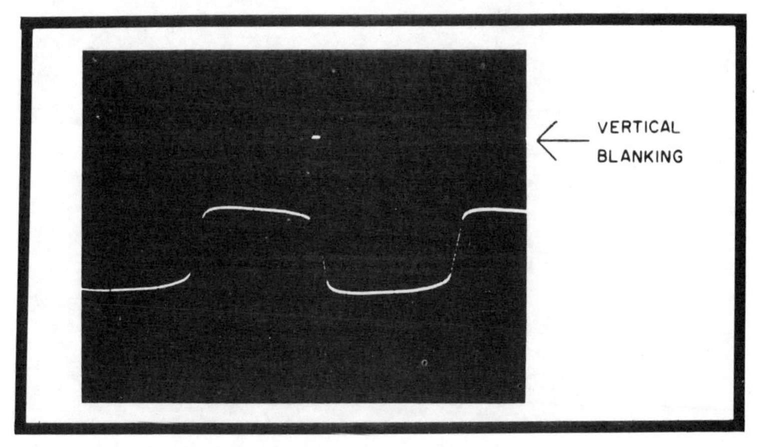

Fig. 5-11. Waveform, fixed reference input test signal with added blanking pulses.

VIOLET LEAD C1 PULLED

DC gains =

Voltage variations at Maximum Brightness – Minimum Brightness

At emitter 2nd video	+ 4 V DC	+ 5 V DC
At delay line input or output	+13 V DC	+ 22 V DC
At emitter driver 3rd video	+14 V DC	+ 17 V DC
At emitter R.G.B.	+14 V DC	+ 17 V DC
At picture tube cathodes	+120 V DC	+220 V DC

Fig. 5-12. Tabulation of signals, DC gains during testing.

From the foregoing it can be seen that the injection of a known fixed amplitude signal will help to isolate video circuit faults. Remember that the blanking amplitude changes when the brightness setting is changed and the AC waveform tracks with the contrast control setting.

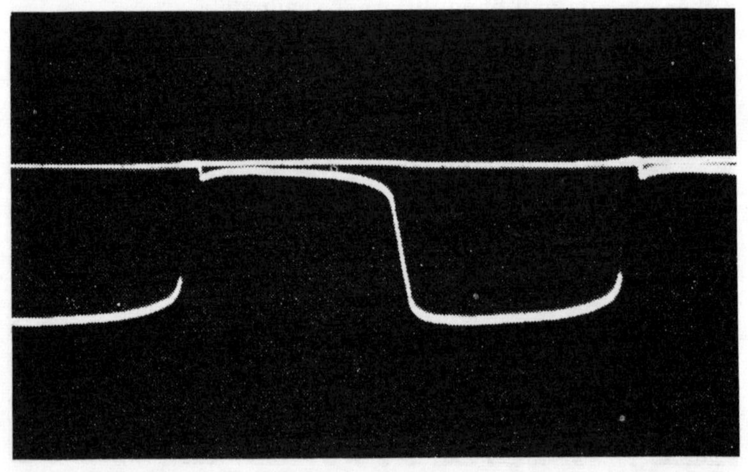

Fig. 5-13. Constant AC waveform when blanking pulse level is varied (brightness change).

Input at violet lead C1		1.4 V P/P	
Violet lead disconnected from C_1			
Square wave	Amplitudes	Contrast	
		Minimum	Maximum
At delay line		.6 V P/P	1.4 V P/P
At emitter 3rd video		.6 V P/P	6 V P/P
At emitter R.G.B.		.3 V P/P	3 V P/P
At picture tube cathodes		12 V P/P	120 V P/P

Fig. 5-14. Tabulation of amplitude changes for AC gains during testing.

SOME VIDEO CIRCUIT QUICK CHECKS

As we know, a very small change in voltage at the video detector will result in a large change of voltage at the CRT. For this reason a voltmeter is not too successful for troubleshooting these video circuits. Refer back to Fig. 5-5 while using the following procedures:

1. Place the "set up" switch in the set-up position and note if a horizontal line can be produced on the screen by adjustment of the G2 controls. If a line of sufficient brightness can be obtained, this indicates the HV is present, the voltages

Blanking (horizontal and vertical) Purple lead disconnected contrast control at maximum - No Input			
	Minimum	Brightness	Maximum
At emitter 2nd video	2 V P/P		.2 V P/P
At delay line	2 V P/P		12 V P/P
Emitter driver 3rd video	0 V		5 V P/P
Emitter R.G.B.	0 V		3 V P/P
Picture tube cathodes	0 V		100 V P/P

Fig. 5-15. Tabulation of blanking amplitude gains.

on the CRT are close to normal, and the color video output transistors are operating normally in the "set up" position. If no line is seen, this indicates that a defect exists with the color video output transistors, their circuitry, or the operating voltages of the CRT. Note that IC902 (the color demodulator IC) sets part of the base bias for the three output transistors. A fault in this IC can cause a brightness of individual red, green, or blue fields.

2. Connect a 680-ohm, 1-watt resistor from the junction of L204 and the delay line to ground. If normal brightness can be restored, the fault occurs prior to the third video stage. If brightness is not restored, the fault is in L204, the third video stage, or the set-up switch.

3. Connect the 680-ohm resistor to the other end of the delay line. If brightness was restored in step 2, but is not now restored, the delay line is defective.

4. Connect the 680-ohm resistor to the base of Q205, the second video transistor. If brightness is restored, the fault is in the IF module, the brightness limiter stage, or the components associated with the contrast control and plug. If brightness is not restored, the fault lies in the second video stage, the vertical blanking stage, the horizontal blanking circuitry, or the brightness control circuit.

Service tip: Should the brightness limiter transistor Q204 develop a short from emitter to collector (and it has happened) the CRT screen will be blacked out. A good case history point to remember.

6 Television AGC Systems

Because the signal strength varies from one TV station to another and sometimes signal fades on a station, some type of automatic gain control (AGC) system is needed. AGC is used to minimize signal variations so they are not visible in the displayed picture. A simple AGC circuit that uses voltage developed from the detector (like AVC for a radio) is not satisfactory. Also, a peak detector is just not practical for the modern color TV receivers. The only practical color TV gain control is that developed by a keyed AGC system.

THE KEYED AGC SYSTEM

In the keyed AGC system, the gain control voltage is developed only during the sync pulse time. This avoids the long time constant of the peak detector, and any noise between sync pulses will have no effect on this system at all.

In the keyed AGC system, the control voltage is developed by the AGC amplifier. The control grid of this stage is fed a positive-going composite video signal and a variable DC voltage which depends on the RF signal level. The AGC amplifier plate has no DC voltage applied, but receives a positive-going pulse from the horizontal sweep output stage at 15,750 Hz. The proper timing of these two signals (Fig. 6-1), is very critical.

In operation, the keying pulse and horizontal sync pulse must appear at the amplifier at the same time. Thus, the amplifier only conducts when the horizontal keying pulse appears at the plate of the AGC amplifier. The AGC voltage developed is dependent both on sync pulse amplitude of the composite video signal and the DC bias on the control grid. If there is any noise pulses riding on the video signal between sync pulses, they cannot put the AGC amplifier into operation, because there is no keying pulse on the plate at this time. Thus, the AGC action of the TV set is immune to noise pulses that occur between horizontal sync pulses.

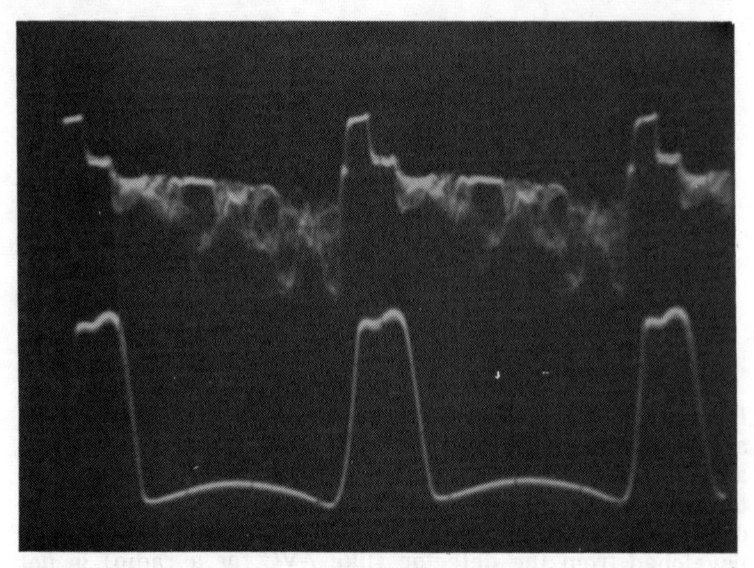

Fig. 6-1. Waveforms of correct keying and blanking pulses and pulse timing. (Scope at approx. 5 kHz.)

An actual keyed AGC circuit (Fig. 6-2) employs the pentode section of a 6GH8A tube (V703A). The control grid (pin 2) is returned through a 2-watt, 6800-ohm resistor (R727) to the +270-volt supply. The cathode (pin 7) is also returned to this same voltage through the AGC control (R136, 50K). Under normal operating conditions the potential on the cathode will be greater than that at the grid, thus the tube will be biased well below cutoff. The signal which appears at the grid is a sync-positive composite video signal (TP3). The bias on the tube is set so only the sync pulses are of sufficient amplitude to drive the tube out of cutoff. The instant the sync pulse appears on the grid, the plate (pin 6, TP4) is driven positive by a pulse obtained from the horizontal output transformer, and the tube conducts. As the tube conducts, a negative charge relative to the amplitude of the sync pulse is developed on the plate side of the pulse coupling capacitor. As the sync pulse passes, cutting the tube off, the negative capacitor charge dissipates through the AGC circuits, providing the negative AGC bias through terminals D to IF and E to RF. Since the AGC voltage is developed during retrace time, the circuit is virtually immune to extraneous noise pulses. The triode section of V703B (not shown) functions as a sync separator.

KEYED AGC CIRCUIT ANALYSIS

The keyed AGC system is more sophisticated and costly, but most of the disadvantages of the other type circuits are eliminated. Because the keying or gating is at the horizontal rate, filtering is simplified and requires a much shorter time constant. This results in a fast acting system which eliminates airplane flutter and reactions due to rapid shifts in the received TV signal level. The keyer tube conducts only during a portion of retrace time (less than ten percent of total horizontal sweep) which establishes effective noise immunity.

Referring to the keyed AGC circuit in Fig. 6-3, positive-going information from the video amplifier is fed to the grid (pin 2) of the keyer tube V6B. DC coupling is maintained between the video detector and the video amplifier and between amplifier and keyer grid so that at keying time the potential appearing at the keyer grid is operated by signal strength only. The plate is supplied with a positive-going pulse from a winding of the horizontal sweep transformer. During retrace (flyback), the pulse is formed and causes V6 to conduct, thus developing a negative charge with respect to ground on C251. The keying pulse must have enough amplitude to make the plate current independent of the plate voltage. Tube gating time is determined by pulse duration so that this pulse must be wide enough to not miss the station sync pulse, yet not

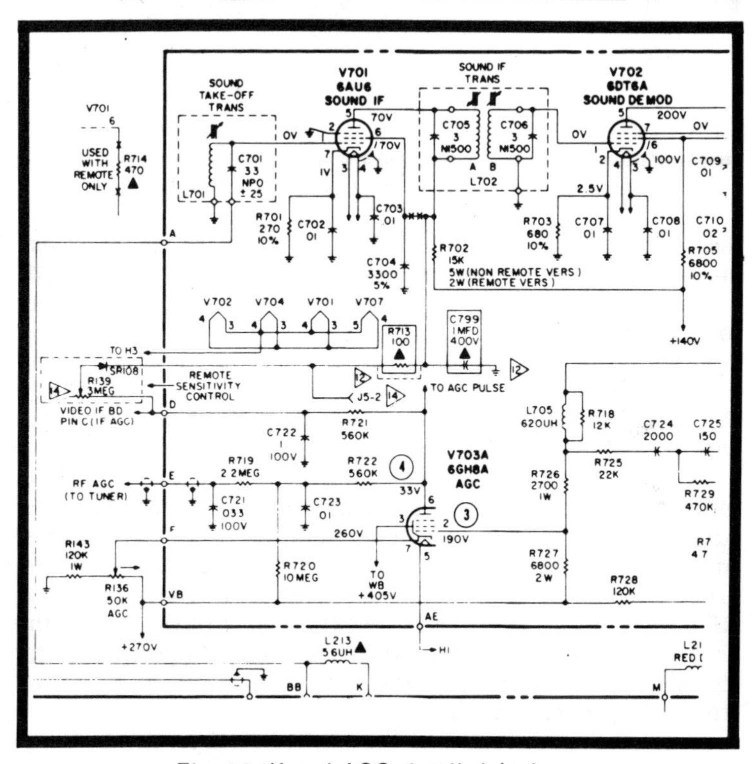

Fig. 6-2. Keyed AGC circuit, tube-type.

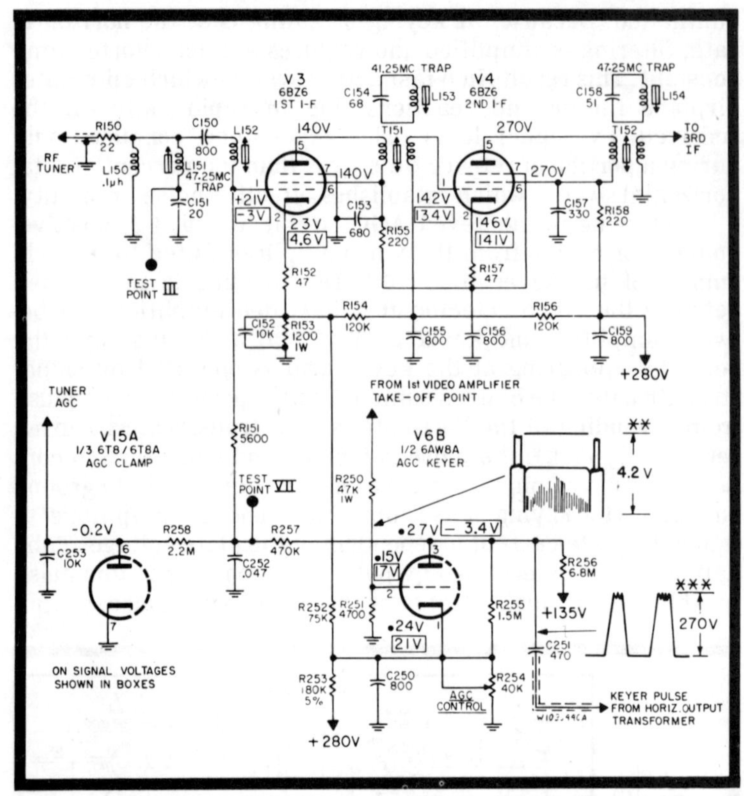

Fig. 6-3. GE "U4" keyed AGC circuit, tube-type.

too wide or conduction time is increased, which will decrease noise immunity. A positive potential is applied to the keyer cathode from the B+ supply through the voltage divider formed by R253 (180K) and the AGC control R254. When the keyer cuts off, some of the charge on C251 leaks off to ground through R255 (1.5 meg). However, since R257 is smaller than R255, a greater portion of the voltage charges the filter capacitor, C252. The AGC bus is supplied from this point.

A wide swing in control voltage is desirable to obtain maximum gain under weak signal conditions and still have sufficient control under medium and strong signal conditions. An improvement in signal to noise will result if the RF section is operated at full gain when the signal input is low. Since the RF gain is at maximum, it is necessary to limit the gain in the IF stages. The opposite is true when operating on stronger signals. Under these conditions, the tuner gain must be controlled to prevent overload.

To increase the swing of the control voltage, the AGC bus is supplied a positive voltage from the +135-volt supply. In

Fig. 6-3 the bus is connected to the voltage divider of R256 and R255. The voltage at this point is about +25 volts with the tube removed. On a weak signal approximately +20 to 25 volts is applied to the AGC bus line. The voltage swing in the negative direction is controlled by the conduction of the keyer tube which in turn is controlled by the amplitude of the sync pulse applied to its grid. The DC grid bias for the keyer tube is established by the divider network R250, R251. Because this network is DC coupled to the plate of the first video amplifier, any sudden changes in conduction of that stage will also cause a bias change on the keyer grid.

Note in the schematic of Fig. 6-3 that these IF amplifiers are stacked. Plate current flow through the first stage (V3) is controlled by the AGC voltage. Since the second stage (V4) is in series with the first, gain of this stage is controlled by the current flow through the first IF amplifier. Additional AGC loop gain is accomplished by returning the cathode of the first IF stage to the keyer cathode, thus a change in IF gain is reflected back to the keyer. Recall that an electronics loop consists of a continuous signal path through two or more circuit elements. Therefore, **each circuit in this path depends on all of the others for proper operation.**

When checking the effect of the AGC voltage on the gain of the IF amplifiers, the voltage developed across R153 in the first IF cathode must be considered. When high conduction is required, the DC potential on the grid of this tube is close to that at the cathode, but when near-cutoff is desired, the voltage applied to the grid must be considerably more negative than the cathode. With the tube near cutoff (strong signal), the voltage across R153 will be a low positive value. As an example, with +2 volts and the grid at ground potential there will be a comparatively high negative value of about -6 volts. When the signal strength is reduced, these two voltages increase in a positive direction **and the difference between the two is reduced.** As the received signal increases, the AGC circuit produces a more negative voltage at the keyer plate which is in turn fed to the AGC controlled sections to regulate the receiver gain.

AGC FAULTS AND DEFECTIVE TUBES

Do not overlook the possibility of faulty tubes anywhere within the loop circuit as they will upset the operations of these systems. When checking the tubes in these receiver circuits, such as delayed and keyed AGC, don't overlook the other tubes included in the loop circuit, as shown in the block diagram of Fig. 6-4.

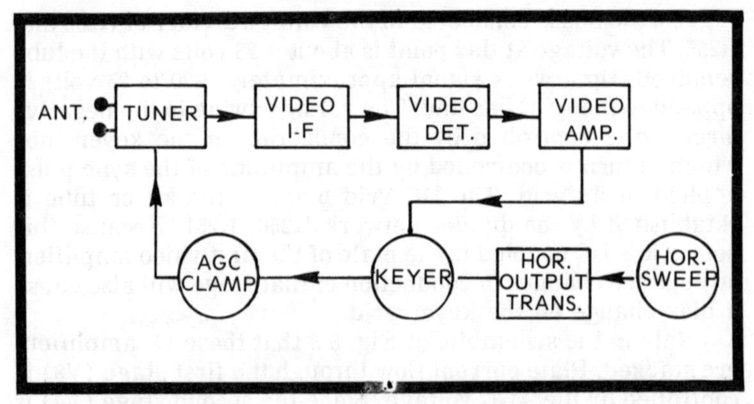

Fig. 6-4. Block diagram of keyed AGC loop. For any tube-type keyed AGC circuit, this drawing can be a helpful troubleshooting aid.

A weak tube, while affecting the AGC voltage, can be recognized by symptoms of a weak or snowy picture or of a "soft" sync lock. A partially shorted tube, or one with some leakage may also cause poor sync lock: however, some picture overload is usually present. At other times a gassy tube will cause the same troubles. In regards to the IF section, this condition may occur in more than one of the tubes. Thus, it is a good practice to replace all of the IF tubes at one time. If this clears the trouble then use the elimination process in order to determine the defective tube(s). If the tube is gassy, excessive grid current will be developed.

CHECKING THE STACKED IF SYSTEM

A stacked IF system is illustrated in Fig. 6-5. These circuits have a different AGC control voltage and bias condition. Voltage is applied to the plate of the second IF tube and the cathode is returned to the first IF plate. The second stage grid voltage is raised to a potential that is high enough in respect to the cathode to properly bias this stage. In effect, the first and second IF stages are in series with the B+ voltage supply. As the control voltage varies the gain of the first stage, the second stage cathode voltage varies; thus the gains of both stages are automatically controlled. It is easy for some confusion to arise while troubleshooting this type circuit or making measurements of the grid and cathode voltages in the second stage where there are very high positive potentials.

The design of an AGC system is predicated, to a great extent, upon tube cutoff characteristics. The proper system operation actually depends on the tube types utilized. When replacements are installed, always make sure that the correct tube types are used.

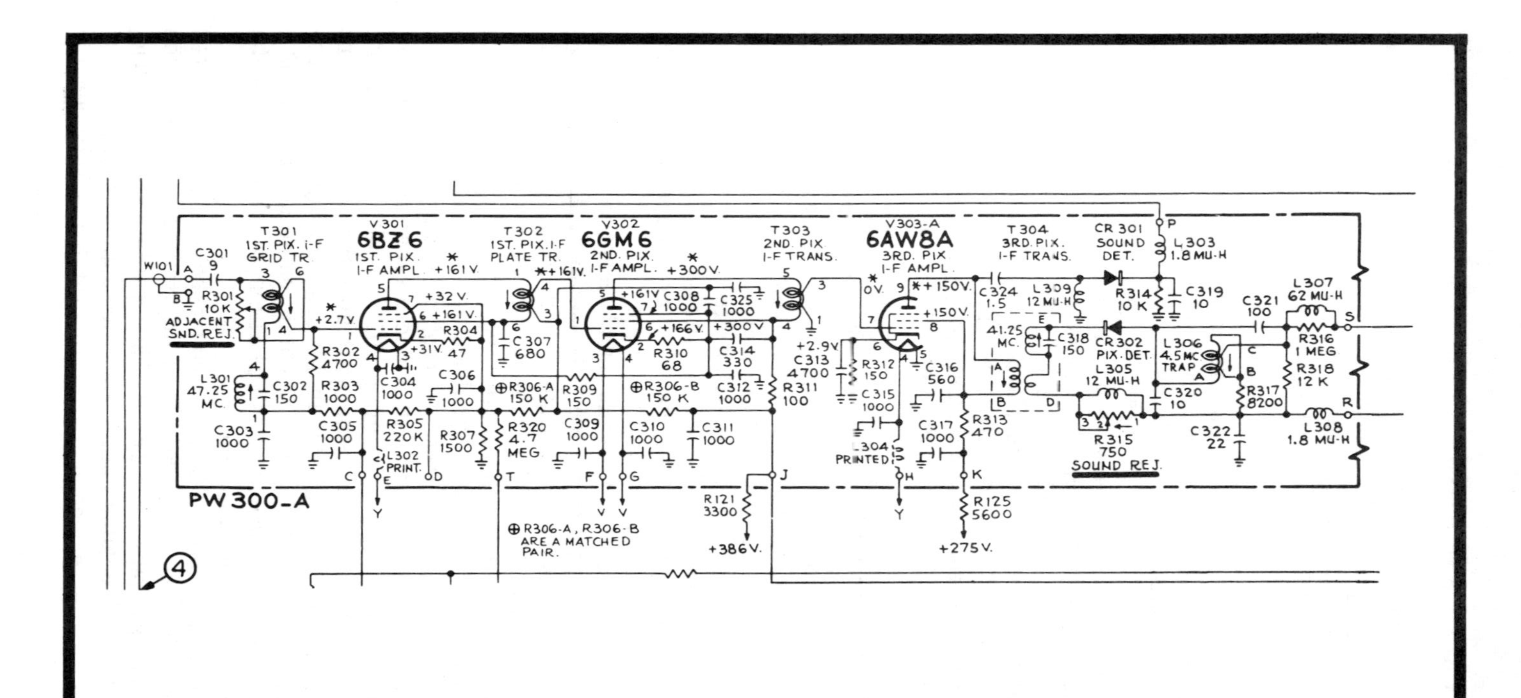

Fig. 6-5. Stacked IF system.

AGC FAULT SYMPTOMS

The following is a list of symptoms that may occur when the AGC circuit malfunctions:

Weak picture but has good sync lock.

Weak picture with bending and some flagwaving.

A very snowy picture.

Picture will look like a negative.

Excessively contrasty picture, with or without poor sync (bending and rolling) and may or may not have some buzz in the sound. This is referred to as an AGC overload condition.

Weak, washed-out picture with good sound.

Blank, white screen with or without sound.

As you can see by the symptoms listed above, AGC faults can be caused by troubles in the antenna system, tuner, IF amplifiers, picture detector, sync circuits, and video amplifiers. The AGC circuits affect the operation of the tuner, IF amplifiers, picture detector, and video amplifiers and at the same time will affect the operation of these same circuits.

QUICK CHECKS OF THE AGC SYSTEM

As with any electronic circuit, logical troubleshooting procedures are needed with the AGC system.

Make a quick visual inspection of the chassis for any obvious troubles. Adjust the AGC control to its correct setting as well as other controls such as fine tuning, horizontal hold and contrast controls. Also, check out or replace all suspected tubes or transistors in the loop stages.

To have proper picture and sound reception, all circuits in the loop system must be "go." To isolate faults in the AGC circuits, a bias-box substitution is needed. As previously stated, the trouble could be in the AGC keyer stage, tuner, IF amplifiers, or the AGC control voltage. The AGC line is clamped or restored to normal voltage with the bias box in order to isolate the defective circuit in the loop chain. With correct bias applied and with video present at the detector, proceed with scope checks in the AGC circuits. If video is not present at the detector, chances are good that there will be no sync pulse at the keyer tube. In this case, try injecting a pulse from a TV analyst in order to obtain some semblance of a picture. If partial operation can be attained, you can scope the circuit to pinpoint the fault.

Let's now look at some circuit checks and actual AGC troubles. Refer to Fig. 6-3. With an oscilloscope, check the waveforms at pins 3 and 2 of AGC keyer tube V6B. At the grid, pin 2 a positive-going composite video signal should be seen. (Refer to the service notes for the set being serviced for correct peak-to-peak voltage amplitude of these pulses.) At the plate, pin 3, a positive keying pulse at a horizontal rate should be observed. Should the waveform or its peak-to-peak amplitude be different from that found on the schematic, look for trouble in that circuit section. Then perform voltage and resistance checks to isolate the defective component.

If coupling capacitor C251 opens, an extreme overload will occur, usually causing both video and audio to block. The normal waveshape (see top trace of Fig. 6-6) found on the keyer plate will be missing or distorted and will appear as shown in the bottom trace. If C251 develops leakage, an overload will occur. However, in this case, the waveshape on the keyer plate may appear normal but the DC voltage will be excessively high because there will be no isolation between the keyer plate and the B+ voltage supply.

An overload condition may be due to a change in the values of the resistors (R250 and R251) which make up the divider in the grid circuit of the keyer. If R250 has increased in value, overload and buzz will be the symptoms.

Overload conditions may be caused by a faulty (open) AGC potentiometer control. Refer to the AGC circuit schematic in Fig. 6-7 for the following checks. The usual test for this control (R66 in this circuit) is to remove all external

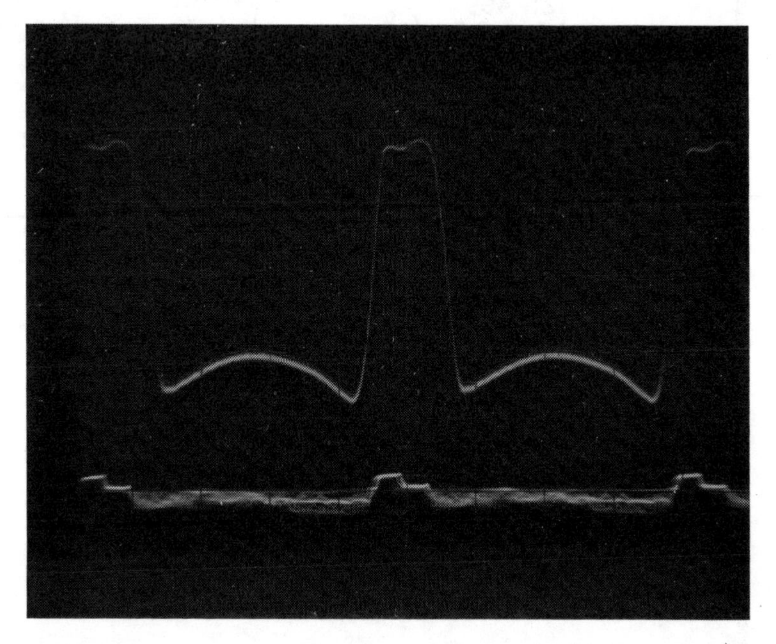

Fig. 6-6. Waveforms, keyer pulses. Top trace, correct keying pulse at plate, pin 3, of V6B in Fig. 6-3. Bottom trace, distorted pulse caused by C251 being open. (Scope at approx. 5 kHz.)

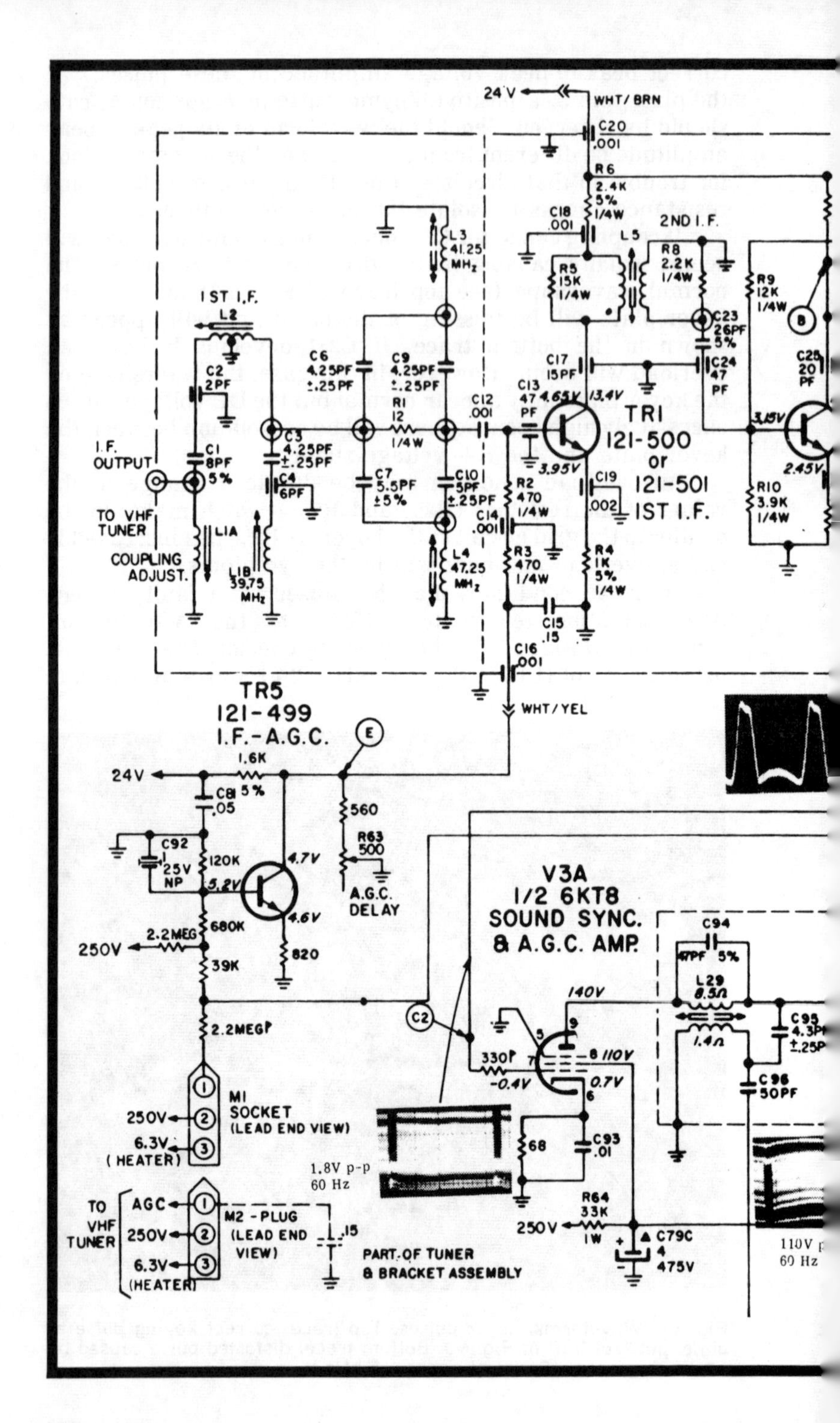
24V
WHT/BRN
C20 .001
R6 2.4K 5% 1/4W
C18 .001
2ND I.F.
L5
R8 2.2K 1/4W
R5 15K 1/4W
R9 12K 1/4W
L3 41.25 MHz
1ST I.F.
L2
C2 2PF
C6 4.25PF ±.25PF
C9 4.25PF ±.25PF
C23 26PF 5%
C17 15PF
C24 47 PF
C25 20 PF
C12 .001
C13 47 PF
4.65V
13.4V
R1 12 1/4W
TR1
121-500
or
121-501
1ST I.F.
3.15V
2.45V
I.F. OUTPUT
C1 8PF 5%
C3 4.25PF ±.25PF
C4 6PF
C7 5.5PF ±5%
C10 5PF ±.25PF
3.95V
C19 .002
R2 470 1/4W
C14 .001
R10 3.9K 1/4W
TO VHF TUNER
COUPLING ADJUST.
L1A
L1B 39.75 MHz
L4 47.25 MHz
R3 470 1/4W
R4 1K 5% 1/4W
C15 .15
C16 .001
WHT/YEL
TR5
121-499
I.F.-A.G.C.
1.6K 5%
560
R63 500
A.G.C. DELAY
C81 .05
C92 1 25V NP
120K
4.7V
5.2V
4.6V
680K
2.2MEG
250V
39K
820
V3A
1/2 6KT8
SOUND SYNC.
& A.G.C. AMP.
C94
47PF 5%
L29
8.5Ω
1.4Ω
C95 4.3PF ±.25PF
C96 50PF
C2
140V
9
5
8 110V
7
330P
-0.4V
0.7V
6
2.2MEGP
M1 SOCKET (LEAD END VIEW)
6.3V (HEATER)
68
C93 .01
1.8V p-p
60 Hz
TO VHF TUNER
AGC
M2 - PLUG (LEAD END VIEW)
(HEATER)
.15
PART OF TUNER & BRACKET ASSEMBLY
R64 33K 1W
C79C 4 475V
110V p
60 Hz

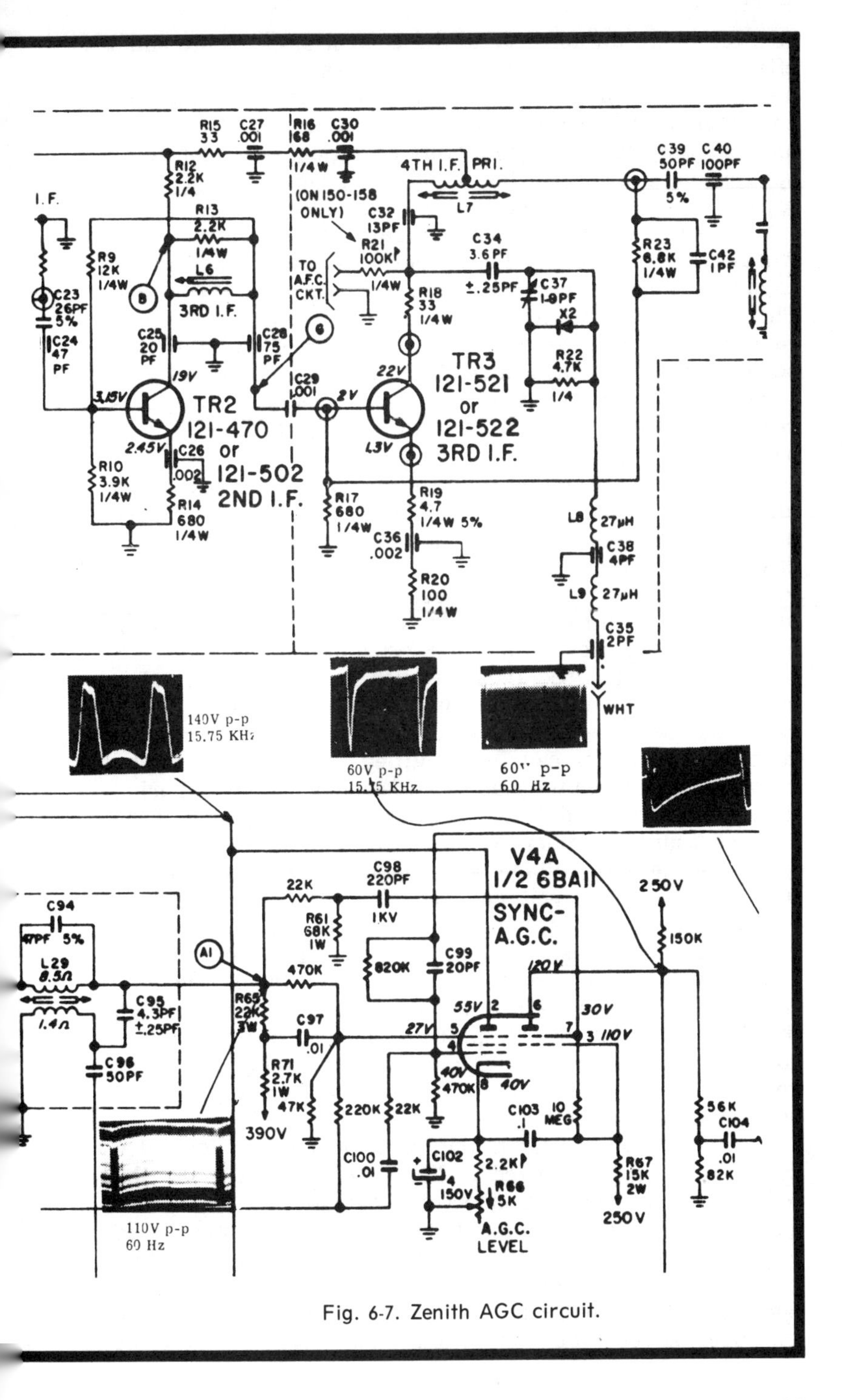

Fig. 6-7. Zenith AGC circuit.

connections to the top side of the control to eliminate shunt effects. Then check the resistance across the end terminals and also between one end and the wiper while rotating the control shaft. Another quick-check is to rotate the AGC control and note any change in the picture condition. If the control is open or has a bad spot, the picture will jump from a weak, washed-out picture to one that is overloaded.

The keyer cathode bypass capacitor C102 (at pin 8 of V4A) will also cause an overload symptom if shorted or leaky. If this capacitor is open, it may result in some type of erratic action. This capacitor is known to be intermittently faulty. The open-condition evidence is that the receiver apparently returns to normal when the AGC control is adjusted to a certain signal strength; when a signal of a different signal strength is received, however, the AGC does not react properly and overload will result.

The video information from the sync amplifier that is fed into the keyer tube grid must be normal in order to produce proper AGC action. If this sync pulse is badly distorted or drawn down into the video information (sync compression), a leaky capacitor should be suspected. This condition (scoped at test point A1 in Fig. 6-7) is illustrated by the scope pattern in Fig. 6-8. Capacitor C97 in Fig. 6-7 should be the number one suspect and be replaced. It can develop slight leakage which is difficult to determine with the normal test gear. Other capacitors to suspect are C100, C99, and C103. Faults in most coupling capacitors can be identified by a scope test, by viewing the pulse on the load side of each one and comparing the results with the display present on the supply side of the suspected unit. (Another good use for the dual-trace scope.)

When the AGC voltage is too high, causing a snowy picture, always check voltages and waveforms in the keyer circuit. Give special attention to the keyer grid voltage and waveform. The resistors in this portion of the circuit are somewhat critical and should be measured if the grid voltage is incorrect. If the waveform is not right, check back through the video amplifier, as the problem may be due to a faulty component that is influencing the sync information fed to the keyer grid. Remember that this information controls the level of conduction for the keyer.

To check the precise timing or coincidence of the video sync and keying pulse fed to the keyer circuit, make the following scope check. Connect one channel of a dual-trace scope to the plate (pin 2) and the other channel to pin 5, grid circuit, for an accurate comparison. These two pulses must be perfectly timed in order to obtain proper overall AGC operation. If one pulse is lagging the other, this will cause improper keyer tube conduction. (The correct dual-trace scope patterns that should be found at this tube are shown in Fig. 6-1.) A faulty capacitor that couples the keying pulse from the sweep transformer winding can cause an out-of-time condition.

Making scope checks on the tuner and IF amplifier AGC line may prove to be a very valuable time saver. You should see a clean, smooth line on the scope like that shown in the top trace of Fig. 6-9. However, should AGC line filters become open or change value, you may see something like the bottom trace, which looks like hash on the AGC line, which will cause all sorts of picture and sound malfunctions. The picture reception in Fig. 6-10 was caused by an open AGC line filter capacitor. These capacitors will also short out, in which case there would be no AGC voltage developed.

SONY SOLID-STATE AGC SYSTEM

This AGC system develops a control voltage that is proportional to the height of the horizontal sync at the video detector. The control voltage is applied to the first and second IF amplifiers while a delayed AGC is applied to the RF amplifier in the tuner. The AGC gate is active only during the sync tip interval so that the control voltage is unaffected by signal brightness level or noise. Also, the AGC filter can be made fast enough to reduce response to aircraft flutter.

Refer to the simplified schematic in Fig. 6-11. Feed for the AGC system is taken from the emitter circuit of 2nd Y amp transistor Q212 and fed through R234 to the base of the AGC gate, Q204, which is turned on by a sweep keying pulse. During this brief period of conduction, the amount of collector current depends upon the amplitude of the horizontal-sync component of the composite signal. A voltage divider in the emitter circuit

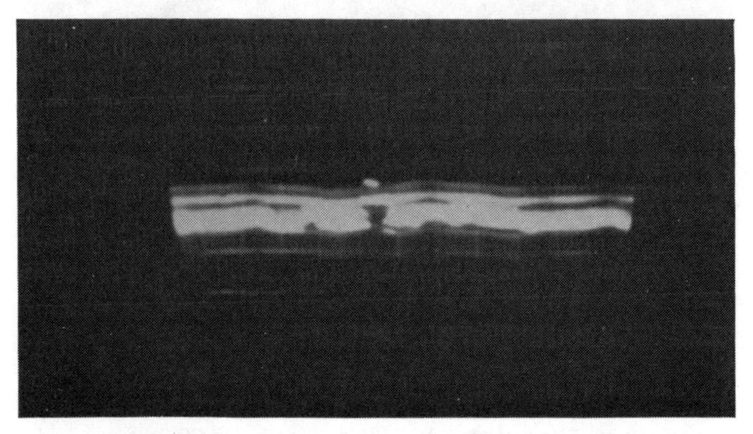

Fig. 6-8. Distorted, compressed sync pulse. (Scope at vertical rate.)

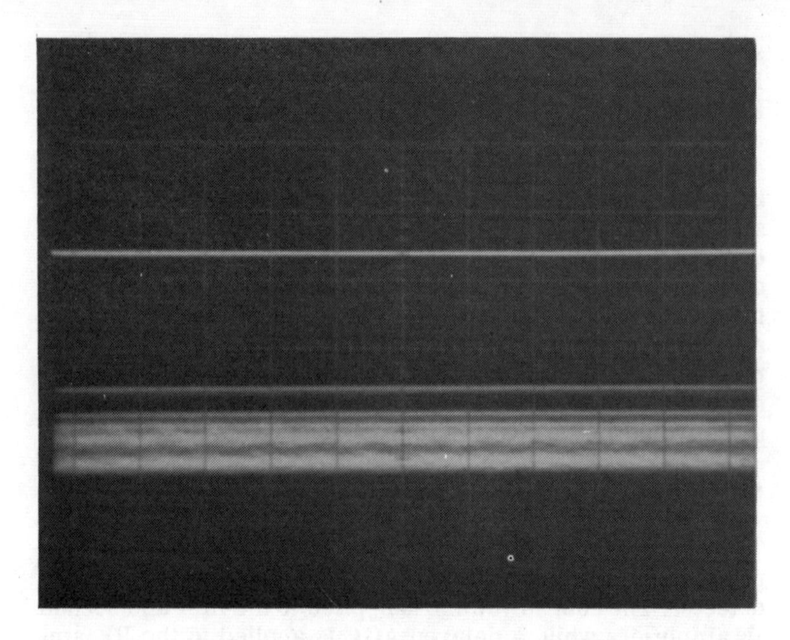

Fig. 6-9. Dual-trace scope patterns during checks on AGC line. Correct waveform top trace. Bottom trace, open AGC filter capacitor.

Fig. 6-10. Effect of open AGC filter capacitor on CRT display.

of Q204 also controls the voltage required to bring Q204 into conduction. This threshold voltage is adjusted by means of VR204 to determine the AGC delay (the minimum input signal at which AGC becomes effective).

During the intervals that Q204 conducts, current flows (as shown by dashed lines) to charge C264, the AGC input filter. Capacitor C264 discharges during the off periods to form a smooth DC voltage that is proportional to signal level. This voltage is further filtered by R238 and C239 before it is applied to the base of second IF amplifier Q202.

Control voltage is distributed from Q202 to Q201 of the IF amplifier stages. Finally, the RF AGC control voltage is developed in the collector circuit of Q201, amplified by Q205 and Q206, and applied for tuner control.

A few troubleshooting hints for this solid-state AGC system are as follows:

Weak Picture Conditions

1. Noisy sound with some buzz. Caused by open D202 diode at AGC gate located on S board.

2. Picture is smeared. Transistor Q210, first video, located on S board, is shorted or has high leakage between base and emitter.

3. Snowy picture and check ascertains weak sync. Probable cause: Q205, AGC amplifier, is shorted.

4. Snowy picture or no picture. Probable causes: Transistor Q201, first IF amplifier or L204 is open.

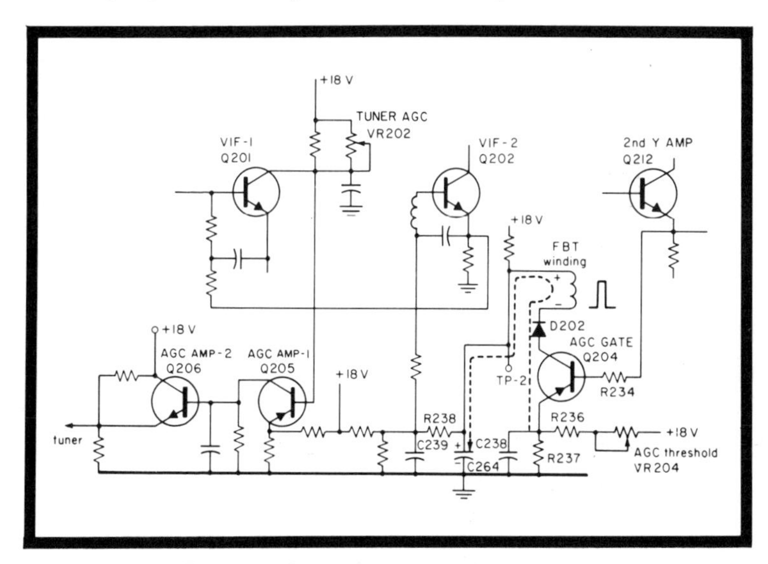

Fig. 6-11. Sony solid-state AGC circuit.

Following Faults are for Picture Overload. (AGC Inoperative)

1. Q204, AGC gate, located on S board has become faulty (open).

2. Transistor Q205, AGC amplifier, has become defective.

3. Diode D202, at AGC gate on the S board, shorts out.

4. Resistor R224, 330 ohms, located on the S board, opens up.

If picture is overloaded and has loud buzz in the sound:

1. Capacitor C265, 33 uf in the Y amplifier circuit, is shorted. Set develops loud buzz in the audio.

2. Resistor R227, 150 ohms, located in the Y amplifier circuit, opens up. Symptoms are a negative picture with loud buzz in the sound.

Chapter 7

Video IF Amplifier Circuits

The transmitted TV signal leaving the television station has an AM RF picture and a FM sound modulated carrier. This composite signal is picked up by the antenna and fed into the tuner. The tuner converts the channel frequencies into a lower frequency IF signal, that is, the "intermediate frequency." This lower IF signal contains all of the modulated RF information, which is amplified and then demodulated by the video detector. The output of the video detector is now a composite video signal which is amplifed by the video amplifiers and is then fed into the CRT control grid (or cathode) circuits.

The video IF amplifiers are designed to pass a broad range of frequencies which will include both picture and sound IF carriers and their associated sidebands. This video IF bandwidth is tuned to approximately 40 to 48 MHz. Shown in Fig. 7-1 is a typical IF amplifier frequency response curve. The picture IF carrier is located at the midpoint of the upper-frequency skirt response at the 45.75 MHz point. The sound carrier is located at the 41.25 MHz point on the lower frequency side of the curve. FCC standards state that the picture and sound carriers must be 4.5 MHz apart.

Many of the modern color receivers have an IF curve that has a smooth appearance which is sometimes called a "haystack" response curve. The center frequency of this curve is at around 44 MHz. The center of the higher-frequency slope is set at 45.75 MHz for the video IF carrier. Traps are usually included to attenuate the sound IF frequency of 41.25 MHz. The adjacent channel sound is also held down to a minimum. Many receivers use traps tuned to 47.25 MHz to reduce this signal. Other TV receivers also have traps at the adjacent channel picture IF carrier at 39.75 MHz. These traps are used for two purposes. The main use is to reduce or eliminate the undesired adjacent channel signal. These traps are also used to shape the skirts of the response curve to produce the desired response curve characteristic. When a complete IF alignment is to be performed, these traps must be adjusted first.

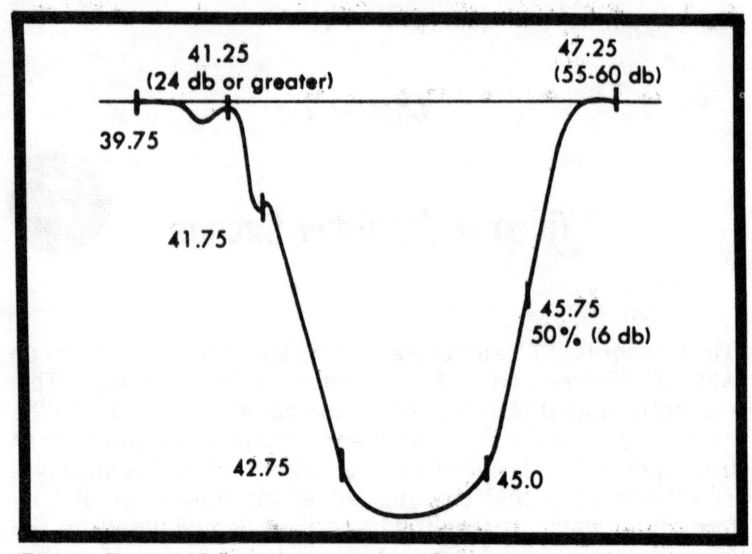

Fig. 7-1. Typical IF amplifier frequency response curve. (All frequencies in MHz.)

TUBE-TYPE IF CIRCUITS

You will usually find two types of IF transformers used in TV receivers which utilize transformer-coupled circuits. Some use the conventionally wound transformer while others use transformers with bifilar wound windings. The larger percentage of the receivers use bifilar windings.

The primary and secondary of these transformers are wound in a process where the two insulated wires are wound together. An IF circuit using bifilar-wound coils is shown in Fig. 7-2. A variable core used for tuning the frequency of this type transformer replaces the single-tuned coils. The IF amplifiers which utilize bifilar coils are also stagger-tuned. Coils T301, T302, and T303 are peaked up at 43.0, 45.5, and 44.5 MHz, respectively. Conventional transformers are used in some receivers that only have two IF stages because the overcoupled transformer has a greater gain than the bifilar single-tuned circuit.

High-gain pentode tubes like the 6GM6 (shown in Fig. 7-2) are used for IF amplifiers. Voltage decoupling resistors are used in the screen grids and plates. The screen grids are then bypassed for the IF signal to be amplified by these tubes. The plate circuits of the IF amplifiers are also decoupled from the B+ supply by suitable capacitors and resistors. Because these amplifiers operate at high frequencies, capacitor lead length can become critical for proper bypassing. Component and lead dress may also become critical. Ground return for each

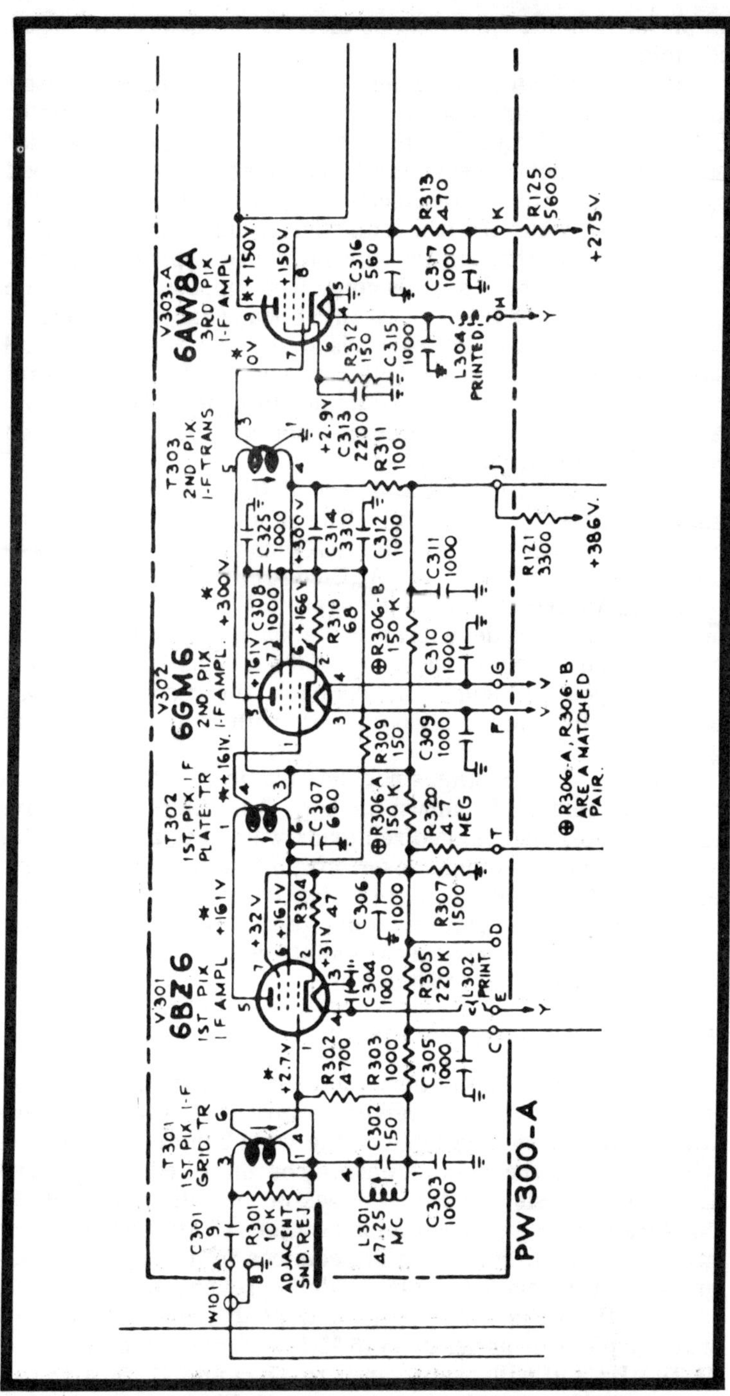

Fig. 7-2. IF circuit showing bifilar wound coupling transformers.

amplifier is made short and as close to the active circuit as possible.

In many designs the screen "bypass" capacitance is used for the neutralizing capacitor. For an example, the value of C312 in Fig. 7-2 (from V302 pin 7 to ground) is selected so that the signal (which is very small) on the screen grid will feed back through the interelectrode screen- to control-grid capacitance, while the same amount of signal to the control grid is fed back through the interelectrode capacitance from the plate. Since these signals come from opposite ends of the plate coil, they are 180 degrees out of phase and will cancel. A point of service to remember is that the value, type, and physical shape of the capacitor is very critical. An exact replacement must be used so as not to upset neutralization. Always check the IF alignment after any parts have been replaced in these circuits.

The grids of the first and second IF amplifiers are normally returned to an AGC bus. The IF gain is then controlled by a DC bias derived from the AGC control circuits. Usually the cathode resistors of AGC-controlled tubes are not bypassed in order to obtain negative feedback and induce degeneration. This degenerative feedback is used to minimize changes in the input capacitance of the tube as the gain of the tube changes. Without such feedback, the response characteristic of the amplifier would vary with changes in circuit gain.

Many of the IF stages now use frame-grid tubes such as the 6EH7 and 6EJ7 shown in Fig. 7-3. These tubes have specially designed ruggedized grids welded to a frame. This structure permits much closer spacing between control grid and cathode, resulting in very high transconductance and a greatly increased amplification gain. One of the characteristics of frame-grid tubes is that the gain at zero bias is actually less than at - 1 volt of bias.

TRANSISTOR IF AMPLIFIERS

The picture IF amplifiers (note block diagram in Fig. 7-4) in a transistorized TV receiver require that sufficient bandwidth and gain be available so the wide band of frequencies which make up the video signal may be amplified. This picture IF system consists of three stages of amplification that use NPN transistors in the common emitter circuit. Consideration must be given to matching impedances from stage to stage, and some provisions must be made to prevent self oscillation.

The IF in Fig. 7-5 uses 44 MHz at the center frequency, with the picture carrier at 45.75 MHz and the sound carrier at 41.25 MHz and with traps tuned to 47.25 and 41.25 MHz (adjacent sound and accompanying sound). The collector circuits

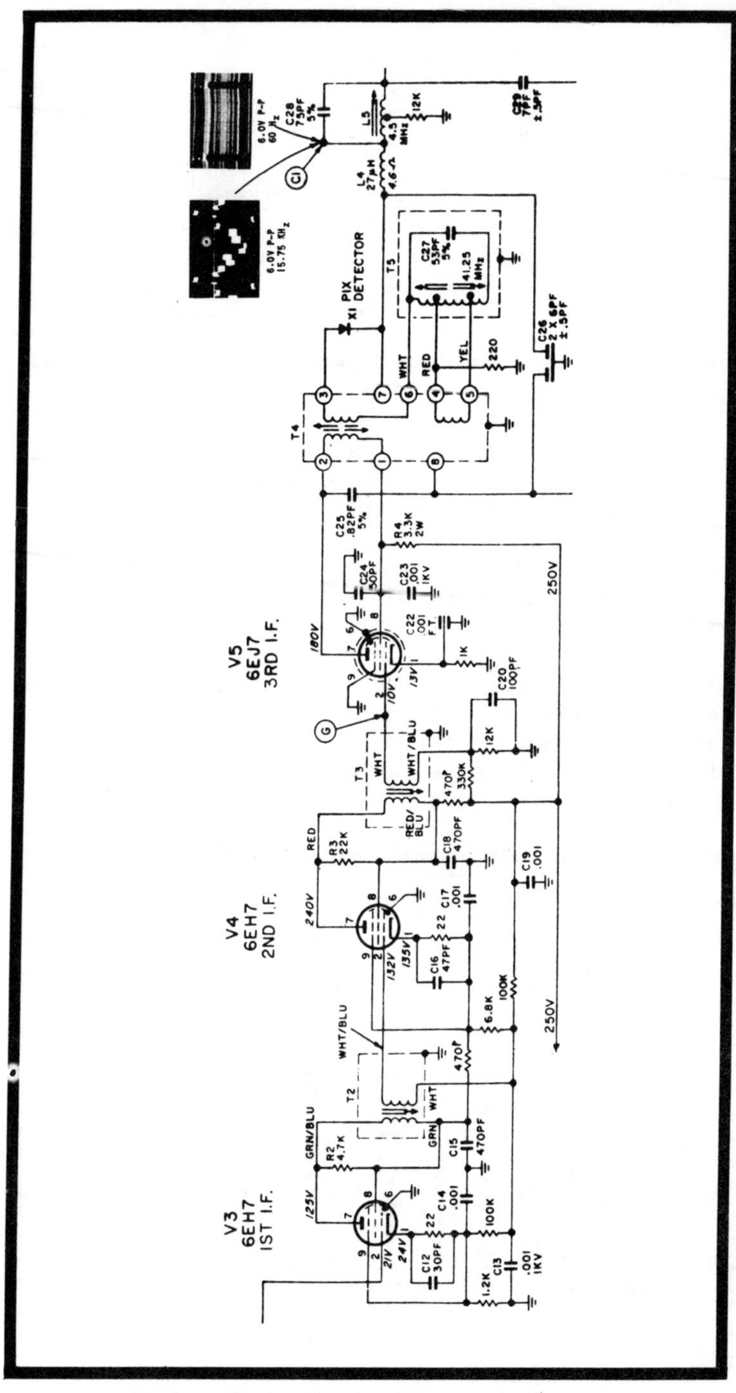

Fig. 7-3. IF circuit using frame-grid tubes.

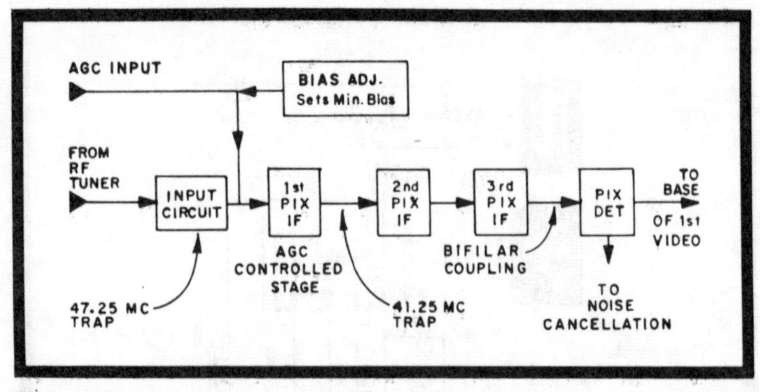

Fig. 7-4. Block diagram, picture IF amplifiers for transistorized TV set.

are tuned and impedance-matched to the input of each succeeding stage. At the second detector, approximately 1.5 volts of video is developed for the video amplifiers.

A special input circuit is used which provides for coupling and matching of the tuner output to the first IF amplifier. Adjacent sound (47.25 MHz) is attenuated in this circuit by means of a "T" notch filter. At the collector of the first IF, a tuned circuit is used to couple to the base of the second IF transistor. This circuit is used to couple to the base of the second IF transistor. This circuit is parallel resonant and matches the base of the next stage by resistive "tap" which is referred to as an "Rx transformer." A 10 to 1 step down is achieved in this transformer.

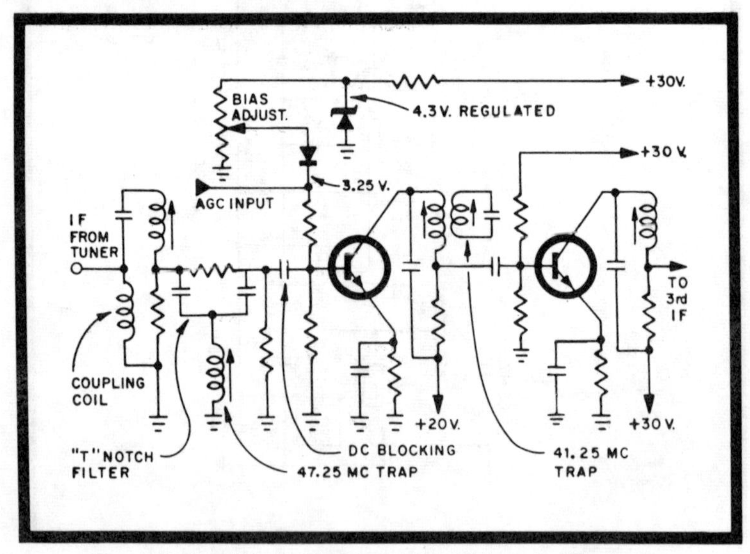

Fig. 7-5. Simplified schematic, transistor pix IF stages.

Coupling circuits between the second and third IF are similar to that found for the first and second IF stages. The output of the third IF is tuned by a transformer with separate primary and secondary windings giving a 2 to 1 step down to match the second detector to the third picture IF collector. The amplifier design also provides for self neutralization.

The second detector circuit, shown in Fig. 7-6 is "elevated" above ground to satisfy the base bias condition of the first video amplifier. An additional feed from the +30 volt supply insures that the second detector is not forward biased by this elevating voltage.

The AGC control voltage (Fig. 7-5) is applied to the first IF amplifier by way of the RF amplifier operating as a DC amplifier. This bias is reverse bias, which makes use of the cutoff characteristic of the first IF amplifier. About 35 db of attenuation is possible by means of the AGC on the first IF and an additional 35 db is introduced on strong signals in the RF amplifier. A bias adjustment on the first IF is set for 3.40 volts with no signal being received.

SOLID-STATE IF MODULE OPERATION

Now let's look at the operation of the Zenith solid-state IF module that is shown in Fig. 7-7.

This IF amplifier input incorporates a series tuned circuit that has good selectivity, input response shaping, and is easy to align. The input circuit consists of C111, C101, and L101A (the first IF coil assembly), and C102 and L101B (the bandwidth adjustment) which compensates for tuner and coax lead variations. L102B, C103, C104 and C105 comprise the 39.75 MHz

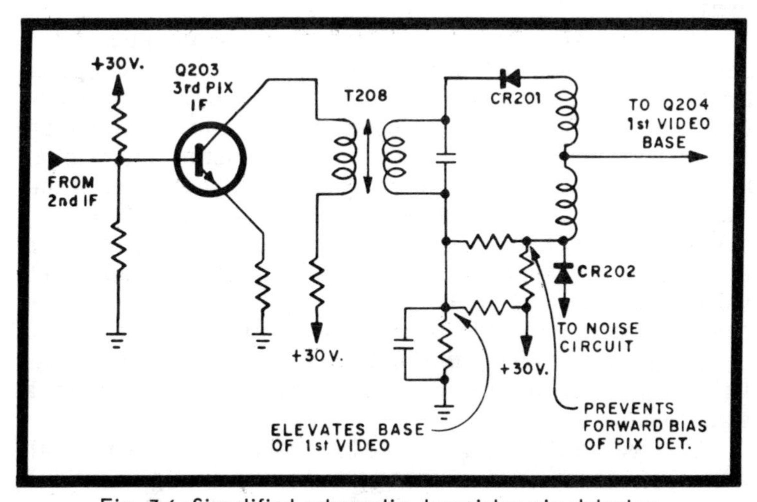

Fig. 7-6. Simplified schematic, transistor pix detector.

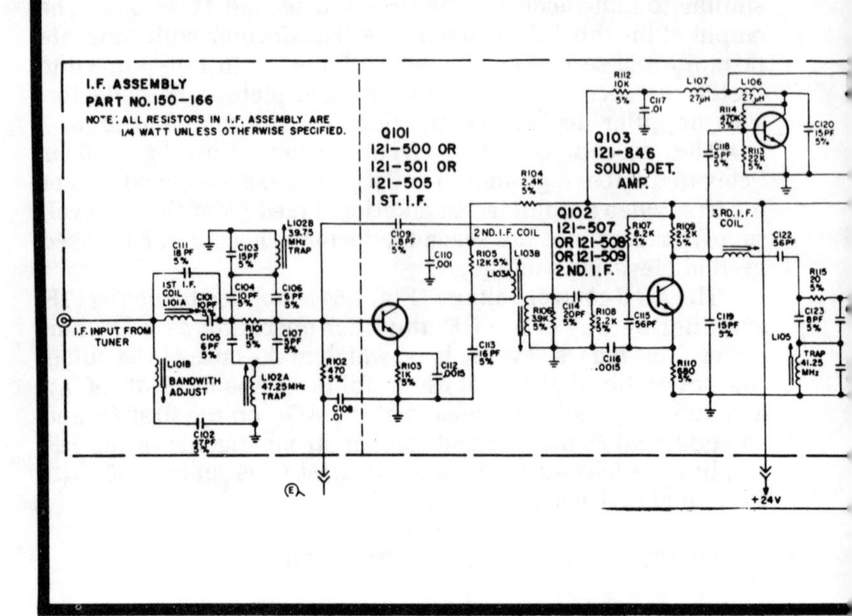

adjacent channel video trap. Coil L102A and capacitors C105 and C107 form the 47.25 MHz adjacent channel sound trap.

It isn't evident from the schematic, but two cores in L102A are provided to assure maximum attenuation in this circuit. During the IF alignment procedure one core (lower Q) is adjusted for proper frequency, and the other core (higher Q) is adjusted for maximum attenuation at the trap frequency of 47.25 MHz.

The first IF transistor (Q101) is a common-emmitter amplifier operated in the forward-bias-AGC mode. Coupling from Q101 to Q102 is by a double-tuned interstage transformer (L103) in the Q101 collector circuit. This circuit has a unique design to provide a "boost" of the sound carrier at weak signal levels. Under conditions of minimum gain the bandpass of the stage appears as shown in Fig. 7-8A. This is the "normal" bandpass to which the IF system is aligned. When the receiver is tuned to a weak station, the AGC voltage drops to about +5 volts, causing Q101 to be biased to maximum gain. The collector impedance that is presented to the IF coil causes a detuning effect, raising the sound shelf (at the arrow, lower left of Fig. 7-8B) to enhance sound performance under weak signal conditions.

The second IF transistor (Q102) is coupled to Q104 through a low-impedance network that permits ease of alignment and a convenient point for the 41.25 MHz sound trap. The double-

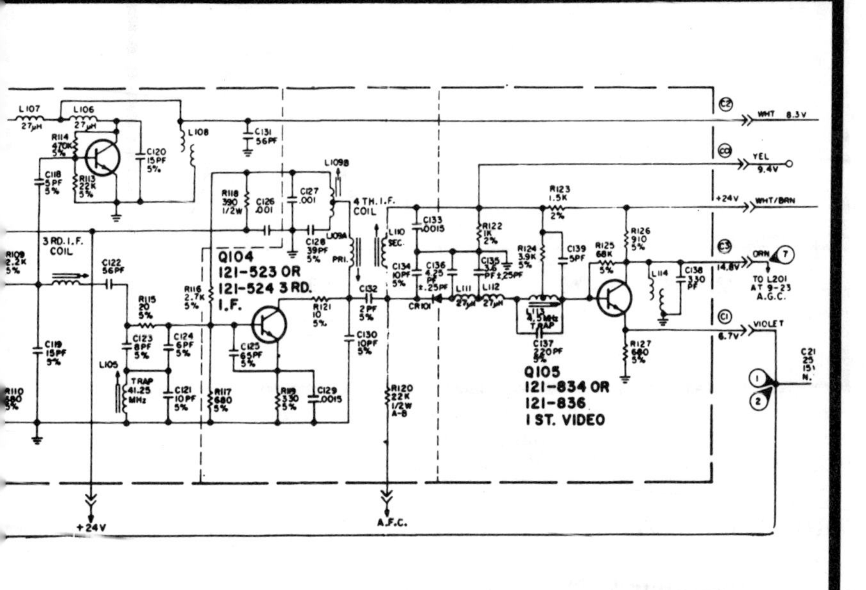

Fig 7-7. Zenith solid-state IF module.

tuned coil between Q104 and the video detector has an additional trap (L109B) to reduce the response between 39.75 MHz and 41.25 MHz to eliminate 920 kHz beat.

Further bandpass shaping is accomplished by L109A and L110, the fourth IF transformer primary and secondary. Video is detected by diode CR101. Coils L111 and L112 together with C134, C135, and C136 form a filter network to prevent harmonics of the IF carrying a "tweet" type interference from being coupled to the first video stage. The detected video is coupled to the base of Q105 through L113, the 4.5 MHz trap. This trap removes 4.5 MHz sound information from the video signal. Video is available from the module at test point C3 and C1. The two signals are of opposite polarity and have a fixed 2 to 1 amplitude ratio. Capacitor C138 in conjunction with L116 functions to keep certain harmonics within the IF module, thereby preventing channel 6 and channel 8 interference.

The second IF (Q102) collector is also capacity-coupled (C118) to the base of sound detector transistor Q103. This sound detector-amplifier stage is biased as a mixer, producing the 4.5 MHz sound signal with sufficient gain to drive the sound system module (not shown). A trap in the collector output circuit is provided to eliminate channel 6 and 8 interference while the low impedance network formed by L107, L106, and C131 is used to couple the 4.5 MHz signal to test point C2 and thence to T1102 the 4.5 MHz sound input coil (not shown).

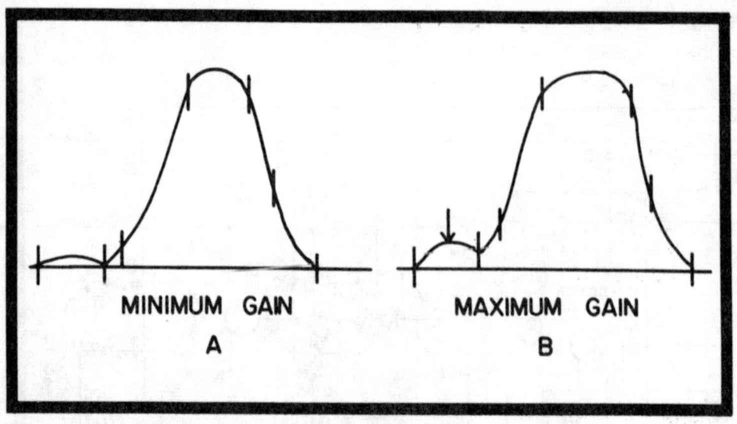

Fig. 7-8. In A, the normal IF bandpass. In B, how the AGC drop under weak signal conditions causes not only maximum gain but also detuning of the response to enhance sound. (See text and Fig. 7-7 for circuitry.)

TROUBLESHOOTING THE SOLID-STATE IF MODULE

This solid-state IF module (refer back to Fig. 7-7) contains not only the usual three IF stages, but also a sound detector transistor and a first video amplifier stage. Thus, it should be noted that complaints of poor or no sound or weak or missing video can be caused by defects within the IF module. Once the fault is isolated to the IF module, use the following service tips. This is a simplified procedure in order to cover most of the expected module failures.

1. Remove IF module from main chassis. With clip leads, connect the module case to chassis ground and connect +24V from the receiver to the +24 volt terminal on the IF module (no tuner input is required).

2. Using a VTVM, measure the emitter voltage of the third IF transistor, Q104. If the voltage is not close to that specified by the manufacturer, replace the third IF transistor.

NOTE: The third IF transistor Q104, will usually have the highest failure rate in these modules. The reason for this is that Q104 is driven very hard, as most of the gain in this video IF system is obtained from the third video stage. The first and second stages are used for shaping the overall bandpass.

3. Measure the emitter voltage of the second IF transistor. If the voltage is not near normal, replace the transistor.

4. Now check the emitter voltage of the first video transistor. If this is incorrect, replace the transistor.

5. Failure of the sound detector transistor is usually accompanied by symptoms of no sound, poor sound (garbled) or poor sound quality when the set is fine tuned properly. A check of the collector voltage of this transistor will usually, but not

always, indicate proper operation. Thus, sound troubles will not always be found in the sound module.

6. Use an ohmmeter to check the resistance ratio of the video detector diode CR101. The diode should have a high front to back ratio.

7. Unsolder one end of L108 and of L114. Check the resistance of these coils. On the high ohmmeter range there should be infinite resistance from one end to the other. That's right, **infinity** ——DC-wise, these are open circuits! Take a closer look at Fig. 7-7.

8. If any defective components have been replaced, always check the IF module after replacement in the TV chassis for correct alignment.

NOTE: Should you have a loss of picture information, don't forget to check out the first video stage Q105 which is also located in the IF module. The emitter bias can be checked at C1 and the collector voltage is checked at C3 plug (orange lead) of the module. No need to go into the IF unit to make these checks.

"SCOPING" THE IF AMPLIFIERS

When a color receiver is operating properly the composite video signal waveform (see Fig. 7-9) will appear at the output of the video detector and can be viewed with the oscilloscope,

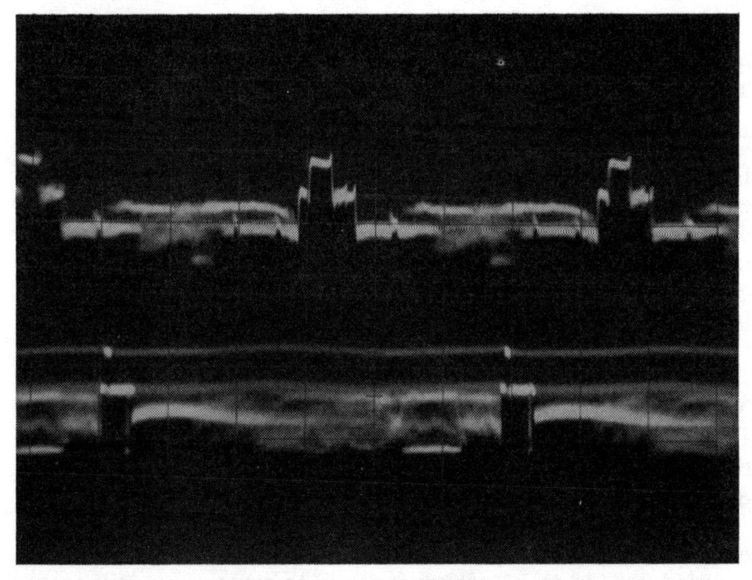

Fig. 7-9. Composite video signal at video detector. Both traces are normal. The top trace was taken at the horizontal rate, bottom at vertical rate.

using a direct probe. The peak-to-peak amplitude of this signal for average signal strength is usually specified in the service information. This amplitude usually measures from 2 to 6 volts peak-to-peak.

The IF signal can be traced through each of the IF amplifier stages with an oscilloscope and a demodulation probe. If your demodulator probe is not compensated for wideband video, the detected scope waveform may appear to be distorted. The waveform will be either positive- or negative-going, depending on the detector probe diode polarity.

The IF signal can now be followed from input to output (control grid to plate for tube stages, base to collector for transistor stages) of each IF amplifier, up to and including the video detector. The polarity of the waveform will remain unchanged as you progress through each stage. In most cases the peak-to-peak amplitude of the scope trace will increase as the test point is moved from the first IF amplifier toward the detector. This will not always be so, because of loading, stagger-tuning, AGC voltage, and other individual circuit design characteristics.

The signal at the first IF amplifier input may be so weak that the scope wavetrace signal may not be observed. In this case it may help to adjust the AGC gain control to give greater IF amplification.

IF AMPLIFIERS ON A "CHIP"

Some of the late model RCA color sets make use of integrated circuit "chips" in the IF amplifier systems. Now for a quick look at how they operate. See Fig. 7-10.

The IF signal from the tuner is coupled into the "chip" via L4 from capacitor C12. L4 is used to tune the input of the first IF amplifier located inside of IC1A. The adjustment of these components is similar to the alignment of the link circuit of the older receivers. The input circuit is tuned to the center frequency of the IF passband (about 44 MHz); L4 principally controls the tilt of the response curve, and C12 establishes the bandwidth.

The schematic in Fig. 7-11 shows the remaining section of IC1 and those external circuit components which connect to it. Notice that L5, C19, and C20 are shown in both of the preceding IF circuit schematics. Signal from the second amplifier appears at terminal 9 of the IC and is coupled to the third IF amplifier input, terminal 13 (Fig. 7-10).

The collector of the second IF is tuned to about the center of the IF passband (44 MHz) by L5 and C19. The signal is coupled via C20 to L6, which is tuned to remove any tilt from the response curve. The 4.25 MHz sound-carrier energy is removed from the IF signal before it reenters the IC at ter-

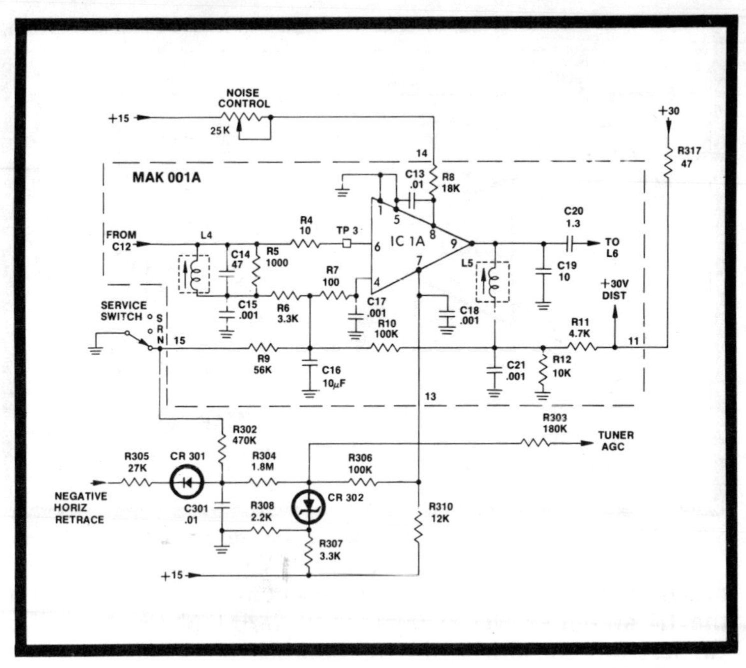

Fig. 7-10. RCA IF amplifiers on an integrated circuit "chip."

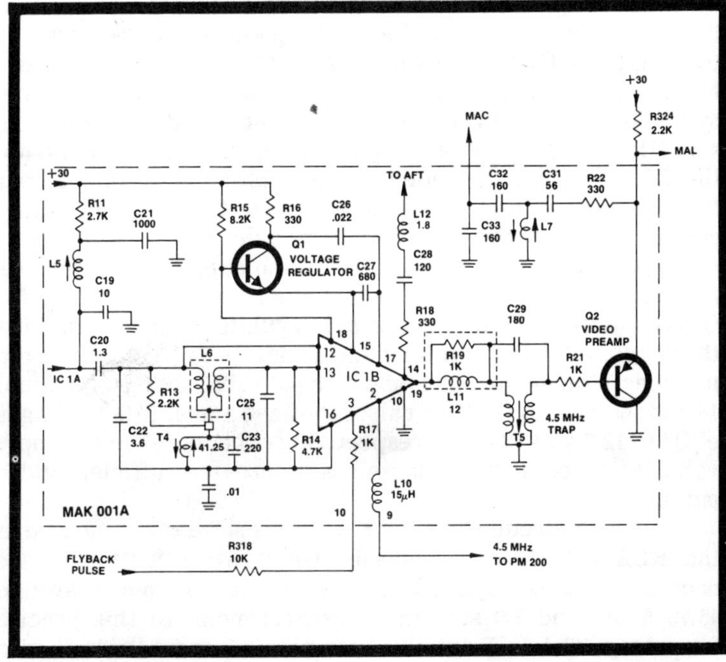

Fig. 7-11. RCA IF output and video preamplifier on an IC "chip".

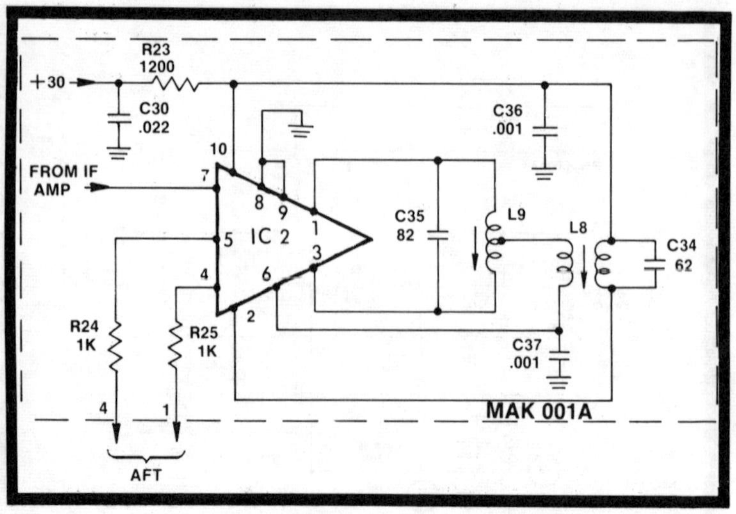

Fig. 7-12. AFT IC on MAK 001A module (RCA).

minal 13, but this energy is still present at the take-off point to the IF amplifier whose input is at terminal 12. As stated earlier, the circuits in the IC between terminals 12 and 2 amplify the IF, produce the 4.5 MHz intercarrier sound signal, and amplify it for injection into the sound module.

The IF signal fed into terminal 13 is amplified, detected, then amplified again, and the resulting video signal appears at terminal 19 (IC1B). The video signal at this point has a peak-to-peak amplitude of about 6.3 volts with a negative-going sync. The sync tip level is about +0.7 volt. Passing through the 4.5 MHz trap and the preamplifier, Q2, the video is fed from the IF module to the video module, MAL. A second output from the preamplifier is developed across the chroma peaking coil L7, and fed to the first chroma module, MAC.

Transistor Q1 is a regulator which supplies voltage to a number of devices in the integrated circuit. Base voltage for Q1 is established by a 12 volt zener regulator in the IC, making the regulated voltage at terminal 15 nearly 11.3 volts. In some modules, a diode may be found connected between IC terminal 18 and the base of Q1. This raises the base and emitter voltages of Q1 to 12.7 and 12 volts, respectively. The increase in supply voltage is necessary for some ICs, to provide sufficient video output.

The AFT circuit shown in Fig. 7-12 is similar to one used in the RCA CTC-42 color chassis. Only two adjustments are necessary, L8 is adjusted for symmetrical response around 45.75 MHz, and L9 sets the crossover point to this precise frequency. The AFT output at terminals 1 and 4 "rides" on a DC level of about 6.7 volts.

Color CRT Diagnosis, Beam Control, and the Trinitron

8

Let's start out by making some comparisons of the Sony Trinitron(TM) with the conventional 3-gun shadowmask color picture tube. The neck of the trinitron is smaller and the electron guns are made differently. However, both tubes still use the same red, green, and blue phosphors, but they are applied to the faceplate differently. The conventional CRT has a shadowmask in back of the screen, while the Trinitron has an aperture grille. Convergence components and adjustment procedures are completely different. Video drive circuits differ because the tube requirements are not the same. In the shadowmask CRT, three electron guns supply three streams of electrons to light up three phospor dots. The trinitron has only one gun, but this single gun emits three streams of electrons.

APERTURE GRILLE

The Trinitron CRT has its primary color phosphors arranged in parallel, vertical stripes (Fig. 8-1). An aperture grille placed behind the phosphors has one slot for each group of three primary-color stripes. Note that the diameter of the single beam spans about two slots, and the electrons strike the centers of two phosphor stripes. The grille intercepts a smaller part of the total beam current than a shadow mask, so most of the electrons strike the phosphor.

The one-beam action for only one of the primary-color phosphors is illustrated in Fig. 8-2. This simplified drawing actually shows the center beam striking green phosphors at approximately screen center. Now let's look at the other two colors.

The beams intended for red and blue phosphor excitation approach the aperture grille from either side of the center (green) beam. Fig. 8-3A shows the electron path for the red beam. This beam emerges from the gun assembly to the right

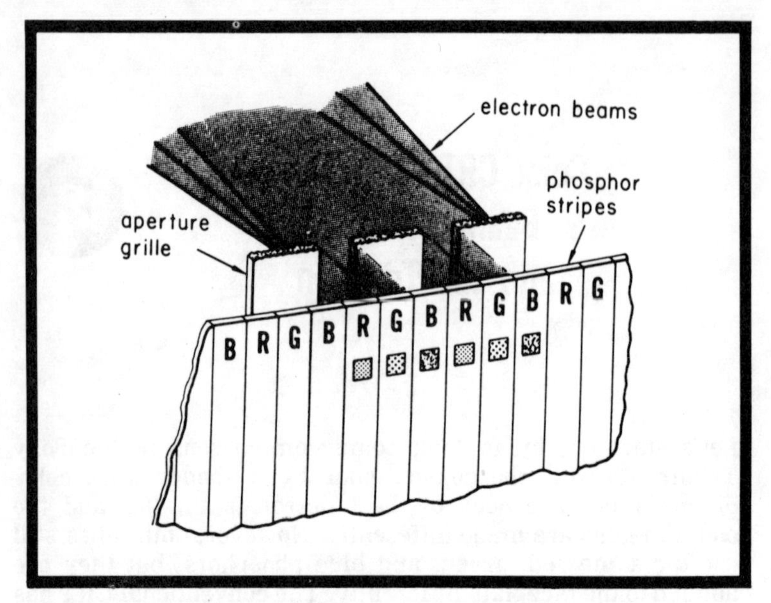

Fig. 8-1. Sony Trinitron CRT aperture grille and phosphor stripes.

of the green beam (as viewed from above) and approaches the aperture grille as shown. The beam is directed into two narrow spots at the center of the red phosphors.

In the same way, electrons intended for blue phosphors approach the aperture grille from the left side of the green beam as shown in Fig. 8-3B.

The beams in a conventional CRT pass through magnetic fields that bend them together (Fig. 8-4).

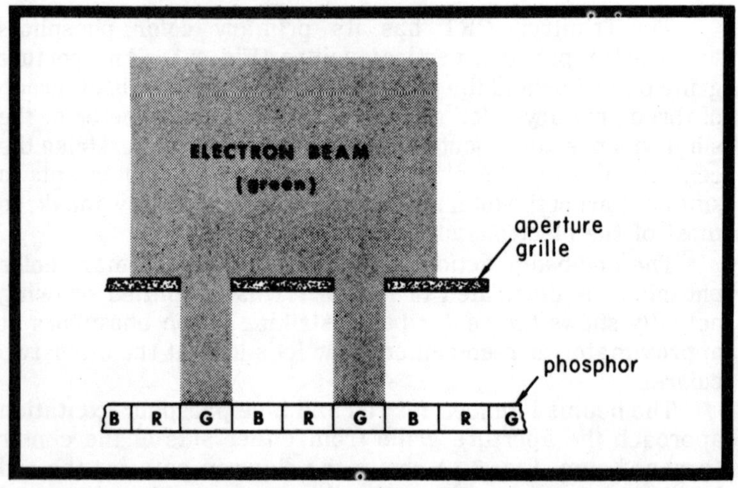

Fig. 8-2. One-beam, one-color action of Sony Trinitron (green).

THE ELECTRON GUN STRUCTURES

The sketch in Fig. 8-5 is a comparison between the gun assembly used in most color CRTs and the Trinitron gun assembly. Note that the beams from the conventional gun assembly produce beams that are displaced horizontally, and in fact this type of gun assembly can be used with the aperture grille / phosphor stripe screen. However, there are certain advantages to having all three beams "in line." For example, it simplifies dynamic convergence corrections. The Trinitron gun assembly is such because of the common lens assembly and thus has a much smaller neck diameter. Since the gap

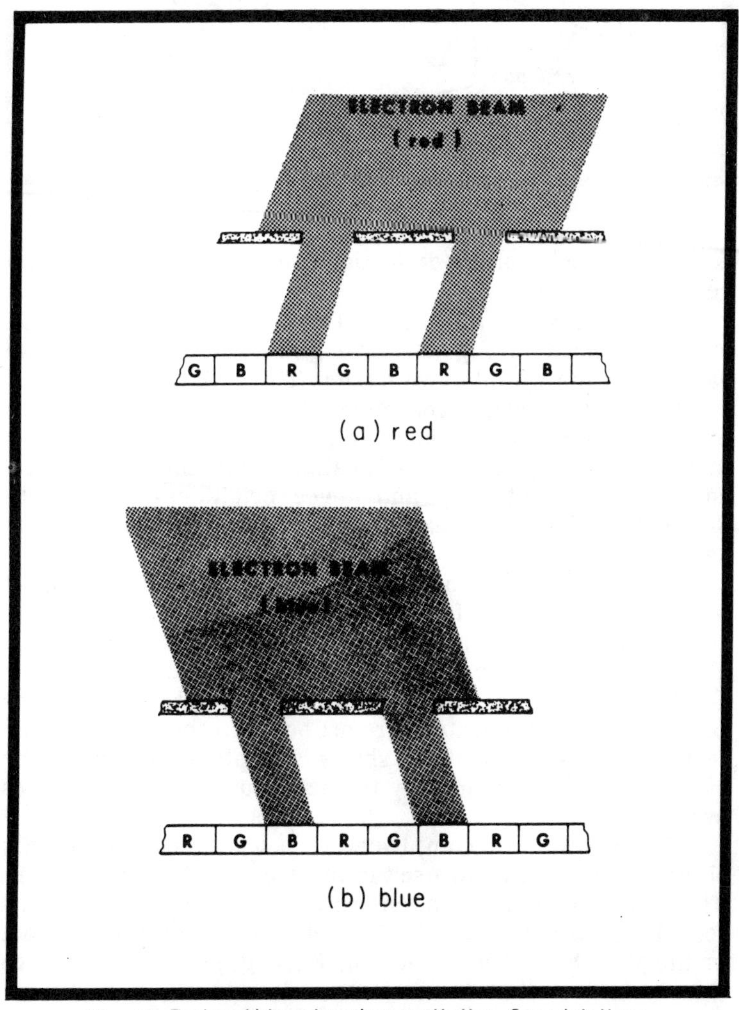

Fig. 8-3. Red and blue phosphor excitation, Sony trinitron.

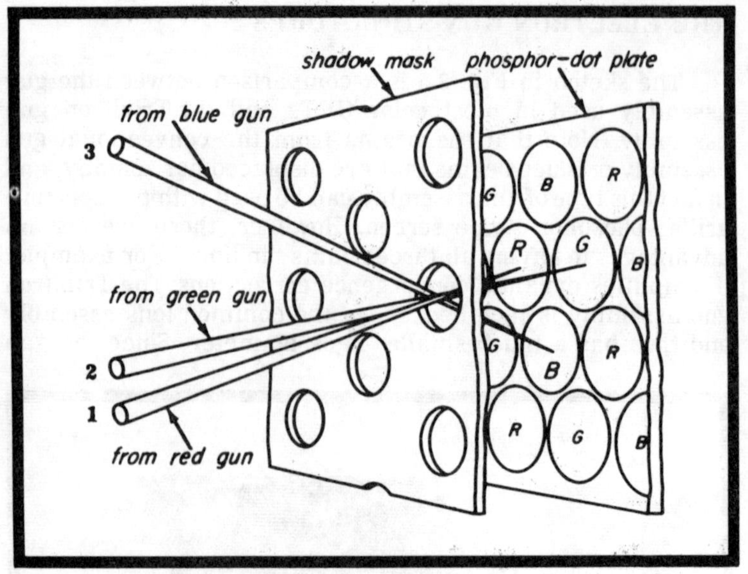

Fig. 8-4. Conventional 3-gun shadowmask color CRT operation.

between the poles of the deflection coils is smaller, deflection current is far less.

A simplified drawing of the Trinitron gun assembly (top view) is shown in Fig. 8-6. The gun uses three individual cathodes, each with its own heater. The cathode assembly is enclosed in a single G1 (control grid) cup with three apertures for the electron beams. Signal voltage, applied as a difference in potential between individual cathodes and the common G1, controls beam intensity and hence brightness for the individual beams. All control signal voltages are applied to the cathodes.

Next along the beam path is a common G2 (screen) cup. It has three apertures for beam passage and provides accelerating potential. The beams now enter the focus assembly. The first electrostatic lens is formed in the space between the G2 cup and the focus assembly. This lens bends the two outer beams to cross the center (green) beam in the middle of the focus assembly. Fig. 8-6B shows an optical model of the electrostatic lens to simplify the action of the electron optics system.

Beam focus is accomplished by a focus assembly that is similar to the guns now used in most B&W TV picture tubes. This assembly acts like a lens with a large diameter, and its focal length is selected to make each beam focus at a fine spot on the phosphor screen. Note Fig. 8-6B again. A good point for this system is that the first lens diverts all three beams through the center of the large focus lens. This gives the same advantage as a larger f-stop for an optical lens. In effect it is the equivalent of a larger depth of field.

The three beams emerge from the forward edge of the focus assembly in slightly divergent paths. They then pass between a set of four static convergence plates as shown in Fig. 8-6A. There is no difference in potential between the two center plates so the green beam passes through unaffected. A static voltage is applied between the two outer and two inner plates, however, and the action of these fields is to turn the

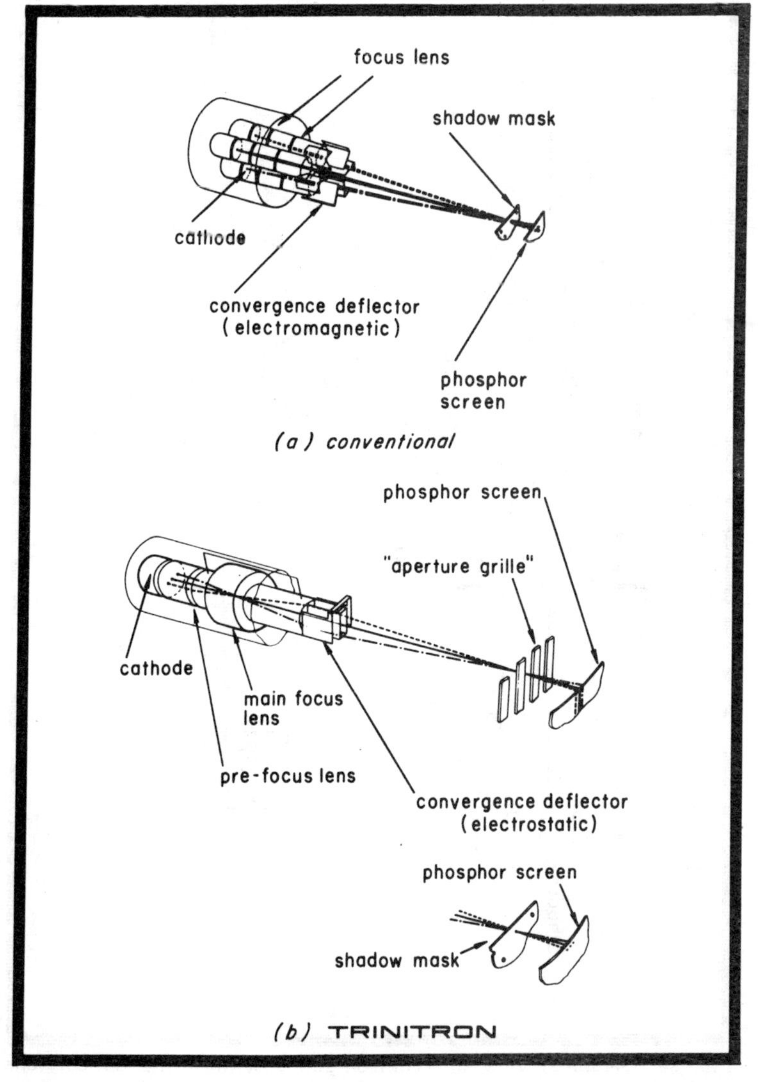

Fig. 8-5. Direct comparison of conventional and trinitron gun structures.

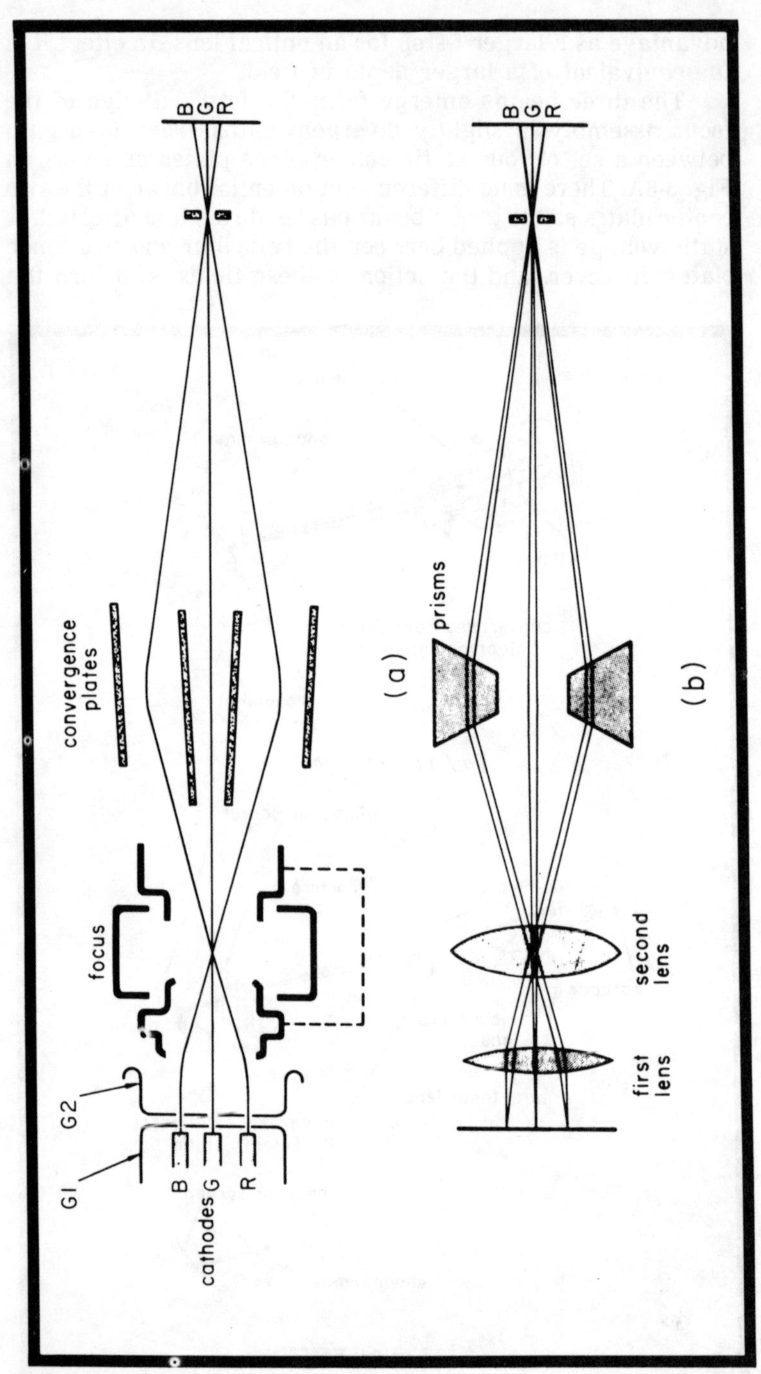

Fig. 8-6. Sectional view of Trinitron gun assembly and optical lens analogy.

outer beams to converge at the aperture grille. This DC voltage is adjustable and is set to effect beam convergence at the aperture grille.

BEAM LANDING CONTROLS

For correct beam landing, the beams must approach the aperture grille at the correct angles. This means that the three beams must come from specific locations with respect to the aperture grille/phosphor plate assembly. In center screen, for example, the green beam must ride right down the axis of the tube.

There are two adjustments required to make electrons land on their designated phosphors. One adjustment is to effect beam landing at center of the screen. This is accomplished by the purity magnet assembly mounted on the rear edge of the yoke. The second adjustment, to correct beam landing at the sides of the screen, is simply a yoke positioning procedure. This is similar to the purity adjustments for a conventional color CRT.

Vertical displacement of the beams does not result in the beams hitting the wrong phosphors. For this reason the earth's magnetic field has no effect upon the beam landing. In addition, beam landing adjustments are simplified somewhat because only horizontal correction is needed.

NECK-TWIST COIL

A beam alignment yoke is positioned on the neck of the CRT, close to the socket, to aim the beams for correct beam landing. This is called the neck-twist coil. The action of this coil is to twist the outer beams around the center green beam. Coil current and polarity are adjusted to put the three beams in a horizontal plane relative to the gun assembly.

DYNAMIC CONVERGENCE

The distance between the deflection centers and the aperture grille increases as we move the beams away from the screen center. Thus, the beams converge short of the aperture grille at the edges. This condition is also found in the shadowmask CRTs. Correction is achieved in the conventional color tubes by applying dynamic convergence correction that is in sync with the scanning signals.

Now let's look at the Trinitron dynamic convergence system. Vertical correction can by symmetrical because all beams are in the same horizontal plane. For horizontal convergence, a parabolic waveform is synchronized with the

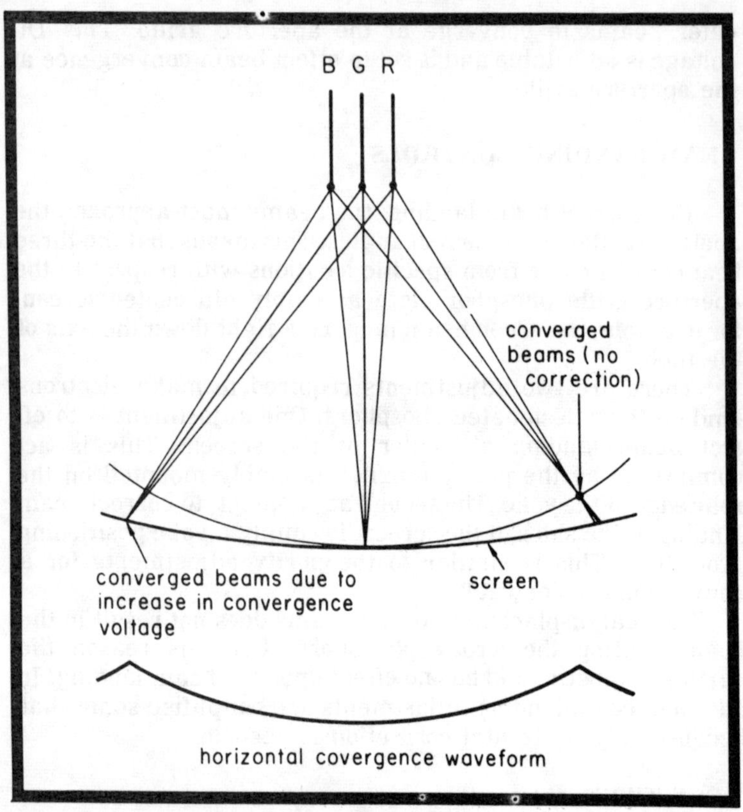

Fig. 8-7. Horizontal dynamic convergence of a Trinitron. Vertical convergence is unnecessary because of the construction.

horizontal scan and is developed at the output circuits of the horizontal output stage and added to the DC voltage applied to the outer deflection plates. The outer plates are normally negative with respect to the inner plates to divert the outer beams. The dynamic convergence voltage acts to reduce the difference in potential between inner and outer plates as the beams sweep toward the sides of the screen. The effect is to "straighten out" the three beams and make them converge at a greater distance from the deflection center. See Fig. 8-7. In other words, the beams spread out slightly to converge at the sides of the screen.

SIMPLIFIED CONVERGENCE

For correct color reproduction with any three-color simultaneous color CRT, a process is needed to position each of the three respective color images at the same position without any slippage. This process is called convergence.

Dynamic as well as static convergence is required for a good quality picture in a color TV receiver.

In the ordinary delta-shaped three-electron gun system, the adjustment of the electron beam direction is very complicated because the beams emitted by the electron guns are not on the same plane. Four adjusting controls are necessary to achieve static convergence. Dynamic convergence is all the more complicated. There are four adjustment controls needed for each red, green and blue beam, which makes twelve controls in total that are necessary to bring all three color images into convergence.

In the trinitron system, convergence has been simplified. Three electron beams are ejected on the same horizontal plane (in-line positioning), therefore, the beam direction can be easily adjusted by shifting the left or the right beam sideways in the plane to meet with the beam emitted from the center cathode on the phosphor screen (or color defining mask).

To do this, a pair of symmetric electron prisms (deflectors) is used. Since the electron deflector works on the electrostatic principle, it is simply required to adjust the voltage applied to the right and the left deflection plates. In fact, very little dynamic convergence adjustment is required because of the use of a newly developed deflection yoke.

TRINITRON CRT CIRCUITRY OPERATION

The Sony Trinitron CRT has a very simple drive circuitry as shown in Fig. 8-8. Because G1 and G2 are part of a single gun, the complex networks used with other CRTs are not necessary. The Y signal and color difference signals are fed into the demodulators, thus the cathodes of the Trinitron receive both color and video signals. The drive requirements are 70 volts (P-P) per cathode, which includes signal and horizontal blanking.

Phosphor efficiencies are about equal so that the drive ratios are about the same. In addition, the transconductance for all cathodes is kept within close tolerance by the common grid structure so that the guns cut off at about the same bias point. This greatly simplifies white balance (color temperature) adjustments.

The operating voltage for the neck-twist coil (NTC) is derived from the +18-volt DC supply via a voltage dropping network. The V. Stat Control permits exact adjustment of the current through the NTC. The control that adjusts neck-twist current is called V. Stat, because the visible effect of beam twist is misconvergence in a vertical direction. The effect is seen most easily on horizontal lines when a crosshatch pattern is displayed.

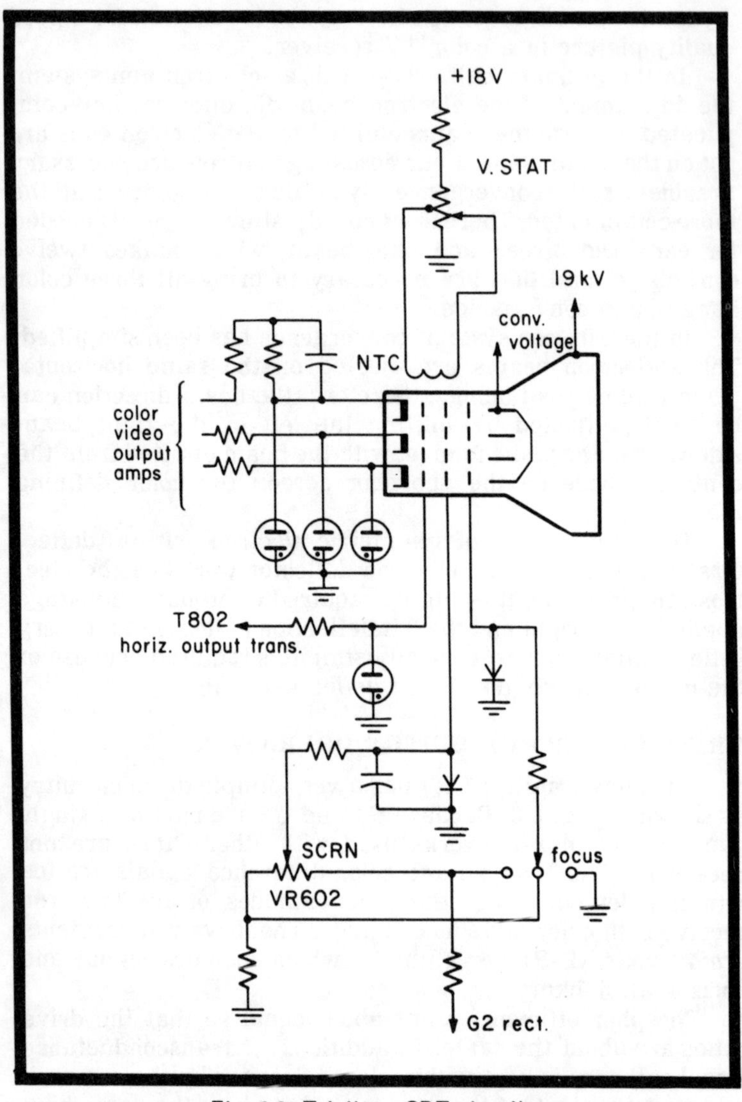

Fig. 8-8. Trinitron CRT circuitry.

The picture tube has neon lamps connected from each cathode to ground. This protects the color video output amplifier (connected to the CRT cathode) from any internal arcover occurring in the picture tube. Similar protection for the circuits connected to the picture tube grids is provided by special spark-gap capacitors for the screen (G2) and focus (G4) grids, and a neon lamp for the control grid (G).

For the conventional color CRTs, matrixing for the Y and color signals takes place within the picture tube. Note the conventional CRT circuits in Fig. 8-9. The color-difference amplifiers drive the control grids of the color guns, while Y information goes to the cathodes. For an equal balance, drive controls set the amplitude of video (Y) signals for each cathode in relation to the other two. These controls will determine the bright screen (white balance) or gray-scale tracking of the CRT.

The no-video gray-scale tracking is determined by settings of the three G2 controls. In most cases you have to work back-and-forth between screen and drive controls to obtain a proper tracking adjustment. However, this is a very simple procedure for the Trinitron tube.

GRAY-SCALE TRACKING FOR THE CRT

The importance of the gray-scale tracking should never be underestimated by the technician. Improper adjustments will

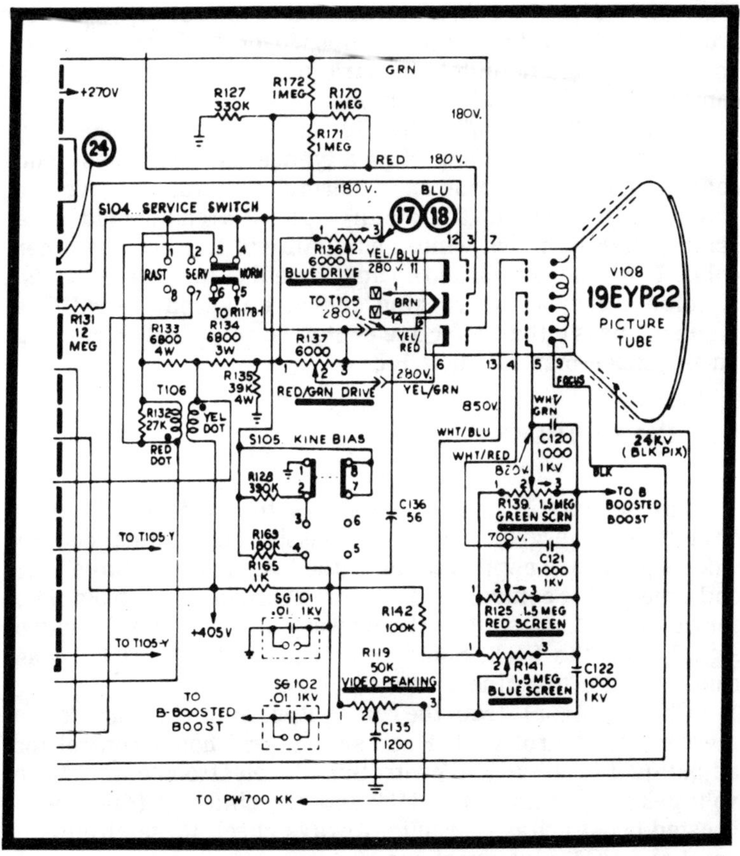

Fig. 8-9. Conventional color CRT circuitry.

cause the color information to be wrong. Make sure the purity adjustments are correct before making any gray-scale tracking adjustments. Refer to Fig. 8-9.

Proper adjustments of both G2 and gain controls is necessary to produce a picture with proper shading or tracking throughout the normal brightness and contrast range. The G2 or screen controls are adjusted to provide correct bias for the CRT and to compensate for minor variations in circuitry and different CRT characteristics. The gain controls (often referred to as drive controls) are used to vary the amount of video signal fed to each individual cathode of the CRT.

In the normal tracking procedure, care must be taken in setting up the screen controls. It's a good habit to always recheck the screen control settings. After placing the NORMAL/SERVICE switch to the SERVICE position and advancing the three G2 controls to produce a white horizontal line of low brightness, return the switch to the NORMAL position. Adjust the brightness and contrast controls to produce a dimly lit picture composed of dark gray areas. This area should definitely be dark gray and not be shaded a combination of red, green, or blue.

A minor "touch up" of the G2 controls may be necessary to achieve this condition. Now advance the brightness and contrast controls to produce a normal picture. Observe the bright white areas of the picture. These areas should be a bright, slightly bluish white and again not tinted some other color. This is accomplished by adjustment of the gain controls. If correct gray-scale tracking cannot be accomplished, the picture tube is either defective or some circuit fault is causing an improper bias for the CRT.

CRT BIAS CHECKS

The schematic in Fig. 8-9 shows the typical CRT bias circuits for a large-screen color chassis. The luminance (Y) video signals are applied to the cathodes. Most sets have blue and green drive controls and the three screen G2 controls. The service switch stops the vertical sweep and removes the video signal from the CRT. Some color chassis have a CRT bias control and a brightness range control.

Chroma signals from the color output amplifiers are fed to the correct control grid. Each screen grid has a control for adjusting its voltage and controlling the electron beam. These voltages vary from 350 to 1100 and are obtained from the B boosted-boost voltage circuits. Always check these circuits if you have CRT bias problems.

Correct gray-scale tracking cannot be obtained if one of the color amplifier tubes is defective. And this will also affect the monochrome picture reception. Use the scope to check for correct signals at the color amplifier plate circuits to help localize faults. And do not overlook a defective color picture tube.

Again, if gray-scale tracking cannot be obtained, check voltages, resistance, and component values in the cathode and screen-grid circuits. Turn the set off and check the variable resistance for each of the CRT bias controls. Also, don't overlook any of the spark-gap or spark-gap capacitor combination components in these circuits. They can cause intermittent flashes across the screen and improper gray-scale tracking.

When the CRT screen changes color in an intermittent to erratic manner the screen (G2) controls are prime suspects. These controls may become pitted or develop burnt spots. If they prove to be defective, always **replace** the control and **do not try to clean them.**

To check the CRT grid bias, use a VTVM and read the voltage between the grid and cathode. This bias voltage will depend on the setting of the brightness control and for normal operation will vary about 40 volts. If you have bias trouble, check the plate voltage of the associated video output tube.

The trouble symptom that is discussed below points out the necessity for looking beyond the CRT control circuits for "odd ball" picture problems.

The trouble was found in a Zenith chassis that used the 6ME8 color demodulator tubes (Fig. 8-10). The picture was grainy and color looked faded. The CRT brightness could not be extinguished even with the screen (G2) controls turned down (CCW).

Looking at the demodulator circuit for this chassis, the plate voltages of V12 and V13 were found to be 100 volts too high. Also, there was no voltage present at pins 1 and 2 of the 6ME8 tubes. This was caused by an open R58 27K, 2 watt resistor from the 250 volt B+ supply. Of course, this caused the high plate voltages on the 6ME8s, thus upsetting the CRT bias and producing the poor quality color picture. This was an unusual picture symptom and can be difficult to track down.

SOME COLOR CRT FOCUS CIRCUITS

Most color CRTs use a high DC voltage to focus the electron beam. The small screen CRTs do not. This DC focus voltage is about 20 percent of the high voltage at the CRT anode button (ultor). This means that you should measure between 4500 and 5500 volts at the number 9 focus pin of the

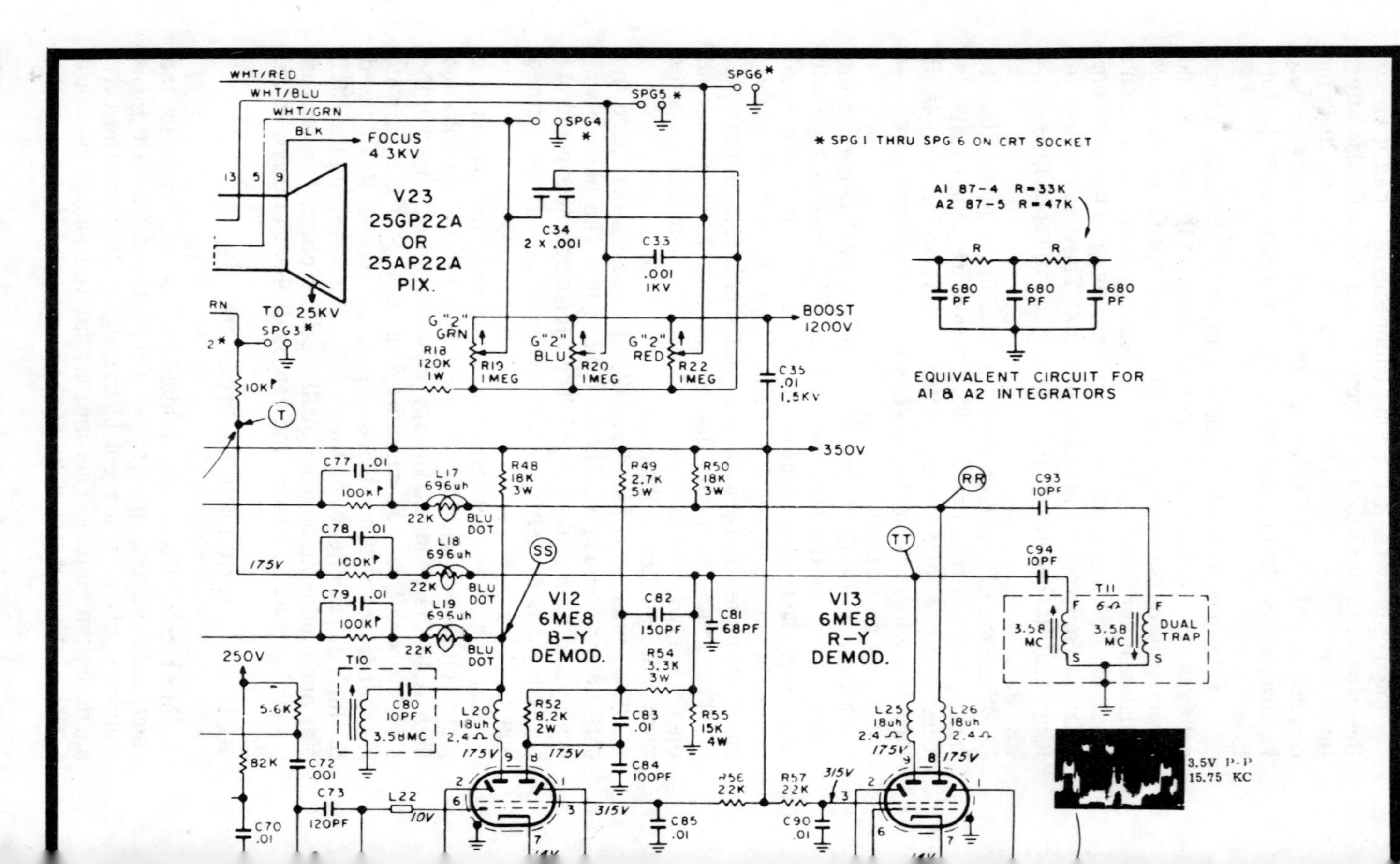

WHT/RED
WHT/BLU
WHT/GRN
BLK
FOCUS
4 3KV
13 5 9
V23
25GP22A
OR
25AP22A
PIX.
TO 25KV
RN
2*
SPG3*
SPG4*
SPG5*
SPG6*
C34
2 X .001
C33
.001
1KV
* SPG1 THRU SPG 6 ON CRT SOCKET
A1 87-4 R=33K
A2 87-5 R=47K
R
680
PF
EQUIVALENT CIRCUIT FOR
A1 & A2 INTEGRATORS
G"2"
GRN
G"2"
BLU
G"2"
RED
R18
120K
1W
R19
1MEG
R20
1MEG
R22
1MEG
BOOST
1200V
C35
.01
1.5KV
10K
T
350V
C77 .01
C78 .01
C79 .01
100K
L17
696uh
L18
696uh
L19
695uh
22K
BLU
DOT
R48
18K
3W
R49
2.7K
5W
R50
18K
3W
RR
C93
10PF
C94
10PF
TT
SS
175V
V12
6ME8
B-Y
DEMOD.
V13
6ME8
R-Y
DEMOD.
C82
150PF
C81
68PF
R54
3.3K
3W
R55
15K
4W
T11
6Ω
3.58
MC
DUAL
TRAP
F
S
250V
T10
C80
10PF
3.58MC
5.6K
82K
C72
.001
C73
120PF
C70
.01
L22
10V
L20
18uh
2.4Ω
R52
8.2K
2W
C83
.01
C84
100PF
R56
22K
R57
22K
315V
C85
.01
C90
.01
L25
18uh
2.4Ω
L26
18uh
2.4Ω
3.5V P-P
15.75 KC

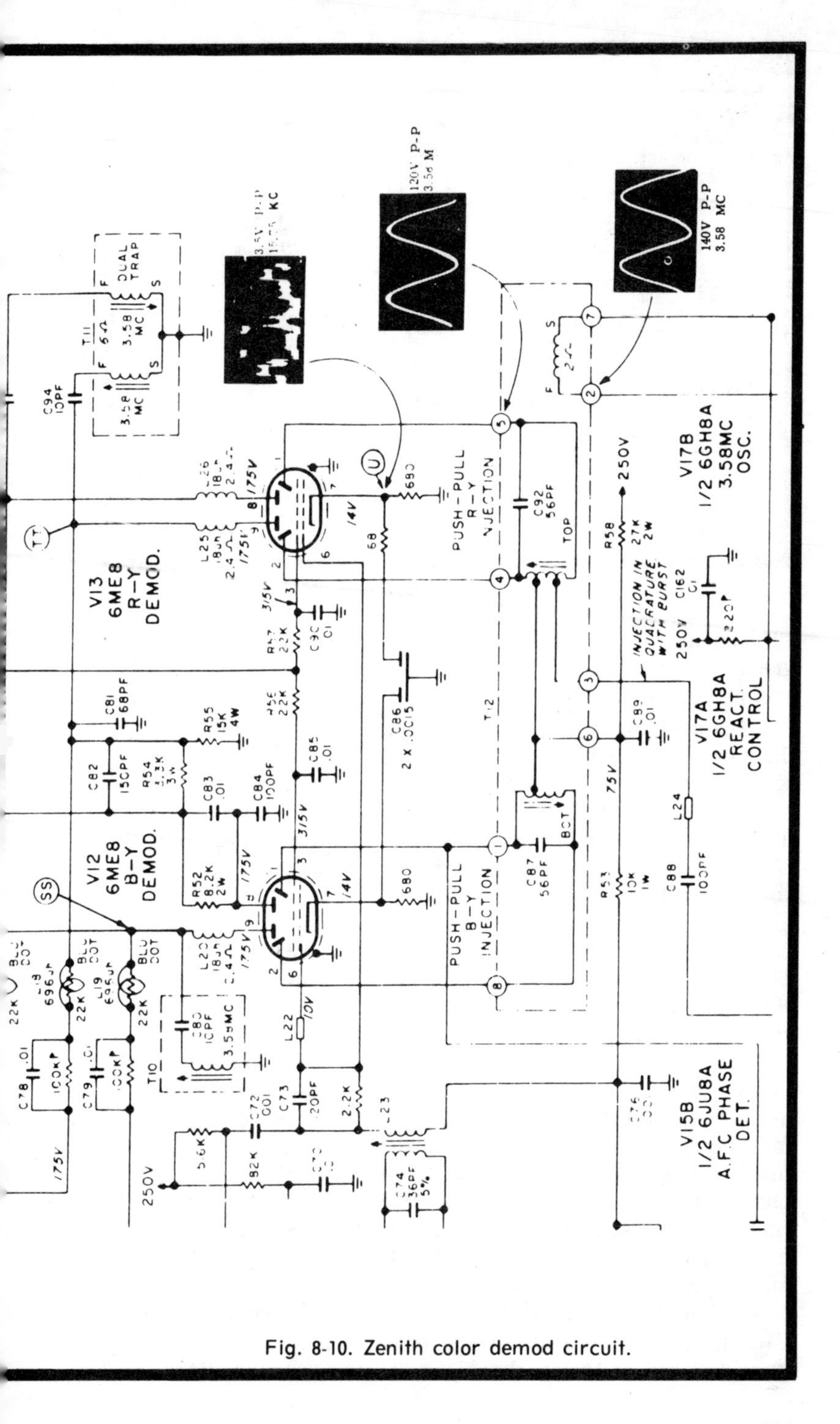

Fig. 8-10. Zenith color demod circuit.

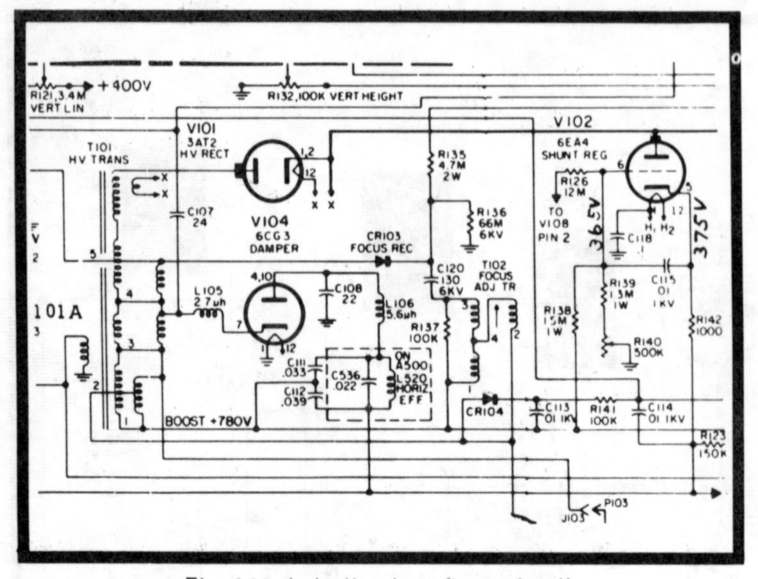

Fig. 8-11. Inductive type focus circuit.

CRT socket. While you are there, check pin 9 and see if it is greenish looking or corroded. Clean it up or replace the CRT socket if it looks this way.

To check for normal focus action, take a very close look at the raster lines while the focus control is varied. You should see a very well defined raster line. If these lines blend together, there is trouble in the focus circuit or color picture tube itself. In order to obtain a sharply focused raster line, the CRT high voltage must also be correct. Measure it with a high-voltage meter probe and adjust if need be.

The focus voltage is produced by rectifying a high-peak pulse taken from the plate of the horizontal sweep output tube. This pulse at about 5000 volts P-P, is fed from the plate of the H.O.T. to a focus rectifier, which can be either a tube or a solid-state diode.

A good many of the "middle-aged" color chassis used the inductive type focus circuit shown in Fig. 8-11. A tube may be used in place of the solid-state rectifier. This circuit features a focus transformer which is adjusted to vary the focus rectifier, CR103, while the focus transformer feeds an out-of-phase pulse signal through capacitor C120 to the cathode of the rectifier. Adjusting the core of T102 determines the amplitude and phase of the signal applied to the cathode and, thus, the amplitude of the focus voltage.

When trouble develops in this circuit check the diode, CR103, and make sure the HV is correct. If the raster wants to go in and out of focus, check the 66 meg resistor, R136, as it may be open. Also, do not overlook a faulty focus-coupling capacitor, C120. Now, if the focus coil adjustment has no effect, it is probably defective. The core could also be stuck.

One of the early focus circuits (shown in Fig. 8-12) had the control pot (R131) across part of the sweep transformer winding. The control slider picks off some of the pulse to develop the correct value of DC at the output for the 1V2 focus rectifier. This arrangement caused R131 to burn out quite often. There aren't too many of these circuits around any more.

Some of the later model color TV - and some portable sets - use the focus system shown in Fig. 8-13. A specially built divider resistor, R340 develops the DC focus potential directly from the high voltage rectifier (V207 cathode). This makes for perfect tracking as the focus voltage is always the same ratio of the high voltage value. A high value pot, R338, at the ground end of the bleeder resistor, provides the manual focus adjustment.

The focus circuit shown in Fig. 8-14 will be found in some of the recent model solid-state color chassis. This is a high-voltage quadrupler system. A 7 KV P-P pulse from the secondary of the high voltage transformer is applied to the "in" terminal of the quadrupler, consisting of eight diodes and capacitors arranged in a full-wave voltage-multiplier configuration.

A tap is provided for application of a 7 KV DC voltage to the focus voltage divider. A 275-megohm bleeder resistor discharges the CRT second anode when the set is turned off.

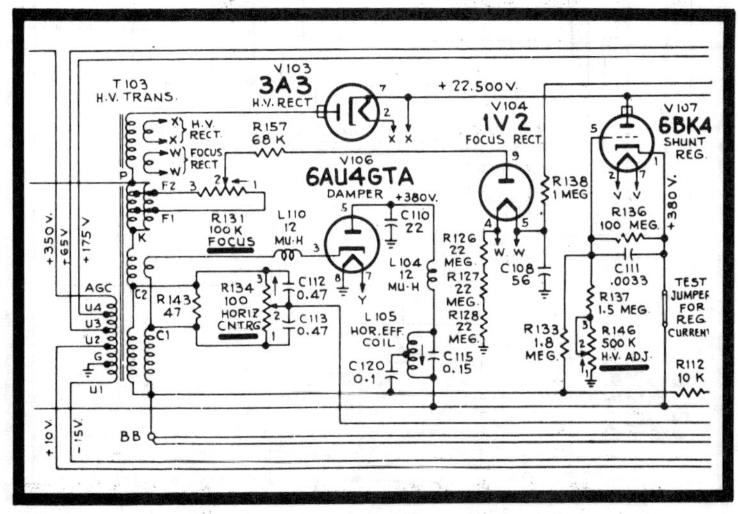

Fig. 8-12. Early-design focus stage with focus potentiometer pickoff from HV transformer.

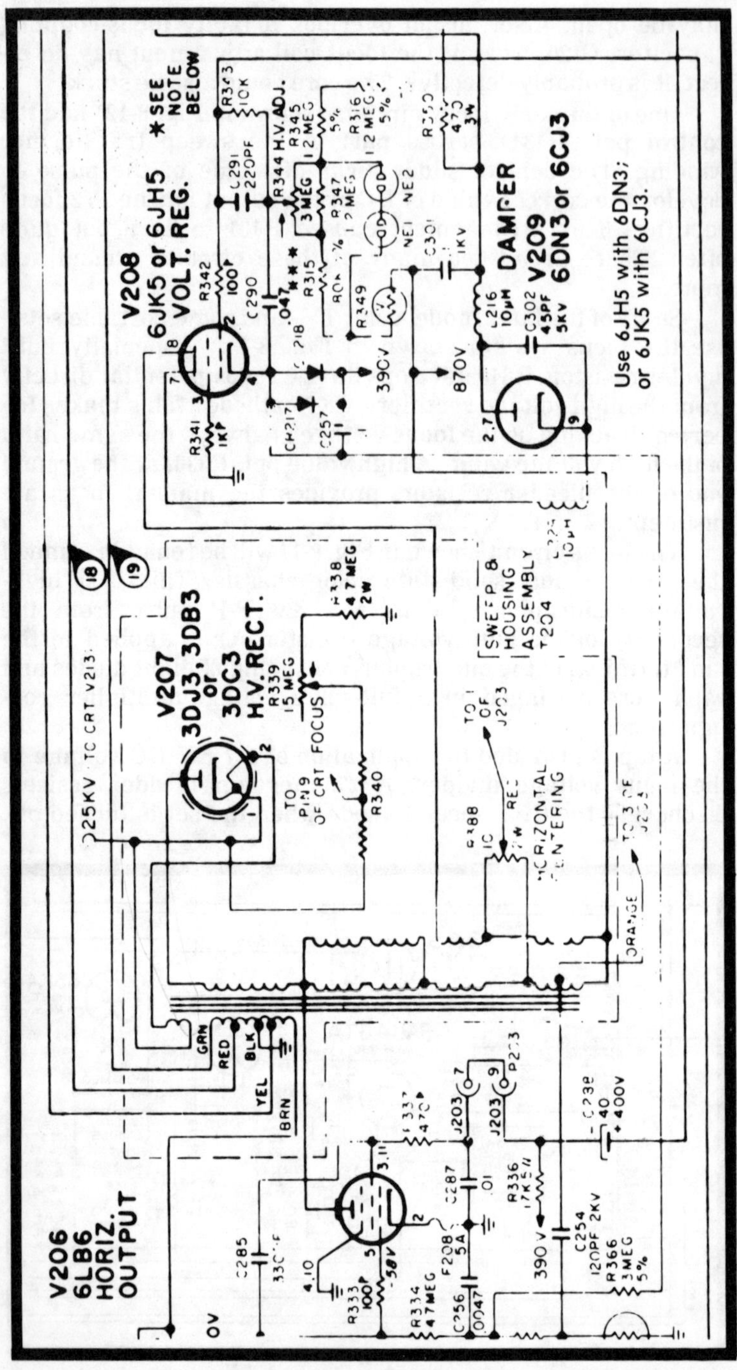

Fig. 8-13. Focus system designed so focus voltage is always the same ratio of the HV.

The output of the quadrupler is 26.5 KV at minimum picture brightness. Divider resistors and a pot set the exact value of DC sent to the CRT focus anode.

A Zenith color chassis using a selenium rectifier diode for focus voltage and using an adjustable potentiometer is shown in Fig. 8-15. This is a popular circuit that Zenith has used for quite a few years. These stacked, selenium focus diodes will cause some troubles at times. Often they will open up (at times intermittently) and cause a blurred or streaked picture. Some will short and burn out other focus circuit components. When these diodes become defective, a faint arcing sound may be detected. A close look at the individual scanning lines will reveal that they are very ragged instead of a smooth line. This type of selenium rectifier is actually many diode units stacked one upon another in order to properly rectify the 5,000 volts required for focus control. Some of these defects are caused by an arc developing between these selenium units. Of course, this will cause all sorts of crazy focus problems.

Let's now look at another focus circuit trouble. See Fig. 8-15. After a fuse was replaced in this set and turned on, it was noted that a burning odor was detected coming from the high voltage cage. Another strange symptom was also noted. As the focus control was turned - to try and sharpen up the blurred picture - it was noticed that starting from the right hand side, the picture would start blacking out. Almost all of the screen could be blacked out this way by adjusting the focus control. It sort of looked like this color set had its own special effects

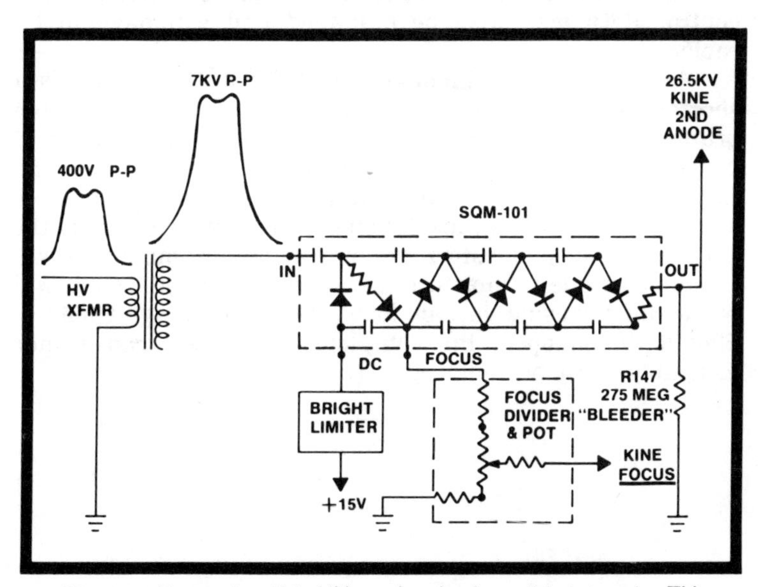

Fig. 8-14. Focus circuit in HV quadrupler for solid-state color TV.

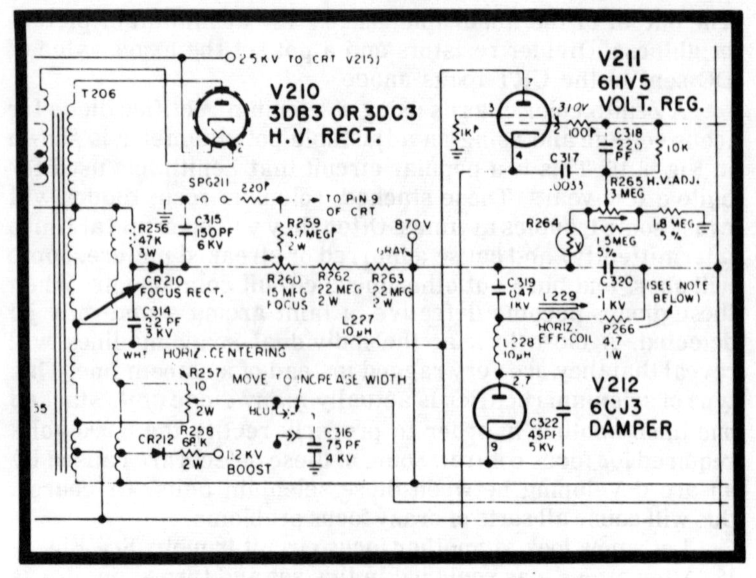

Fig. 8-15. Zenith selenium rectifier focus stage.

studio control board. That is, this "wiper" action would make the screen either dark or light by adjusting the focus control.

This windshield wiping action was caused by leakage in C314, a 22 pf, 3 KV capacitor in the focus circuit near CR210. When C314 develops leakage or shorts out, resistor R256 and the focus rectifier CR210 will start to smoke. Sometimes focus control R260 may also be damaged and will have to be replaced.

Another focus problem to watch for in this circuit is a shorted spark gap SPG211. Also the built-in spark-gap protection device in the CRT socket may break down and cause an intermittent focus problem. When this occurs, the focus control, focus rectifier, or R256 may be damaged. In some chassis, the fiber insulation board (on which most of the horizontal sweep transformer components are mounted) will break down due to moisture- caused leakage and arcing will occur, thus causing various focus and high voltage troubles. Should this happen, the complete HV sweep transformer assembly must be replaced.

Color Sync and Associated Circuits

9

Color sync circuits are necessary in order for the correct color to be reproduced on the receiver screen. For correct color to be developed, the color demodulator must be fed the correct reference signal phase. The incoming color burst is used to control the local 3.58 MHz reference signal and this control is performed by the color sync circuits.

The color burst has at least 8 cycles of 3.58MHz subcarrier signal (see Fig. 9-1) which is transmitted at blanking level after each horizontal sync pulse. The color sync process begins by separating the burst from the other transmitted information. And this is accomplished by the burst separator circuit. The burst amplifier is turned on by a pulse from the horizontal sweep section only during the duration of burst. You should only find the color burst at the output stage of this amplifier.

Other essential links in the color sync system are the 3.58 MHz local reference oscillator and the reactance control circuits. In some color chassis the burst is compared to the reference signal with an in-phase detector. This detector output is proportional to the incoming burst amplitude and is used for ACC (automatic color control) and color killer control.

OVERALL COLOR BURST AND ASSOCIATED CIRCUITS

The overall operation system of one color receiver is shown in Fig. 9-2. The burst amplifier (pentode section of a 5GH8A), couples separated burst into a 3.58 MHz injection locked oscillator. The output of the 3.58 MHz oscillator is coupled directly to the "Z" demodulator, then phase shifted approximately 90 degrees and coupled to the "X" demodulator. Also note the associated ACC and color killer circuits.

Burst Separation Circuit

A simplified schematic of a color burst amplifier is shown in Fig. 9-3. Let's look at its operation. A 250 volt peak-to-peak

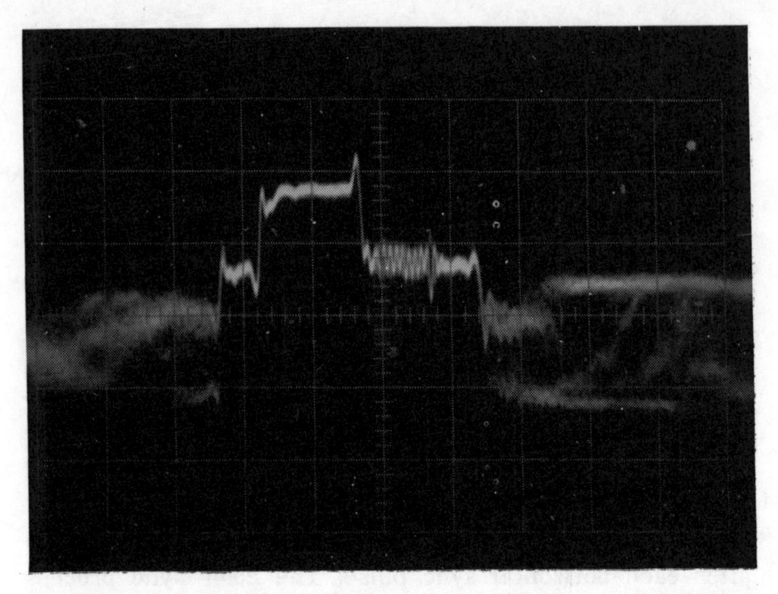

Fig. 9-1. Color burst signal (3.58 MHz) on back porch of horizontal sync pulse. (Scope at 5X.)

horizontal pulse is used to "key" the grid of the amplifier "on" during horizontal retrace time (this is the time color sync is present). The bottom waveform shown in Fig. 9-4 is what the scope should see at the grid-pin 2 when a color bar signal is applied to the receiver. At the top tip of the horizontal keying pulse rides the burst signal while the rest of the chroma information is along the bottom of this wavetrace. The pulse is also coupled through C716 and R716 to the screen-plate circuit of this amplifier. This connection, in conjunction with diode CR707, increases the plate and screen voltage at the 5GH8A during retrace time.

In operation, diode CR707 acts as a switch and clamp. During retrace time (positive pulse is present at C716), the diode cuts off, and the pulse adds to the 280 V B+ on C716 thus increasing the plate and screen voltage to approximately 400 volts during retrace time. During scan time, the diode conducts and clamps C716 to 280 volts. This eliminates scan ripple, and also gives the desired amount of pulse shaping. The combined results are a high-level, clean burst signal available to the 3.58 MHz oscillator.

The correct output waveform (see top trace of Fig. 9-4) should consist of color burst information only. The load for this stage is the burst transformer. This transformer is tuned for 3.58 MHz to give maximum burst output and this reduces any noise present. (Also note that some burst transformers secondaries are center-tapped.)

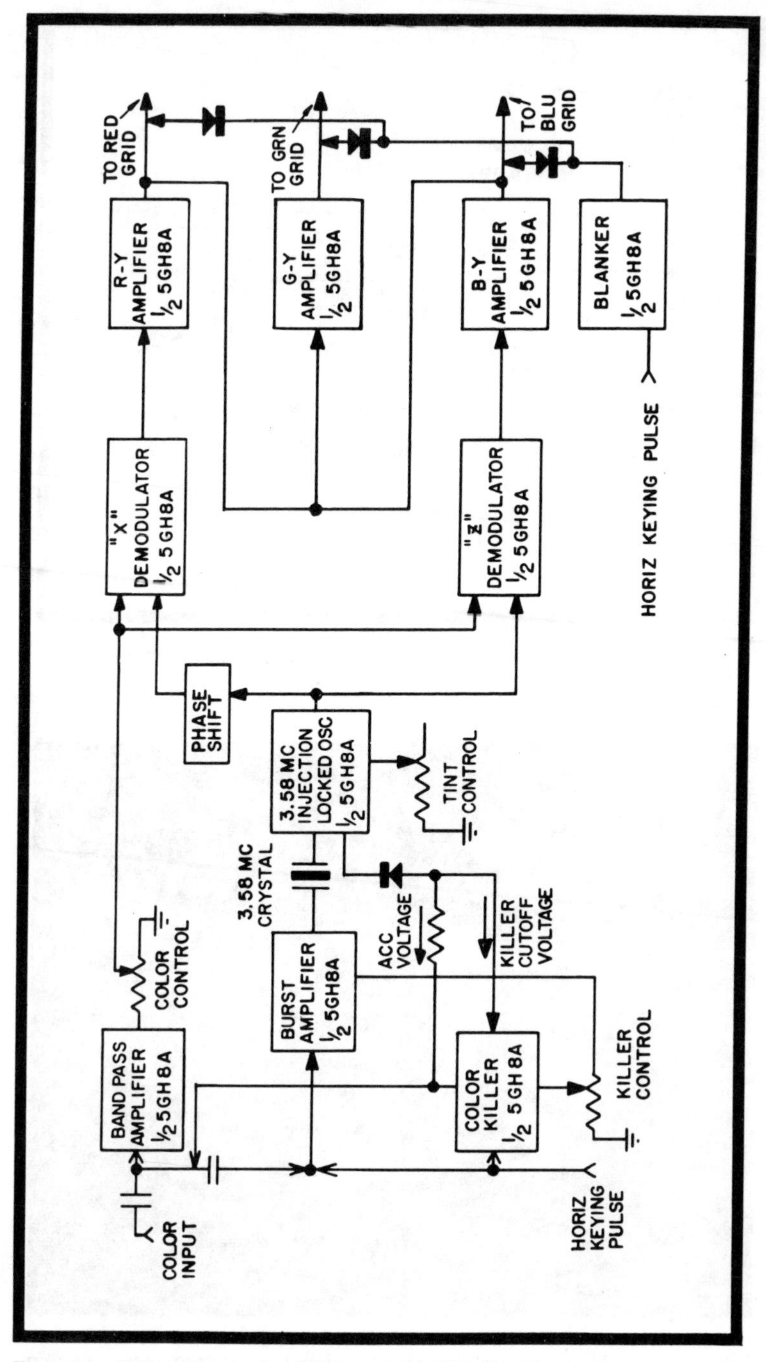

Fig. 9-2. Color stages in CTC-22 RCA TV, to illustrate color sync operation.

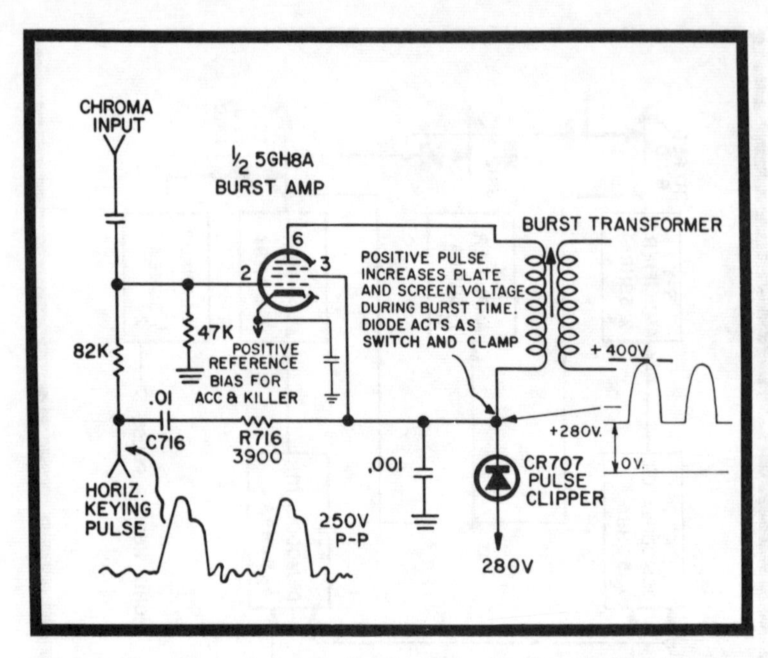

Fig. 9-3. Simplified schematic showing color burst amplifier operation.

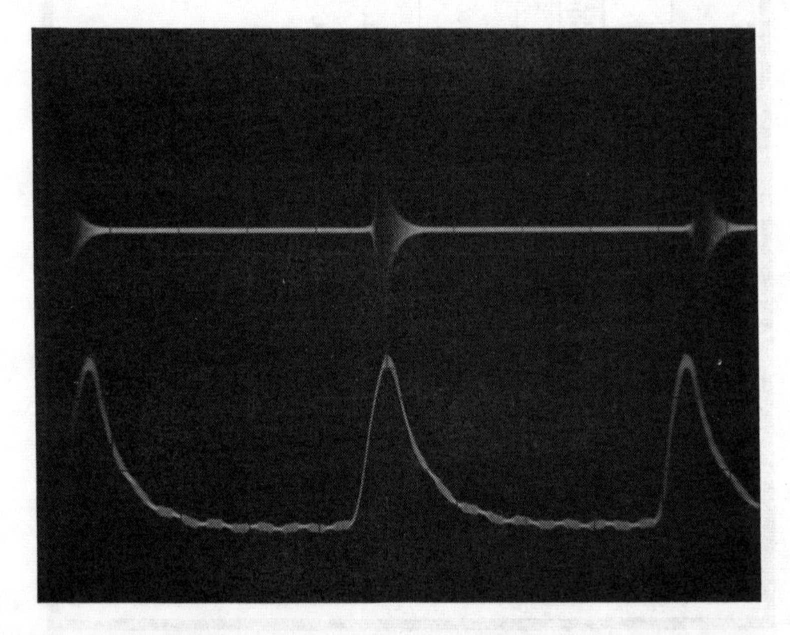

Fig. 9-4. Top trace, correctly color burst pulse. Bottom trace, keying pulse from horizontal circuitry, correctly timed so as to be coincident with color burst. (Scope at horizontal rate.)

Fig. 9-5 shows the 3.58-MHz burst signal riding down on the side of the keying pulse. Of course, this mistiming will cause poor color sync or complete loss of color. This condition could be caused by a faulty coupling capacitor that feeds the keying pulse from the horizontal sweep transformer. However, do not overlook circuit trouble in the horizontal oscillator, AFC, and sync amplifier stages. Proper timing is very important in this circuit as the horizontal keying pulse must coincide with the transmitted color burst signal. With a dual-trace scope the color burst can be simultaneously compared with the horizontal keying pulse. (Additional color burst circuit information will be found in Chapter 10.)

3.58 MHz Crystal Control Circuit

For proper color reproduction the phase of the 3.58 MHz oscillator must be accurately controlled. The local color oscillator must be kept close to the correct frequency because the phase error introduced by this system increases as the oscillator drifts away from the correct frequency. The modern color receivers use crystal controlled oscillators which will provide the required frequency stability. A typical 3.58 MHz crystal oscillator circuit is shown in Fig. 9-6.

The oscillator action takes place between the control grid and the screen grid of V703B. As the oscillator will see it, the

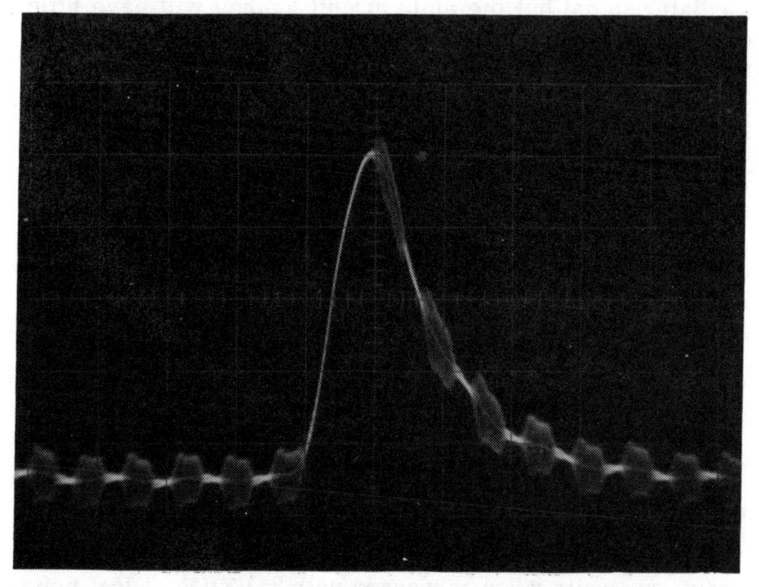

Fig. 9-5. The keying pulse which is sliding to the right from the top should be directly straddling the peak of the keying pulse. This improper timing will cause poor color sync or even loss of color. (Scope at horizontal rate.)

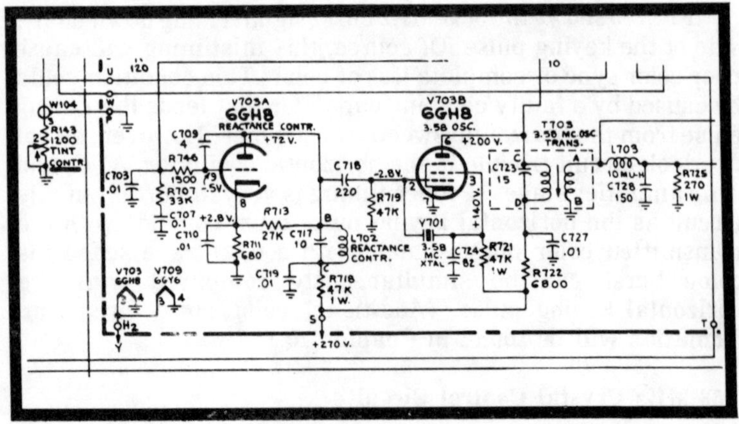

Fig. 9-6. Typical 3.58 MHz crystal controlled tube oscillator and reactance control circuit.

screen grid will act as an effective plate. The crystal is connected between grid and screen and is effectively shunted by the plate-to-ground impedance of the reactance tube V703A which is in series with C724, screen capacitor. The frequency of the oscillator is controlled by the effective capacitance presented by the reactance tube.

The plate of V703B is electron coupled to the oscillator. This will cause the plate current to vary according to the oscillator signal but the plate circuit loading will have little effect on the oscillator action. Thus the oscillator frequency will not be upset by load changes from the demodulators. The oscillator plate load transformer T703 is tuned to 3.58 MHz to produce a good sine wave signal for the demodulators.

The phase-shift network is composed of L703, C728, and R725. This network gives a 90 degree shift to produce the correct phase relation between the R-Y and B-Y chroma signals. Always use the exact replacement should these components ever be replaced. The 90 degree phase-shifted 3.58 MHz signal is fed to the AFPC detector. At the same time an in-phase 3.58 MHz signal is fed to the in-phase killer detector.

SERVICING THE ZENITH COLOR OSCILLATOR CIRCUIT

Refer to Fig. 9-7 for the following circuit checks:

NOTE: If color set has AFTC (automatic fine tuning control) provisions turn AFTC switch to the off position while making these adjustments.

A properly operating Zenith color set will generally read about zero volts when measured with a VTVM at test point "W" under no signal condition (tuner set between channels). Should this voltage exceed one volt (positive or negative),

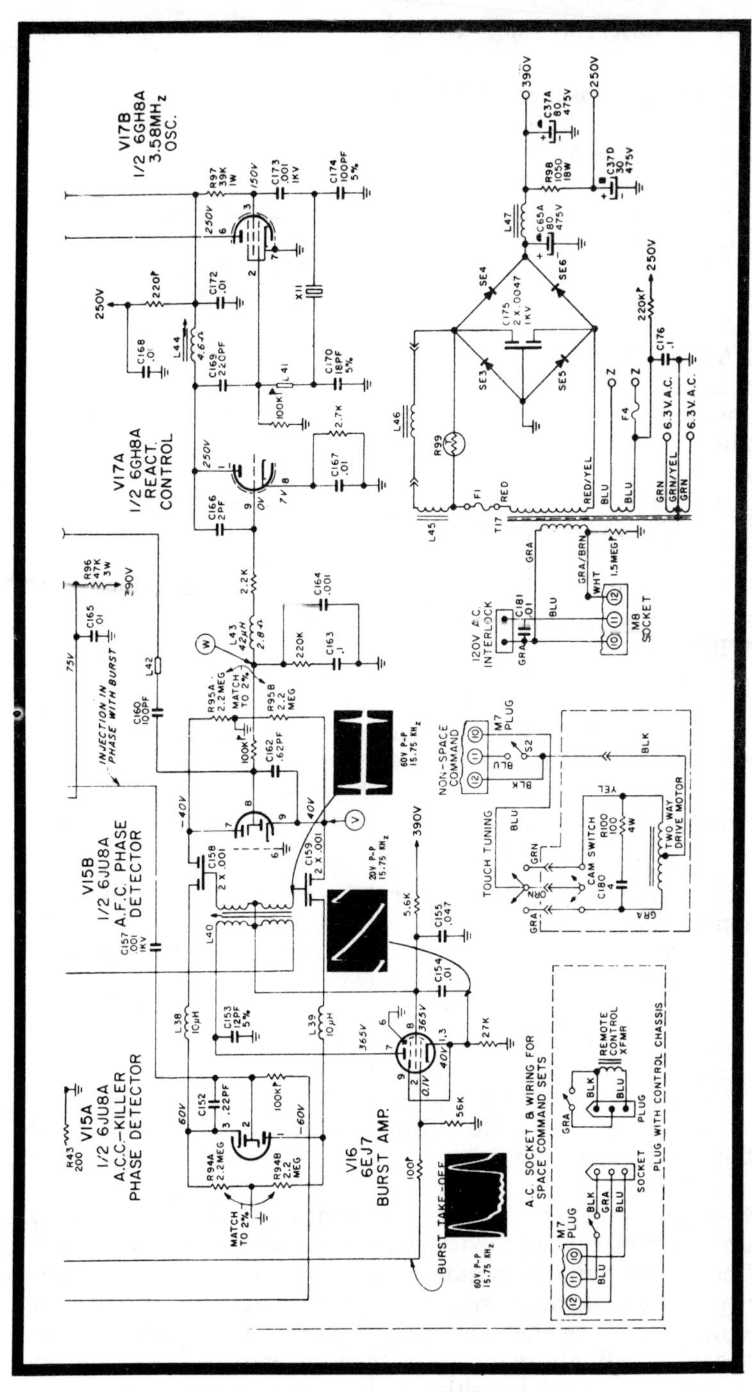

Fig. 9-7. Zenith color oscillator circuit.

several different color picture symptoms may appear as follows:

1. Complete absence of color information
2. Intermittent loss of color sync
3. Color dropout
4. Critical fine tuning range for color
5. Delay in color "lock-in" when changing from one color program channel to another.

Some of the probable causes of voltages in excess of one volt at test point "W" under no signal conditions are as follows:

1. A faulty 3.58 MHz CW oscillator and control tube
2. A defective AFC phase detector (tube or solid-state diode). One diode may be conducting considerably more than the other.
3. An open or leaky section in one of the two dual .001 uf capacitor(s) in the AFC phase detector circuit (C158 or C159).
4. A faulty 42 microhenry choke (L43) at test point "W." This choke is important for stability of the oscillator.
5. Component leakage or value change in the anti-hunt network (from test point "W" to ground).
6. Incorrect setting of the 3.58 MHz color oscillator frequency adjustment, L44.
7. The 2.2 meg resistors have changed value (not matched) in the AFC phase detector circuit. These resistors must be matched to within 2 percent of each other.

Voltage Check at Test Point "W" with Color Signal Reception.

The range of correction voltage at test point "W" to maintain color sync for the 3.58 MHz color oscillator should be approximately 5 volts (positive or negative) for optimum pull-in range. The inability to develop this range of voltage may result in several varied troubles as will now be noted.

The following information can be used to check the pull-in range (voltage correction range) at test point "W."

1. Feed a color signal into the tuner.
2. Connect a VTVM from test point "W" to ground.
3. Ground test point "K." Killer voltage can be measured at this point (not shown in Fig. 9-7). (Ground to open color channel.)
4. Adjust the slug of the color oscillator frequency coil (L44) by turning it clockwise as viewed from the bottom of the chassis. Check the voltage reading on the VTVM. About 5 volts negative should be present with the color pattern remaining in sync. Adjust the slug counterclockwise. The negative voltage should slowly approach zero, then reverse polarity and approach about 5 volts positive with the color pattern remaining

in sync. This range (approximately +5 volts) is considered normal and correct for consistent color sync.

If this range is considerably less than +- 5 volts, circuit troubleshooting is then called for.

A few service tips are as follows:

1. Check or replace the color oscillator and AFC phase detector tubes.
2. Check all components from test point "W" to ground. A prime suspect could be C163, the 0.1uf capacitor (leaky).
3. Check the 42 microhenry choke at test point "W" (L43).
4. With the oscilloscope, look for presence of burst at the phase detector coil, L40.
5. Check out the dual .001 uf capacitors (two are used) in the color AFC phase detector circuit, and also capacitors, or cold solder joints, in screen grid circuit of the burst amplifier tube, V16.
6. The 3.58 MHz crystal may also be defective. (Replace and retune oscillator for zero beat.)
7. Oscillator coil L44 may be faulty.

The adjustment of the 3.58 MHz oscillator must be made correctly. The procedure for adjusting the 3.58 MHz CW oscillator for "zero beat" with an "on the air" color program is as follows:
1. Tune in a color program and observe color on the CRT.
2. Ground test points "K" and "W" on the chassis. Should the color be "running" across the screen at a rapid rate, it is an indication that the 3.58 MHz oscillator requires tuning for a "zero" beat condition.
3. Adjust the core of the color oscillator frequency coil, L44, until color band movement comes to a near standstill. Continue to tune the 3.58 MHz oscillator coil while noting colors on the screen, until the color on the screen becomes a **solid** hue. It is at this point that all color has come to a standstill.

Service Note: At this point of the adjustment an error can easily be made. It is possible to adjust the core until two or more "red" and "blue" wide diagonal bars are seen to be at a standstill. However, this is **not** "zero" beat but can easily be mistaken as such. Proof of "zero" beat can easily be made by removing the ground from test point "W" and connecting a VTVM from this point to ground. A reading of up to 2.5 volts, either positive or negative, may be present. When the 3.58 MHz oscillator is correctly set for "zero" beat, the voltage measured at test point "W" will be virtually zero (or less than one volt positive or negative). Remove the VTVM and again ground test point "W," if further adjustment is necessary.

The proper tuning of the 3.58 MHz color oscillator will solve many color sync and fade out customer complaints.

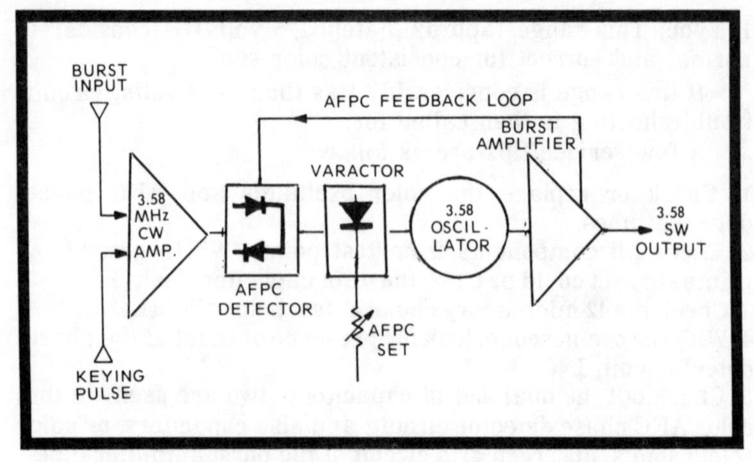

Fig.9-8. Solid-state color sync block diagram, CTC-40 RCA.

SOLID-STATE COLOR SYNC OPERATION

The block diagram in Fig. 9-8 shows the major functions for an RCA CTC-40 color sync system. The color sync signal (burst) is coupled to the burst amplifier from the collector of the second chroma amplifier. The burst amplifier is gated into operation by a keying pulse from the horizontal sweep transformer.

This amplified burst signal is fed to a phase sensitive discriminator (AFPC detector), which generates a DC voltage proportional to the phase difference between that of the burst signal and the 3.58 MHz local oscillator. The DC output varies the capacity of a varactor diode that in turn causes the proper correctional changes in the phase of the 3.58 MHz oscillator. The oscillator output is increased to a desired level by a separate CW amplifier. The resulting 3.58 MHz CW signal (synchronized and amplified) is fed to the color demodulator and automatic chroma control (ACC) circuitry.

COLOR BURST AMPLIFIER

Refer to Fig. 9-9 for the simplified schematic of the burst amplifier. The color sync signal (BURST INPUT) is coupled to the base of the burst amplifier from the collector of the second chroma bandpass amplifier through capacitor C721 A 15 volt positive-going keying pulse, from the horizontal sweep transformer, is also coupled to the base through R773. The integrating action of C721 and R773 provides the proper timing delay for the keying pulse. The transistor (Q701) is gated into conduction by the keying pulses which arrive at the base coincidentally with color burst. Burst is then amplified by the transistor and appears across transformer T701. During conduction, R701 sets the correct Q701 emitter operating point and C701 provides an effective AC bypass. Resistor R702 offers the desired amount of emitter degeneration for amplifier stability and thus determines the maximum voltage gain. Capacitor C703 is a neutralization component which cancels the effect of amplifier internal feedback capacity.

The burst amplifier base-emitter bias is maintained below cutoff between burst keying pulses (during scan time), to keep any information except color sync from being applied to the AFPC detector. The necessary scantime bias is provided by the discharging action of the emitter bypass capacitor (C701) through the emitter resistor (R701). As illustrated in Fig. 9-10, the emitter current flow caused by action of the burst keying pulse places a positive bias on the emitter which effectively reverse-biases the transistor during scan time. Diode CR715 prevents this bias voltage from exceeding the reverse emitter-base breakdown voltage.

AFPC DETECTOR SYSTEM

The purpose of the AFPC (automatic frequency and phase control) detector is to provide a DC voltage which is proportional to the frequency (and phase) difference between the applied color sync signal (burst) and the 3.58 MHz color oscillator. A very close control over the phase of this oscillator is needed for proper color demodulation because the oscillator acts as the reference against which demodulation occurs.

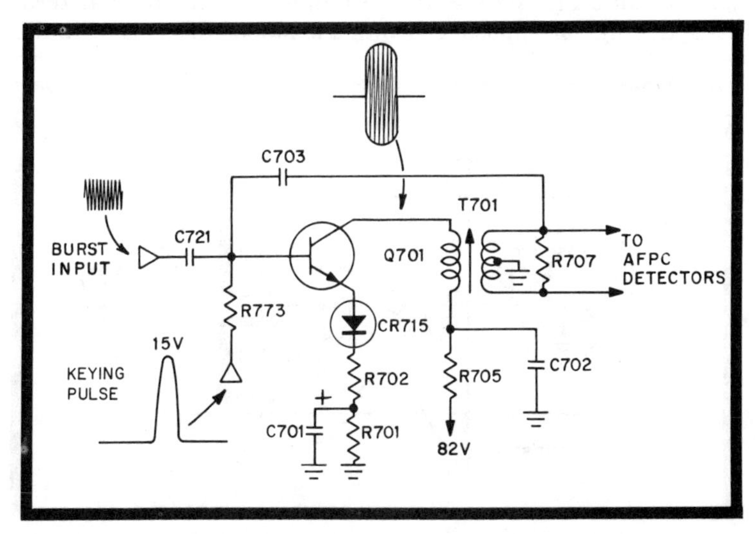

Fig. 9-9. Simplified schematic, solid-state burst amplifier.

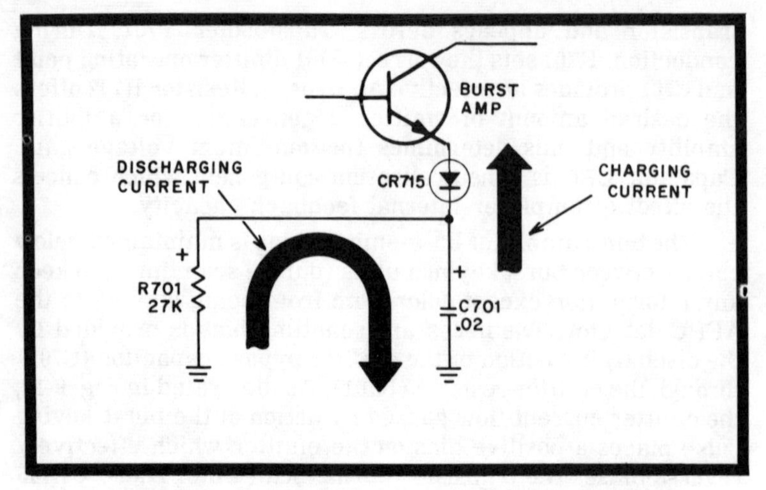

Fig. 9-10. Charge-discharge cycle of burst amplifier.

The AFPC detector uses a phase sensitive discriminator circuit in the basic operational configuration shown in Fig. 9-11. Looking at this schematic, the basic operation of the AFPC detector may be explained as follows: The burst signal is applied at equal amplitude but opposite phase through capacitors C1 and C2, to diodes D1 and D2 respectively. A sample of the 3.58 MHz oscillator output is applied to point C, the junction of diodes D1 and D2. If the oscillator is in sync with the burst signal as shown, the diodes will conduct equally. This can be verified by observing the three waveshapes applied at time T1. The 3.58 MHz oscillator waveform is at zero, thus, adding - or opposing - voltage is placed in series with the

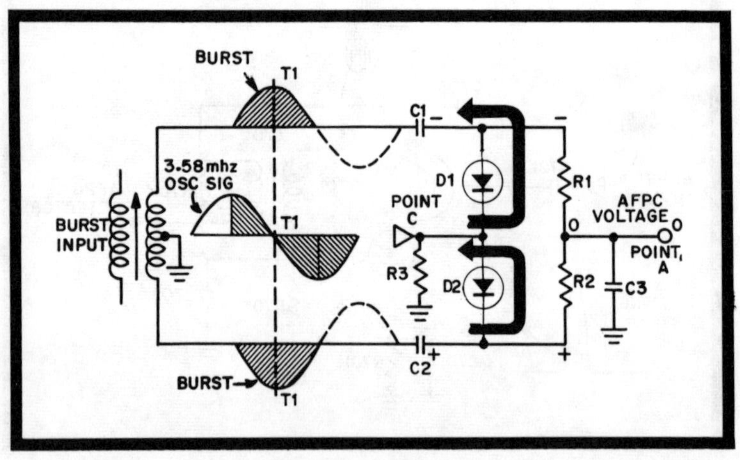

Fig. 9-11. AFPC detector operation. (Oscillator on frequency.)

diode conduction through R3. Therefore the resulting current flow through each diode is equal and opposite in direction. This places equal but opposite charges on C1 and C2. As these capacitors are discharged through resistors R1 and R2, this will cause equal but opposite voltages across the resistors. The voltage at the junction of these resistors (point A) will then be zero. The AFPC correction voltage can also be measured at point A.

The detector action shown in Fig. 9-12 is when the oscillator output is lagging the applied burst signal. At time T1 the oscillator waveform, due to its lower frequency, is somewhat delayed relative to the in-phase signal (dotted line). This results in the positive portion of the waveform being effectively placed in series with the diode conduction path through R3. Diode D2, therefore, conducts heavier than D1, causing the value of the positive charge on C2 to become greater than the value of the negative charge on C1. The resulting unbalance in current flow through R1 and R2 (when the capacitors discharge) causes the potential at point A to become positive. A similar but opposite action occurs when the oscillator frequency is leading that of the applied burst signal.

The actual schematic of the color sync detector is shown in Fig. 9-13. Inductor L707 acts as the diode return impedance (referred to as R3 in the previous figure). Capacitors C730 and C748 form a capacitor voltage divider network to supply the desired level of 3.58 MHz signal to the detector circuit. Resistors R703 and R704 act to minimize the possiblilty of diode radiation. The complete AFPC filter network consists of C707, R714, and C713.

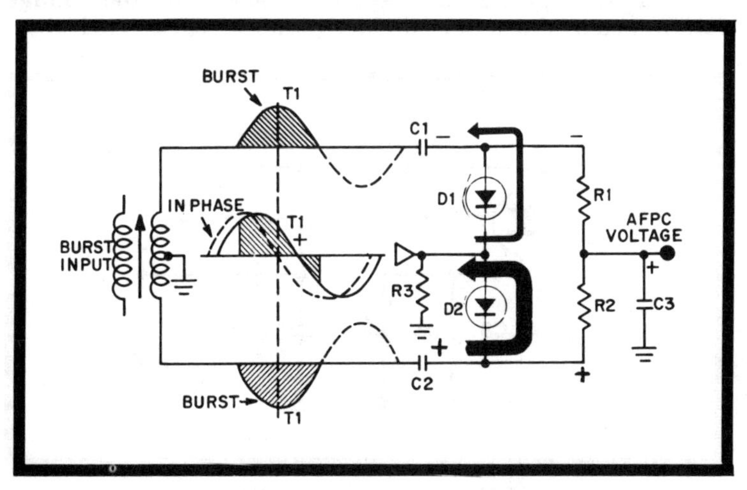

Fig. 9-12. AFPC detector, oscillator frequency low.

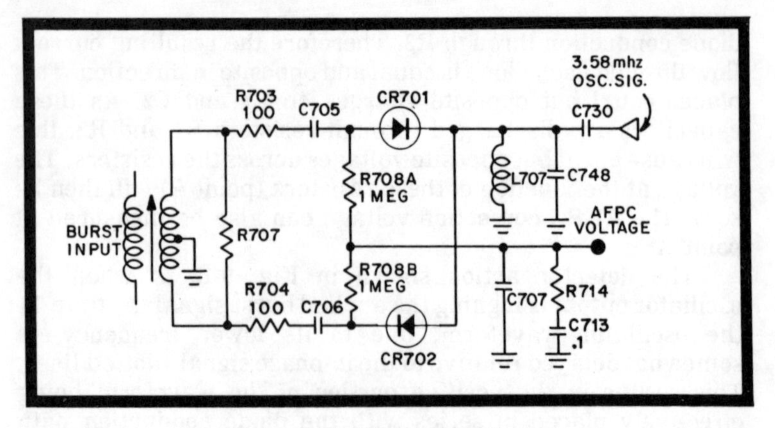

Fig. 9-13. CTC-40 AFPC detector.

SOLID-STATE 3.58 CW OSCILLATOR

The basic schematic of a solid-state 3.58 MHz oscillator is illustrated in Fig. 9-14. This oscillator is a modified Clapp circuit. Feedback energy necessary to sustain oscillation is coupled from the emitter impedance, consisting of R2 and C2, to the base via capacitor C1. Operating frequency is determined by the 3.58 MHz crystal and the capacity totals of capacitors C1, C2, and the varactor diode. The varactor has a specially constructed junction which enhances the normal voltage dependent capacity characteristics of a diode. Therefore, the frequency of the oscillator can be varied (over a limited range) by changing the voltage across the varactor diode. In this manner, the AFPC voltage - and the voltage determined by the voltage divider network represented by R1

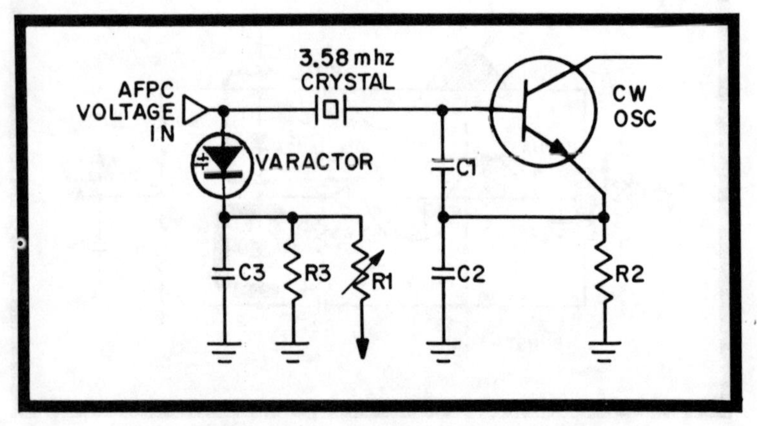

Fig. 9-14. CTC-40 solid-state 3.58 MHz oscillator.

and R3 - will change oscillator frequency. Capacitor C3 acts as a low impedance ground return for the varactor and has no effect on the oscillator frequency.

A more complete schematic of the 3.58 MHz oscillator is shown in Fig. 9-15. Resistors R727 and R728 provide oscillator base bias. Resistor R730 acts as an emitter stablilization resistor while R706 provides the proper collector load.

Also shown in Fig. 9-15 is the 3.58 MHz CW amplifier. This amplifier operates into a high-Q, single-tuned transformer (T702), which develops a sine wave from the output current pulses. Capacitors C748 and C730 act as a voltage divider network to provide the desired 3.58 MHz reference signal level to the AFPC detector circuit. The secondary of the oscillator couples 3.58 MHz energy to the color demodulators.

SOME ACTUAL COLOR TROUBLES

1. The slanting color bars on the screen of this color set gave forth some very misleading information. During a color program it was noted that the picture had slanting color bars across the screen and the hue and color level controls would have some effect on the color picture. This symptom had the appearance of an off-frequency 3.58 MHz color oscillator or a faulty color sync stage, but these circuits checked out fine. A few more visual and voltage checks were made but no clues turned up. Refer to Fig. 9-16. The scope was used to trace the chroma signal through the color stages. The defective culprit was pinned down to C129, a 220 pf coupling capacitor in the base circuit of TR7, 2nd color amplifier stage. This capacitor

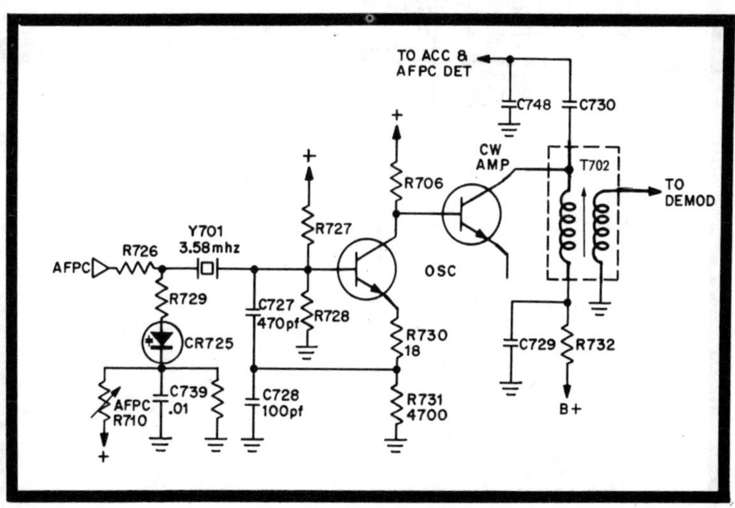

Fig. 9-15. More complete schematic of oscillator in Fig. 9-14, including CW amplifier.

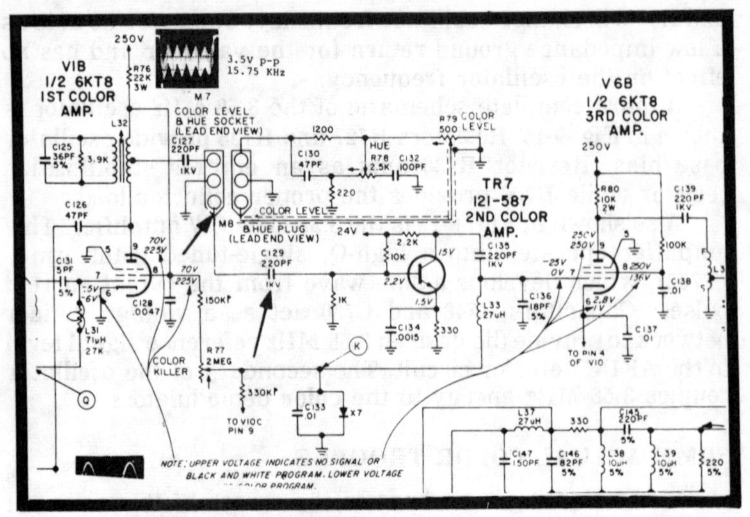

Fig. 9-16. Color troubleshooting example No. 1.

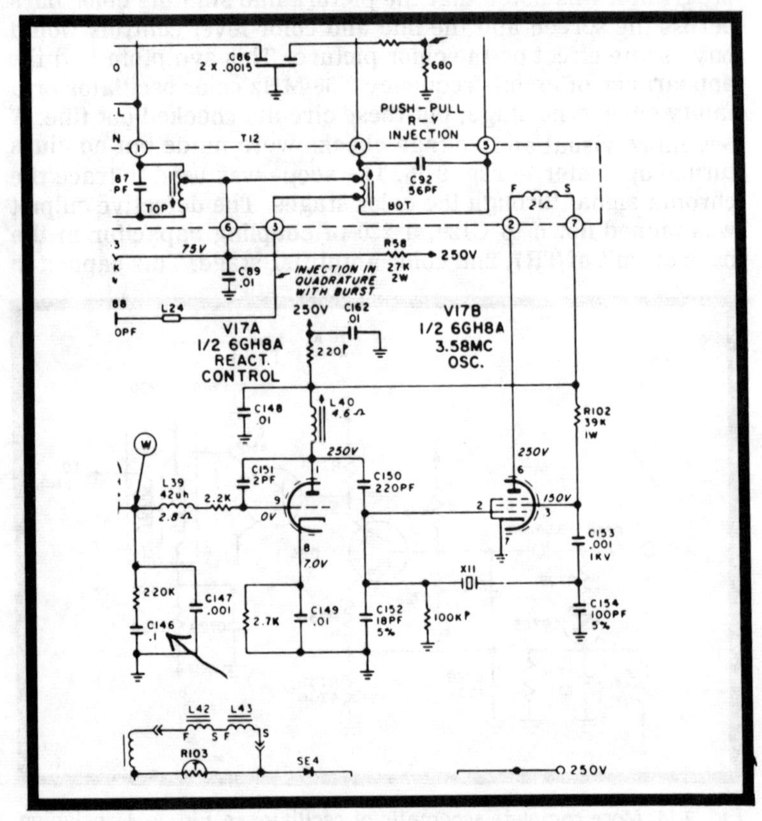

Fig. 9-17. Color troubleshooting example No. 2.

may become faulty intermittently and cause this odd color reception problem. Also check out the color level control plug M8 connections for any type of intermittent color reception problems.

2. Another set had a very discriminating color chassis and would only produce color when a color-bar generator was connected to the antenna terminal. When tuned to a TV station signal the color would not be produced or only very faint colors could be seen flittering across the screen. Also, the fine tuning control had to be set just right in order to catch this faint flicker of color.

This malfunction was tracked down to a faulty (leaky) 0.1 uf capacitor in the control grid circuit of V17A reactance control tube (Fig. 9-17), part of the antihunt network. I replaced the defective capacitor C146, and zero beat the 3.58 MHz color oscillator.

3. This Zenith chassis had a good black and white picture and sound, but no color reception.

All tubes, transistors and the IC chip were checked out in the color section but were found to be good. Voltages measured correct and were of no help.

A scope was then used to trace through the color burst stage and 3.58 MHz signal to the chip (pins 6 and 7). These were also found to be correct. However, there was no chroma signal found at pins 3 and 4 of the 221-62 IC color demodulator. Tracing back with the scope the chroma signal was found (see scope trace in Fig. 9-18) at the color level control side of capacitor C262. This 220 pf coupling capacitor was found to be faulty (open).

A new C262 capacitor restored good color operation.

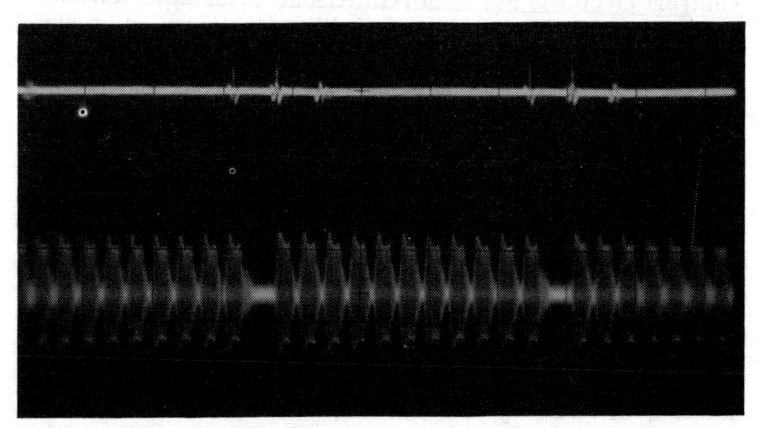

Fig. 9-18. Color troubleshooting example No. 3, scope waveforms. Top trace, no chroma at coupling capacitor. Bottom trace, chroma at other side of capacitor. Open 220 pf coupling capacitor was replaced and operation returned to normal.

10 Modern Color TV Crystal Ringing Circuits

In this chapter we will look at some modern crystal ringing circuits, ways to toubleshoot them and actual problems that have been known to develop. It's always good to have a little insight as to how these circuits should work before we can tackle a faulty circuit condition.

COLOR RINGING CIRCUIT FUNCTION

The burst amplifier stage separates and amplifies the color burst signal, but also provides a 3.58 MHz CW signal as subcarrier reinsertion for color demodulation. The crystal ringing circuit works on the principle that if we place a crystal in the output of the burst amplifier, burst itself will cause this crystal to ring at the 3.58 MHz rate.

The burst signal at the plate of the tube (V208, burst amp) "rings" at high Q circuit (see Fig. 10-1) consisting of a crystal (CR207) and the tuned circuit. Because the incoming burst signal is used to create the 3.58 MHz, a separate oscillator is not required. Since the 3.58 MHz signal is automatically locked-in by virtue of design, an AFC (automatic frequency control) circuit is not required. Also, a color killer circuit is not needed, but may still be used. When the color burst signal is not present, the 3.58 MHz CW at the color demodulators will not be generated, thus no color confetti is seen during black and white picture reception. Automatic color killer action is of course the resultant.

The ringing (oscillatory effect) of the crystal and tuned circuit is sustained over each completed horizontal scan line across the face of the picture tube which has a time duration of approximately 63 microseconds. The crystal must ring for this time period and at the same amplitude in order to produce correct color on the CRT. Since the Q of the circuit is very high, the decay (or fall-off) of the ringing signal (Fig. 10-2) is extremely small and can be disregarded. A circuit having a lower Q, or circuitry where this signal may be loaded down to a high degree, would cause a noticeable degree of decay. This, of course, would be undesirable since it would cause a color phase shift and produce wrong or weak color across the

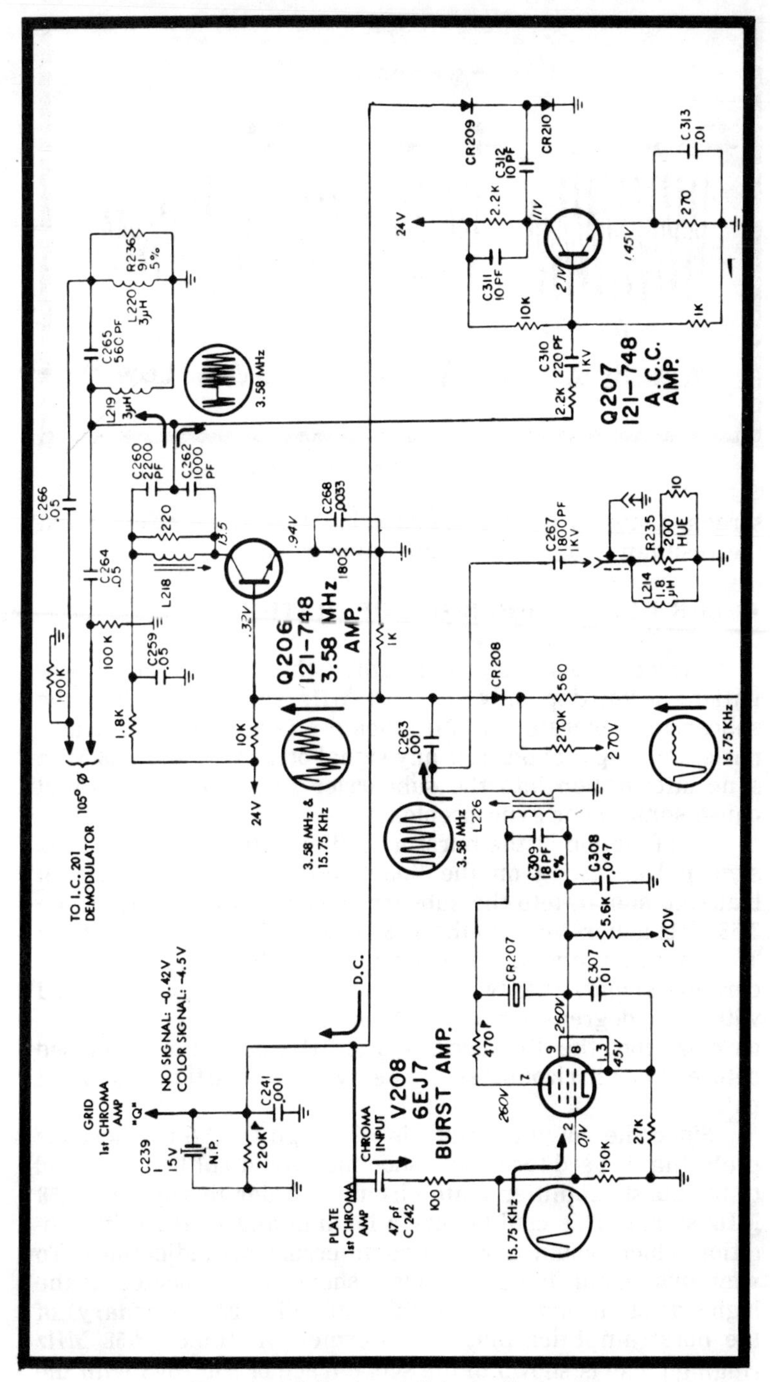

Fig. 10-1. Color burst "ringing" circuit.

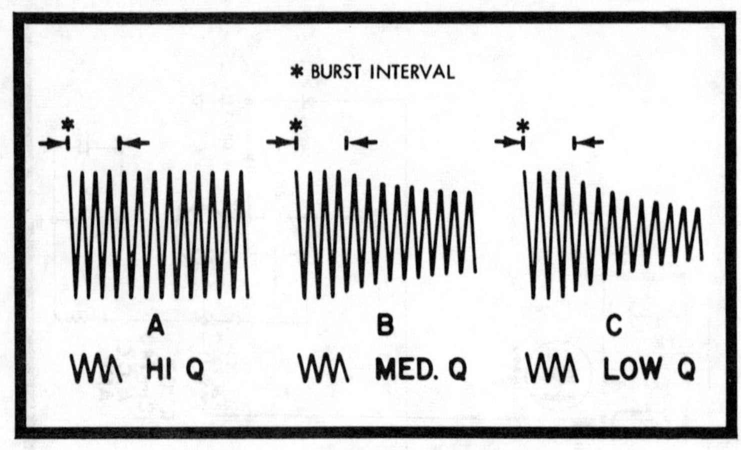

Fig. 10-2. Crystal ringing trace.

screen. Note in Fig. 10-2 the differenv Q of three circuits and the resultant 3.58 MHz signal waveforms.

COLOR BURST AMPLIFIER OPERATION

The color burst signal is gated through the burst amplifier, permitting only the 3.58 MHz sync signal into the subcarrier generator circuit. Should a fault in the burst amplifier develop that permits any signal other than the 3.58 MHz sync information into the subcarrier generator, this would cause some color sync problems.

The function of the burst amplifier is to let the 3.58 MHz sync pulse, sitting on the "back porch" of the horizontal blanking signal, into the subcarrier generator to sync in the 3.58 MHz generator with the 3.58 MHz oscillator located at the TV broadcast studio. To recover color information from the chroma signal, the two oscillators must be synchronized within 10 degrees. In order to maintain this critical synchronization, the TV receiver 3.58 MHz oscillator is pulsed before every horizontal scan line by the 3.58 MHz color burst pulse.

Since the ringing circuit is of a high Q, the oscilloscope probe that is used for troubleshooting should not be connected to the burst amplifier plate circuit in order to view the 3.58 MHz signal. This could cause a loading and phase shift condition which would result in some erroneous indications. To view this signal, the oscilloscope should be connected at the high side or secondary of the take-off coil (L226 secondary) of the burst amplifier plate transformer. A "Line" 3.58 MHz ringing pulse is shown in the scope trace of Fig. 10-3 with the probe at the high side of L226 (Fig. 10-1).

For "touch-up" color alignment the plate transformer (Coil L226) is adjusted to "center" the range of the hue control. When adjusting, the hue control is set to midrange (during a color program) and L226 is then adjusted for proper flesh tones. The hue control is located in the secondary winding of L226.

The output of the burst amplifier (ringing signal) is coupled to the base of a 3.58 MHz amplifier (transistor Q206) through capacitor C263. The ringing signal is less than 1 volt P-P and is amplified to provide approximately 1.5 volts P-P to the I.C. 201 color demodulator.

ANOTHER TYPE OF CRYSTAL RINGING CIRCUIT

Let's now take a look at the General Electric N-1 chassis color burst ringing circuit shown in Fig. 10-4. A positive pulse from the horizontal sweep transformer forward-biases both Y501 and Q501 at the proper time to allow the "burst back porch" reference signal to be amplified by Q501. This same pulse that gates Y501 and Q501 on, also charges capacitor C506. After the gating period, C506 begins to discharge, reverse-biasing Y501 and Q501. The charge on C506 is of enough amplitude (150 volts) that the base-emitter junction of Q501 is forced into zener action and thus becomes a discharge path for C506.

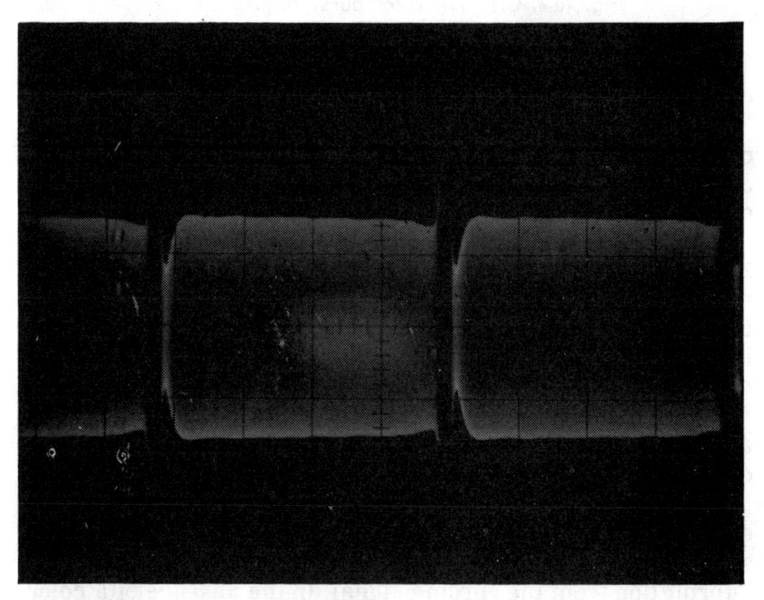

Fig. 10-3. Waveform of 3.58 MHz crystal ringing pulses. (Scope set at horizontal rate; 8 V P-P.)

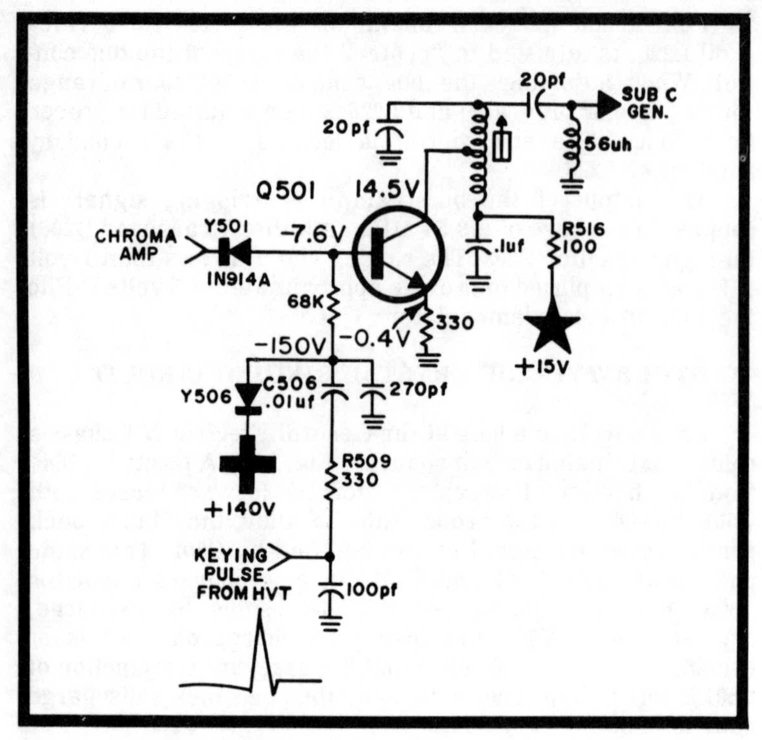

Fig. 10-4. G.E. N-1 color burst ringing circuit.

Diode Y506 is a pulse clamp and limits the positive excursion of the keying pulse to 140 volts.

The amplified 3.58 MHz sync pulse is now in the Hi-Q crystal ringing circuit and the subcarrier generator will continue to generate 3.58 MHz of the correct phase for the duration of the horizontal scan interval.

Now refer to Fig. 10-5 as we examine the function of the 3.58 subcarrier generator. A quick look at this circuitry reveals that V8 is simply a pentode amplifier using cathode bias and having a tuned plate circuit. The control grid circuit is a crystal filter. As the gated reference pulse is applied to the crystal circuitry it is started to ringing at the correct frequency and phase. The voltage of the ringing crystal is amplified by tube V8 and transformer-coupled to the color demodulator system.

The crystal filter is pulsed with the reference signal before each horizontal scan begins, insuring that the subcarrier signal will be in the proper phase for recovering the color information from the chroma signal. In the absence of a color burst reference signal, the crystal circuitry becomes dormant and there will be no output from the 3.58 MHz subcarrier generator. After seeing how these different types of crystal ringing circuits function, let's now put this all together by delving into some actual troubleshooting techniques and circuit troubles.

CRYSTAL RINGING TROUBLESHOOTING TECHNIQUES

In regards to the crystal ringing circuit, let's see how the waveform viewed on your scope looks relative to the color information seen on the screen of the color CRT. As we know, the color burst is located on the "back porch" of the horizontal sync pulse. Shown in the scope trace of Fig. 10-6 are two horizontal sync pulses with one complete line of video information between them which makes up one line of picture information across the CRT. When we pick the 3.58 MHz burst signal off the horizontal pulse at the left the crystal has to ring during the entire length of that scanning at the same amplitude. If it does not ring for the entire time (about 63 microseconds) the 3.58 MHz CW fed to the color demodulators will be weak or lost. So, if the crystal would only ring about half-way across the scanning line time, the color would fade out on the right side of the CRT or incorrect color would be developed. The prime suspect here would be a faulty 3.58 MHz crystal or other components in this circuit.

We will start these color checks by referring to the Zenith 12A1OC15 chassis in Fig. 10-7. Now, should the set have no color, connect a VTVM to test point Q (ACC voltage) then short across test points K and KK, thus defeating color killer

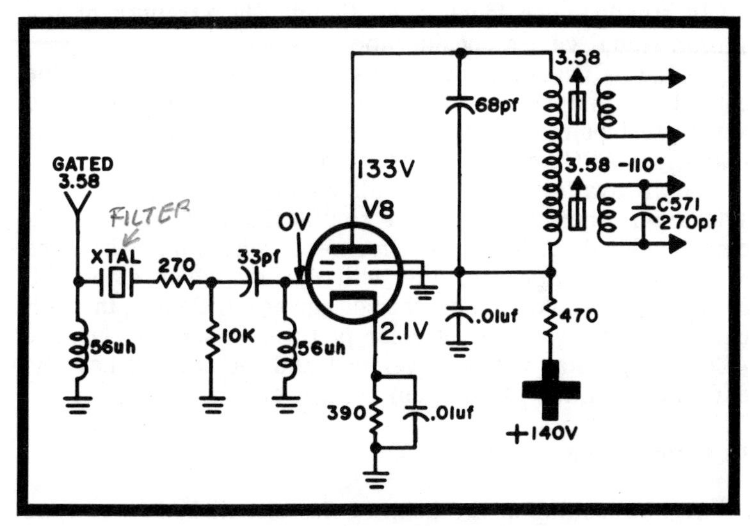

Fig. 10-5. 3.58 MHz subcarrier generator.

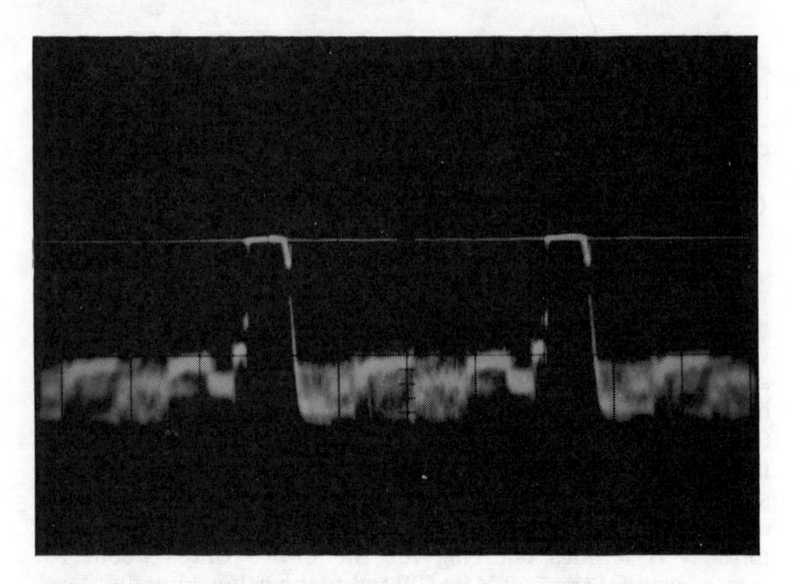

Fig. 10-6. Two horizontal sync pulses with video information between them. The video represents one line (scan) of picture detail. Color burst is barely discernible on "back porch" of the horizontal sync pulse. (Scope rate at one-half horizontal frequency.)

action and letting the chroma stages develop maximum gain.

Many color chassis will have a color killer / ACC detector that senses the strength of the incoming color burst signal. This will then develop a controlling voltage in the ACC stage (automatic color control) which is then applied to the chroma bandpass amplifier and controls the stage gain. The result is a fairly steady color level, even though the strength of the station signal varies considerably.

If the ACC voltage measures too low, a possible trouble could be a mistuned coil L226. To check this, clip a test lead across test points K and KK and then adjust coil L226 for maximum ACC voltage at test point Q.

So always take the ACC/killer control voltage into consideration when color circuit troubleshooting. If the bias developed in the ACC system becomes much too high (because of a faulty component) this will cut the color stage off, resulting in no color. Should the voltage be a little low, this will cause a weak, washed-out color picture. In some cases, the ACC voltage may not be developed and the set will have no chroma amplifier (ACC) gain control at all.

Checking the gating pulse

Again, looking at the Zenith color circuit in Fig. 10-7 we will now consider the horizontal gating pulse that should ap-

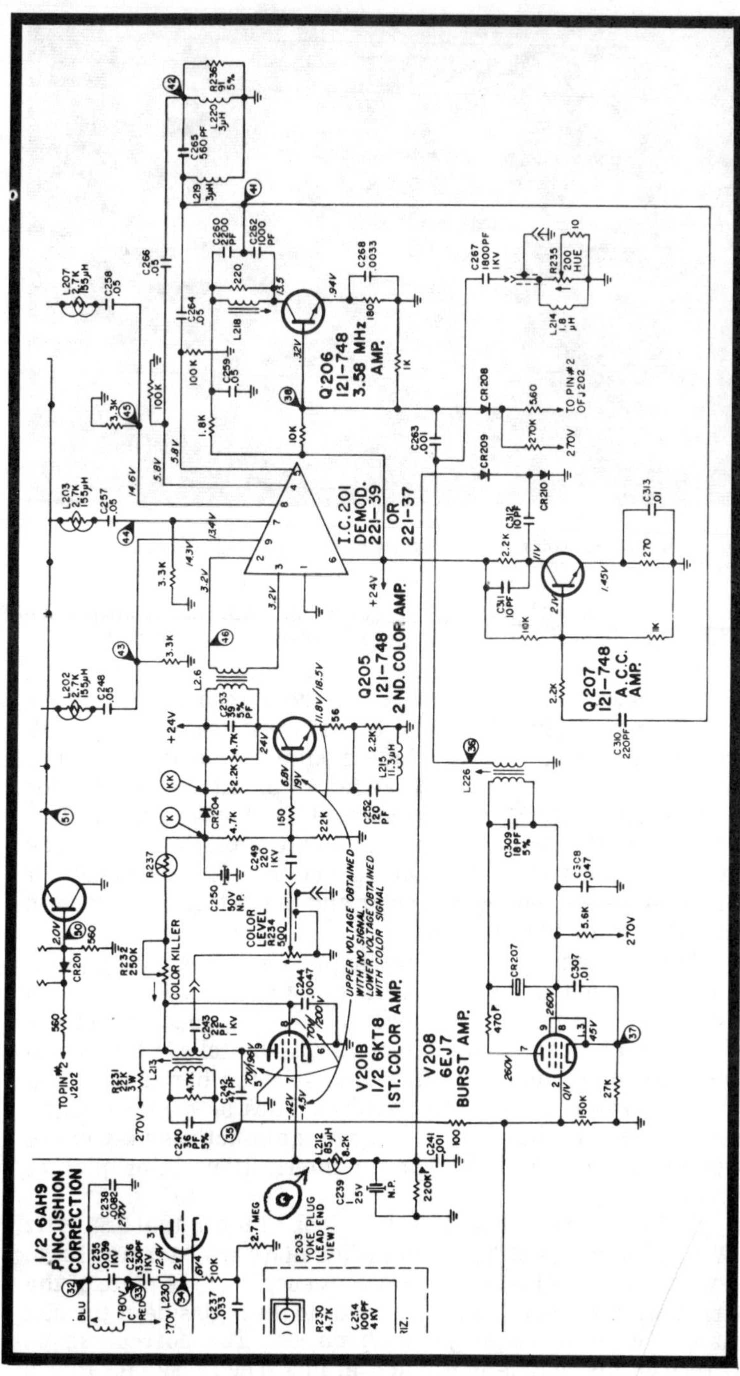

Fig. 10-7. Zenith 12A10C15 partial schematic.

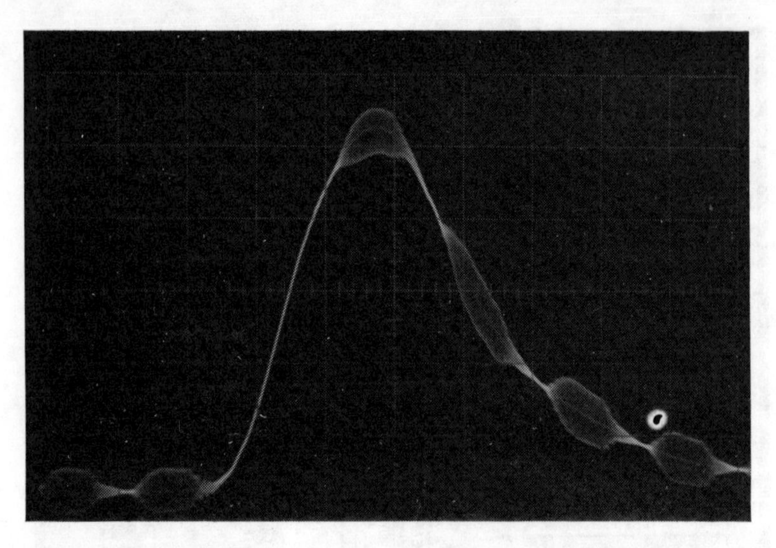

Fig. 10-8. Correct color burst at peak of gating pulse. (Scope at 5X for detail; horizontal rate.)

pear at the control grid - pin 2 - of V208 color burst amplifier tube. This circuit separates and amplifies the 3.58 MHz color burst. The eight cycles or more of color burst should sit right on the peak of the gating pulse in order to perform proper crystal ringing action; when sitting at the pulse peak, this will produce maximum crystal ringing output. Shown in Fig. 10-8 is the correct gating pulse-expanded 5X on the scope trace for more detail - that should be evident at the control grid of V208. If this gating pulse is missing, the burst amplifier will not be turned on and no crystal ringing action will take place. Of course, there will not be any color and the gating pulse system will have to be checked out.

Another symptom that may happen with the gating pulse is illustrated in Fig. 10-9 where the burst appears way down on one side of the pulse. This incorrect timing condition of the burst could be caused by defective (mismatched) horizontal phase detector diodes or other faults in the horizontal AFC system. Also, it may be the wrong diodes or a misadjusted horizontal hold control. These circuit malfunctions may cause poor color lock, no color, or an intermittent color picture condition.

Still referring to Fig. 10-7, let's look at the actual 3.58 MHz CW ringing signal at burst amplifier plate transformer L226. Never measure this signal (with your scope) at the plate of the burst amplifier tube! Always go over to the high side of the coil (L226) for accurate scope information. The correct scope pattern at this point is shown in Fig. 10-3. The reason for this is that the scope probe would be adding capacitance across the

High-Q crystal system and would load down and broaden the circuit response.

The low peak-to-peak signal (only 2.5 volts) shown in Fig. 10-10, 3.58 MHz CW scope trace, was caused by installing the wrong type 3.58 crystal. The correct crystal part number must be used in these sets. The crystals used in the sets with a 3.58 MHz oscillator will not work properly in these ringing circuits. This is a good point to remember.

While on the subject of faulty crystals - refer to Fig. 10-7 once more - the scope probe was placed at pin 2 of color demodulator "chip" I.C. 201 for a look at the chroma signal with the color-bar generator connected to the set. The very distorted chroma signal seen in Fig. 10-11 was caused by a defective 3.58 MHz crystal.

We came across a late model, small portable Zenith color set that had a sharp black-and-white picture, but not a trace of color. This was a 14CC15 chassis (see Fig. 10-12). For a problem of this nature you should defeat the color killer stage (if the set has one) and then measure the automatic color control voltage (ACC). In this chassis, the ACC voltage at test point Q measured about +5 volts and thus cut off the 1st color amplifier stage V201B. Diode CR210 was found to be faulty (open) and was causing an incorrect ACC voltage at the grid of V201B. Should diode CR200 open up this will also cause a no color condition.

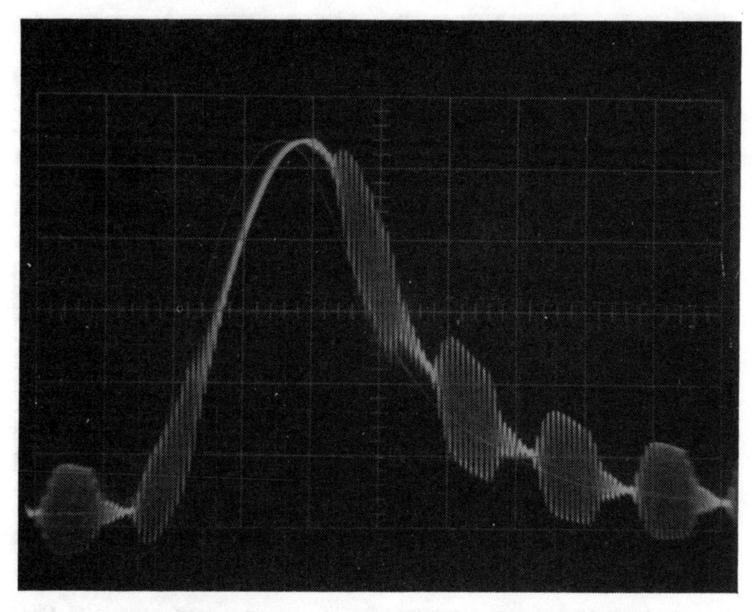

Fig. 10-9. Color burst not at peak of gating pulse will cause poor color lock; intermittent color or no color. (Scope at 5X; horizontal rate.)

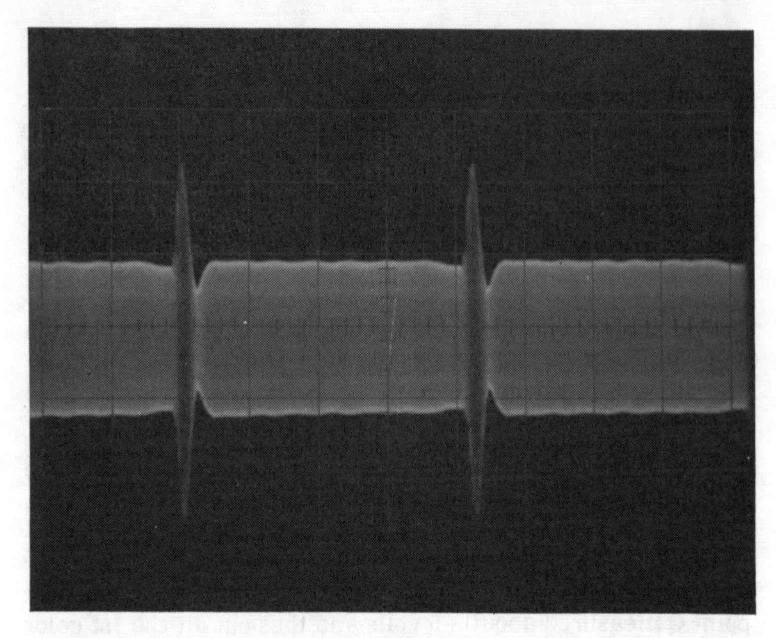

Fig. 10-10. Effect of installing wrong type 3.58 MHz crystal. The p-p voltage here is only 2.5 volts -- not enough to do the job properly. Always replace 3.58 MHz crystals with exact part number. (Scope at 15.75 kHz.)

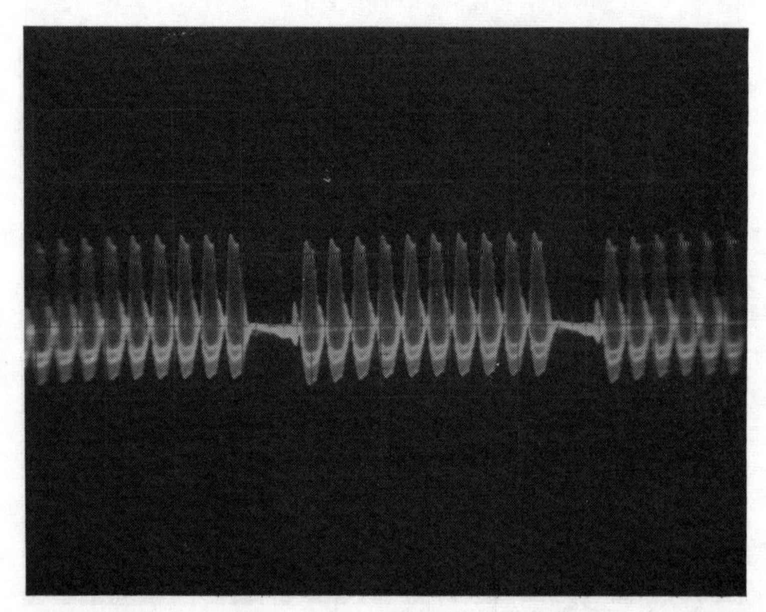

Fig. 10-11. Distorted chroma information caused by defective 3.58 MHz crystal. (Input: color-bar generator. Scope at horizontal rate.)

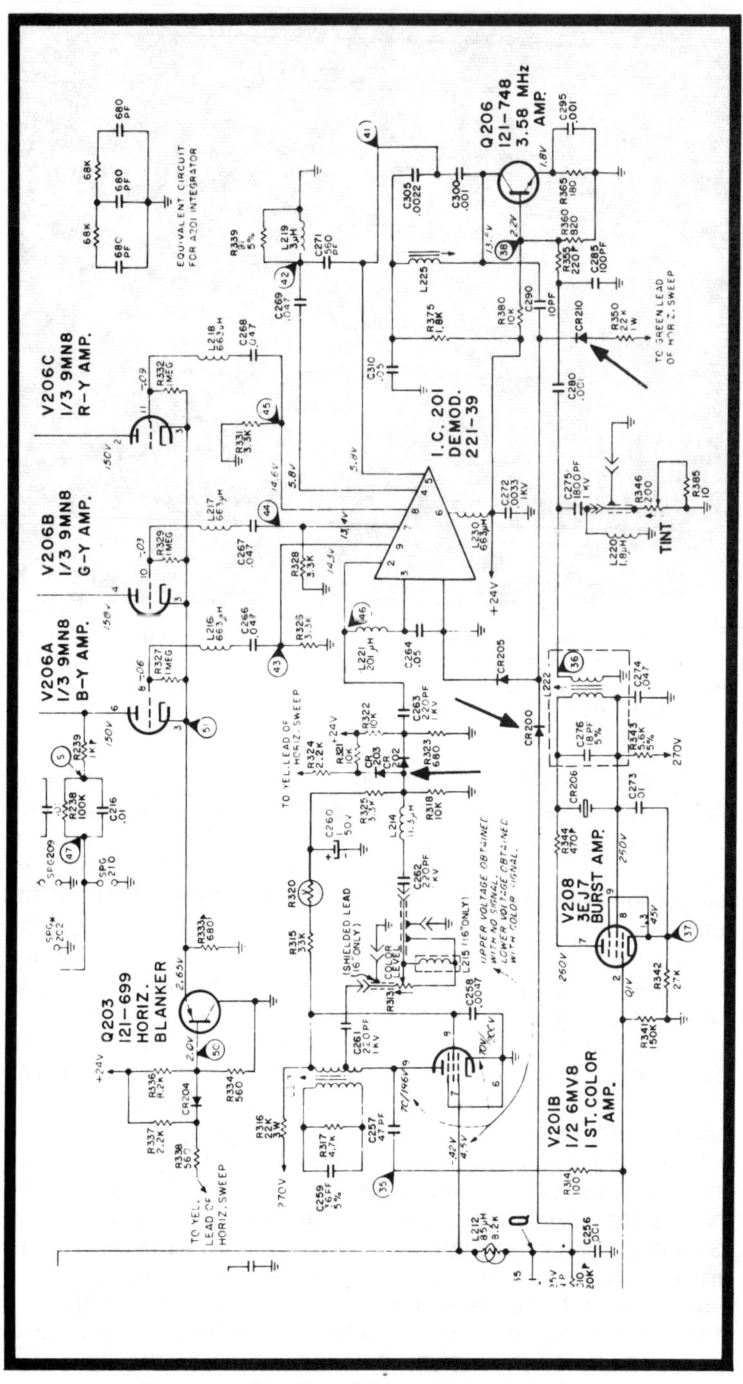

Fig. 10-12. Zenith 14CC15 partial schematic.

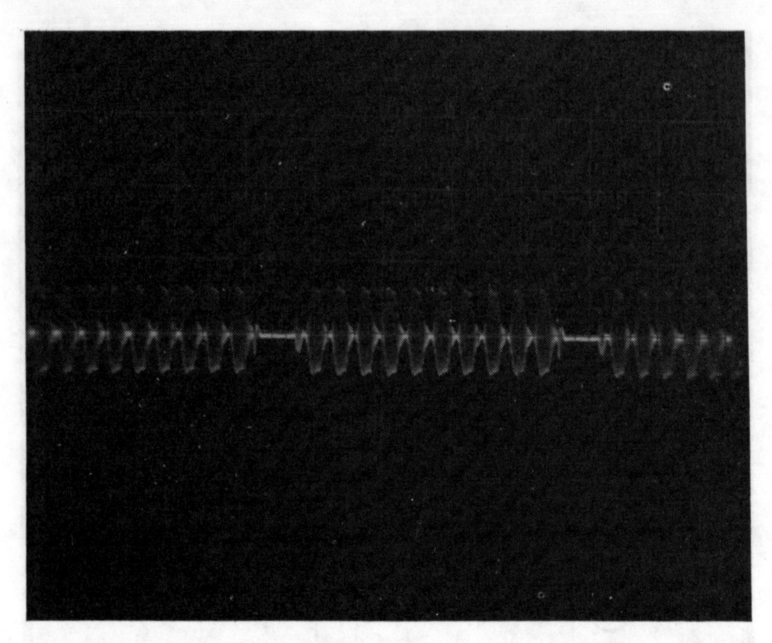

Fig. 10-13. "Weak color" waveform. The chroma information is correct (color-bar generator input) but p-p voltage is too low because the plate load resistor of 1st color amplifier tube had increased in value. (Scope at horizontal rate.)

If diodes CR203 - CR202 were faulty there would be no blanking pulse at the "chip." These diodes also eliminate any ripple voltage between blanking pulses that are used to provide a low amplitude pulse for cutting off the color demodulator chip, I.C. 201, during the horizontal blanking interval. If this blanking pulse is lost, there will probably be some irregular color lines on left or right sides of the picture that will change with video information. Leakage in diode CR210 may cause very strong color with no ACC action at all.

To locate a weak color symptom, use your oscilloscope to signal-trace through the circuits because this is the surest approach to this problem. The usual complaint is weak or tinted colors even when the receiver is properly tuned and the color control is turned full on. The trouble is tracked down the same way as for a no-color condition. Of course, you are now looking for a loss of color signal. The scope pattern in Fig. 10-13 shows a color-bar signal of the chroma information that is much too low, for a color chassis that had weak color. This low chroma signal was caused by the plate load resistor R316 of the 1st color amplifier stage V201B increasing in value to 500K ohms. (Refer to Fig. 10-12.) Misalignment of L213 may also cause this weak chroma signal condition.

Many other stages can cause loss of color or weak color reproduction. Any stage or system that the color signal must pass through can be at fault. The antenna, leadin wire, or CATV system can cause the color problem and they must not be overlooked.

Misalignment of the tuner or video IF amplifiers may also cause a loss of color. A trap in the IF or tuner system that is not tuned properly can cause a suckout effect on the signal response and cause loss of color or picture resolution.

Color Killer Circuit Operation

The color killer is a very useful and important circuit, but it is often overlooked by the TV service technician. Always check into the color killer section should you experience any loss of the color picture information.

The purpose of the color killer circuit is to "shut-down" the chroma amplifier stage during black-and-white picture transmission in order to eliminate color confetti on the receiver screen. Color killer circuit malfunction is a common cause for loss of color during a color program. Check the color killer adjustment and set it during a black-and-white program to a point where the color confetti (speckles) just disappears. For a "loss of color" complaint, always adjust the killer control (for most sets this control is turned counterclockwise) and see if the color will then come in. For a "loss of color" or a "weak color" condition, always make sure the color killer is operating properly.

The color killer in most receivers works in conjunction with the color burst amplifier and the 3.58 MHz reference oscillator. The signals from these two stages are applied to the killer detector (in later model sets, the ACC circuits) for further processing in order to develop the ACC and color killer control voltages. These killer circuits function on the premise that when a color burst is received, the killer circuit turns on the color bandpass amplifier(s), but when no burst is present (black and white program being received) the color amplifier(s) stage is biased off. When burst comes back in, the killer circuit has to receive a control voltage from the killer phase detector in order to bias the bandpass amplifier(s) back on again.

Of course, some of the receivers that utilize crystal ringing circuits (as described in Chapter 10) do not use or need a color killer circuit. In these sets, when there is no color burst to ring the high "Q" resonant crystal circuit there is, in a sense, an automatic color-killer action.

EARLY MODEL COLOR KILLER CIRCUIT

This early model RCA CTC-5 chassis color killer circuit is shown in Fig. 11-1. A horizontal pulse, supplied from a winding of the sweep transformer, is fed to the grid of V704-A, the burst keyer tube. This pulse is delayed by the RC network in the grid of the keyer tube, so as to arrive at the grid at the same instant burst is present on the color signal. Thus, the tube will conduct at burst time only. The burst signal is then fed to the grid of burst amplifier stage V704-B through T-703. The "hue" control, R164, is combined with capacitors in the primary of T703. Hue control adjustment effectively varies the load on the primary winding which in turn changes the **phase** of the burst signal at the burst amplifier grid.

The amplified burst information is fed to the grid of the 3.58 MHz subcarrier oscillator V705 through a filter network and crystal Y701. The oscillator is kept at the crystal frequency by the injection-lock principle.

The color killer circuit is in the tride section of the 6AW8 bandpass amplifier, V701-B. When burst is present in the received signal, an increased DC grid bias is present at the grid of the 3.58 MHz subcarrier oscillator (V705). This grid bias is also used as the color killer grid bias source and is sufficient to hold the killer tube at cutoff.

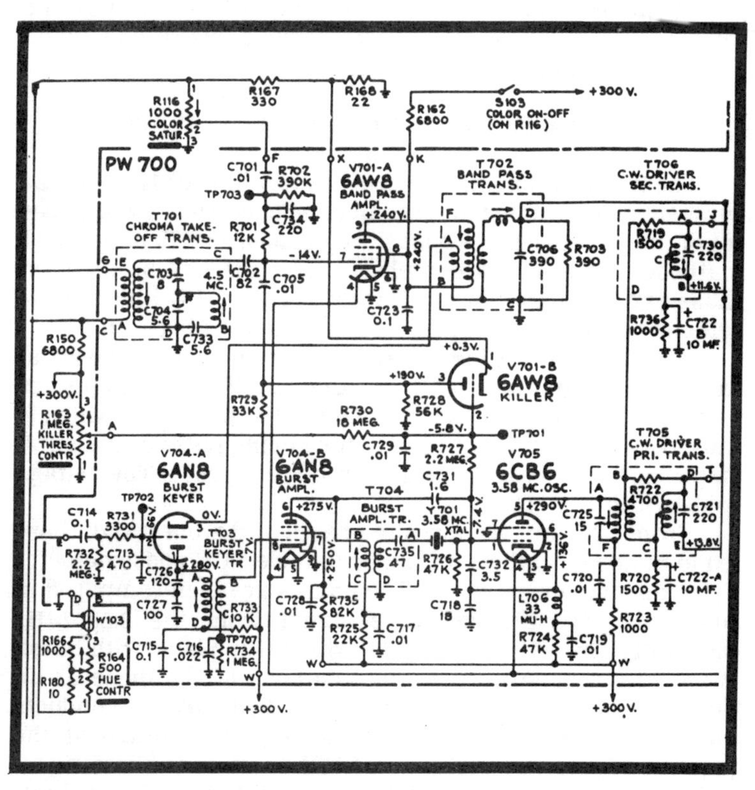

Fig. 11-1. Early model color killer circuit. (RCA CTC-5.)

A positive pulse obtained from the color saturation control (developed at cathode of horizontal blanking amplifier) is applied to the cathode of the killer, and when the killer conducts, the pulse is amplified and coupled to the grid of the bandpass amplifier. This causes the bandpass amplifier to cut off during picture trace time while zero bias is held during the pulse time. When burst occurs during pulse time (coincidence timing) the color killer is biased off and the bandpass amplifier operates normally.

The killer threshold control, R163, is adjusted to vary the level at which the color killer tube conducts. This must be adjustable for various levels from time to time because the color signal strength and color burst pulse being received will vary considerably. Should the color killer circuit malfunction it may bias off the color amplifier during a color program and only a black and white picture will be received. However, if the color amplifier stage is not cut off when receiving a black and white program, noise (colored snow) will develop on the screen due to this circuit operating at maximum gain. This is quite evident in low signal or noisy locations. The color killer circuit (if operating properly) will automatically bias the color amplifier stage into cutoff when a black and white program is being received.

EARLY MODEL ZENITH COLOR KILLER SYSTEM

The Zenith color killer circuit shown in Fig. 11-2 has the burst signal appearing across injection coil L30 at 180 degrees out-of-phase between the opposite coil ends. The signal at pin 3 (cathode of the ACC killer phase detector) will be at burst phase and the signal at pin 3 (plate of the ACC killer phase detector) will be 180 degrees out-of-phase with burst.

When measuring the voltage waveform (with a scope) at test point L (burst take off), only the amplified burst appears with at times a slight positive-going sawtooth voltage, which is caused by a slight integration of the keying pulse at the burst amplifier by the type scope and probe being used. The voltage is shown in Fig. 11-3.

From L30, the two opposite phases of the burst signal are fed to plate and cathode of the dual-diode ACC phase detector V21A, as mentioned before. The phases are then compared in amplitude with the reference signal from the 3.58 MHz oscillator. The reference signal, which is in phase with burst, is fed to the opposite plate and cathode of the diode at all times. Regardless of burst signal, one diode always conducts more than the other to develop a negative voltage at the junction of resistors R55A and R55B (with zero burst signal input, the voltage at this junction is almost zero). The voltage

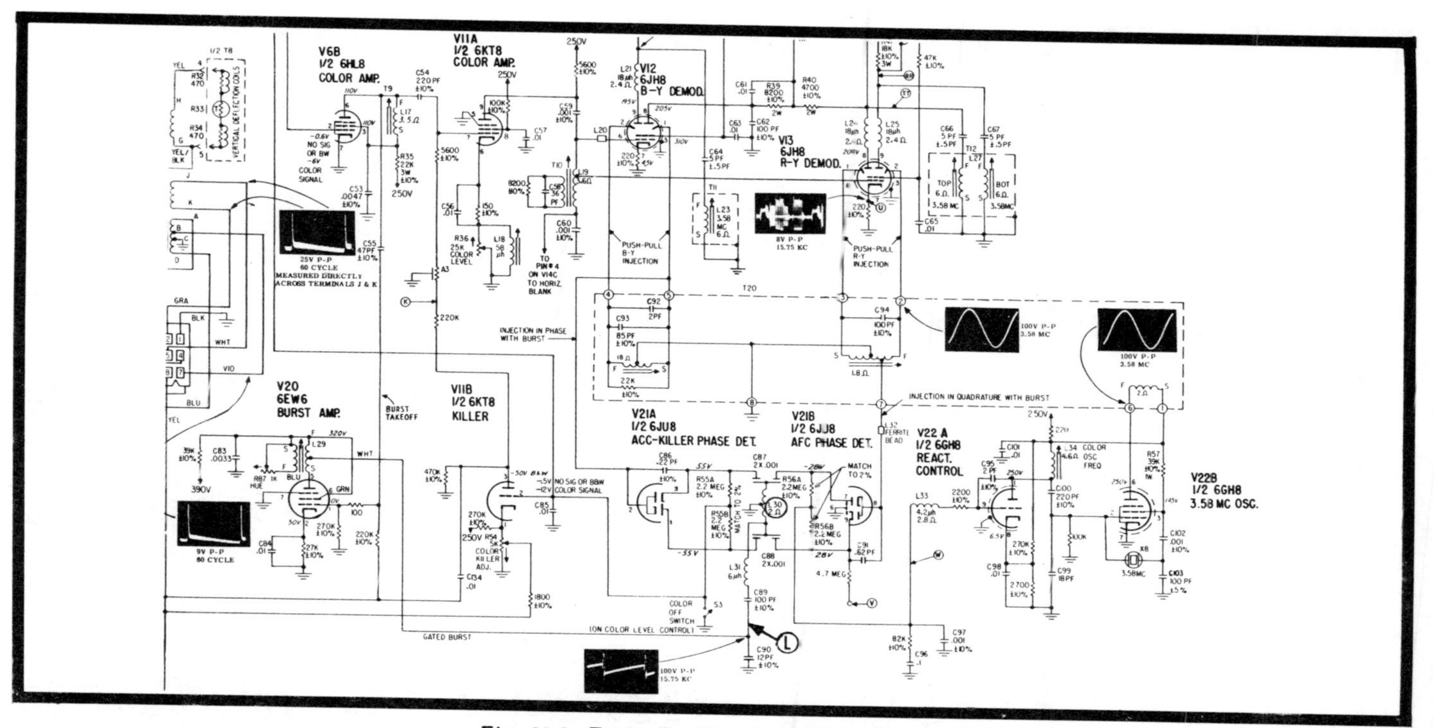

Fig. 11-2. Early Zenith color killer circuit.

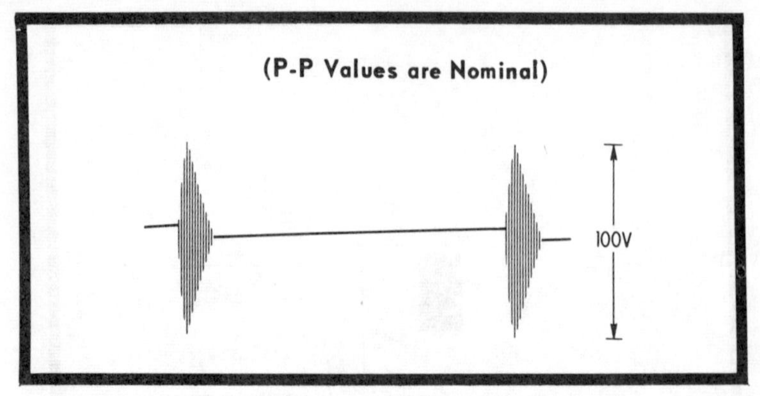

Fig. 11-3. Typical amplified burst signal.

at this junction may vary from a high of 14 volts negative to almost zero depending on signal input. (NEVER POSITIVE.)

The large negative voltage present during burst is used for two purposes: (1) as an ACC (automatic color control) voltage for the first color amplifier, and (2) as a cut-off voltage to disable the color killer stage. A "color-off switch" S3 is connected to the ACC and color killer grid voltage. As noted, in the absence of a burst signal, this voltage is essentially removed. In either case, the color killer tube conducts heavily (during the on-time established by the pulse at the killer plate during the burst interval) and develops approximately 50 volts which is applied to the 2nd color amplifier control grid to disable the color channel.

THE IN-PHASE COLOR KILLER SYSTEM

The ACC system senses the burst amplitude and controls the gain of the chroma amplifier, keeping the chroma output at a fairly constant level, even during signal fading or changes in TV station chroma signal strength. As noted previously the color killer biases the chroma amplifier to cutoff during a black-and-white picture but turns the chroma amplifier on during a color transmission. The negative bias required for the color killer and gain controlled (ACC) chroma amplifier is derived from the combined killer and ACC detector.

The combined color killer and ACC detector shown in Fig. 11-4 is an in-phase rather than a quadrature detector. Which means that the burst signal and 3.58 MHz oscillator signal are fed **in phase** to the ACC/killer detector, rather than 90 degrees out of phase.

Diodes CR401 and CR402 that are oppositely connected from those in the AFPC detector are fed the same reference signal from the burst transformer as are the AFPC diodes.

However, the 3.58 MHz from the oscillator transformer T402 secondary is not phase shifted. When the color oscillator is on frequency and at the correct phase, the two signals fed to diode CR401 will be in phase and those fed to CR402 will be 180 degrees out of phase. Thus, CR401 will conduct heavily during the positive half cycle, charging capacitor C401 negatively, as shown. As a result, the output of this detector will be at maximum negative voltage when the 3.58 MHz oscillator is in sync with the burst. The negative DC voltage level developed in the output of the killer / ACC detector is made to depend on the color burst amplitude. This negative voltage cuts off the color killer and applies a bias to the chroma amplifier. If a large burst signal is present, the high negative bias developed by the in-phase detector will reduce the gain of the chroma amplifier, which in turn lowers the color output level. Similarly, a low-level burst signal will increase the gain of the chroma amplifier, increasing its color signal output.

Since the level of the burst signal is dependent on the level of the chroma signal, ACC is accomplished by reference to burst rather than to chroma. The reason is that the average chroma level changes with variations in color, whereas burst remains at a more constant level.

ANOTHER RCA COLOR KILLER CIRCUIT

Refer to Fig. 11-5 as we look at this popular color-killer circuit in the early production CTC-16. A pulse from the burst transformer (T702) is fed to V705B, another phase detector

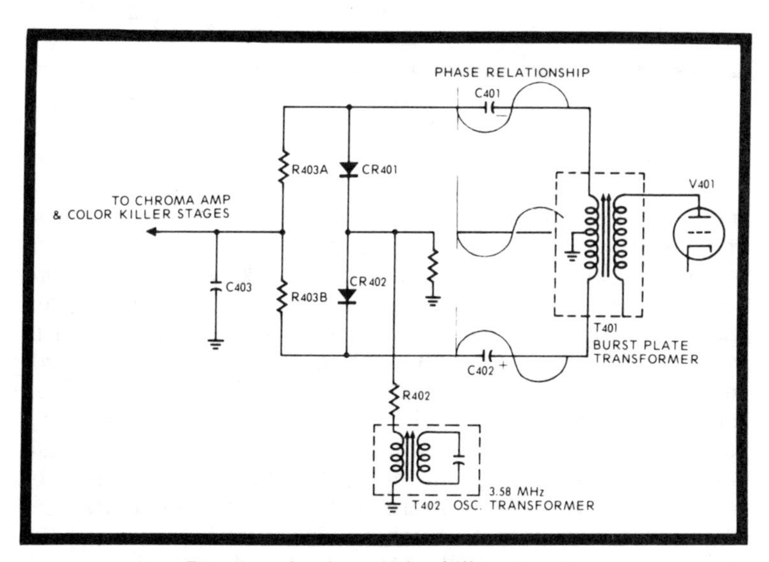

Fig. 11-4. In-phase color killer circuit.

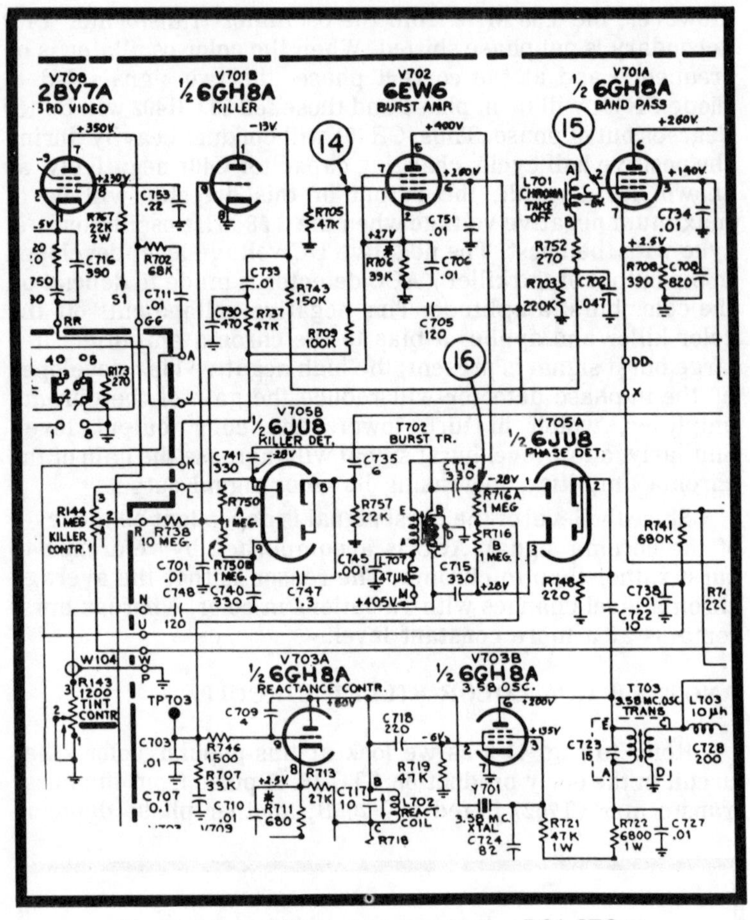

Fig. 11-5. Color killer in early production RCA CTC-16.

that is almost like the one used for locking in the 3.58 MHz color oscillator. Of course, the color oscillator control detector is of another phase, which does not add equally to both rectifiers. The negative voltage at pin 7 is not equal to the positive voltage at pin 9. This phase detector works as a percentage type ratio system. The difference between the two voltages appears across two matched resistors, R750A and B, and is then applied to the grid, pin 9 of the color killer, V701B. A small positive or negative fixed voltage is supplied by color killer control R144 to this same grid to establish the right operating point for the individual tube and burst level. A horizontal pulse is present at the plate, and the grid voltage sets the level of rectification. With -3 or more volts on the grid, plate rectification occurs with as much as —16 volts measured at the plate. The DC voltage at the plate is filtered and reduced with a voltage divider, then fed to the grid circuit of the chroma IF amplifier. About —8 volts is necessary to completely cut the tube off. (Pins 2 and 8 or the 6JU8 killer detector have no DC voltages.)

LOGICAL KILLER CIRCUIT ANALYSIS

Let's now check out the DC voltages of the color circuit shown in Fig. 11-5 with a VTVM. The color picture was way out of lock and the DC voltages were as follows:

		Faulty Operation	Normal Operation
6JU8	Pin No. 1	-48	-28
6JU8	Pin No. 3	-37	+28

If C740, a 330 pf capacitor, were open, this would indicate no burst rectification at pin 3, but some positive rectification of the 3.58 MHz would still be measured at pin 1. The fault in this case was a defective diode section of the 6JU8 tube.

Another trouble symptom was very weak color locking and touchy color killer action. When the killer control was adjusted for color, then there was color snow off channel. And if the color killer control was set for no color snow when off channel, there was no color picture.

The DC voltages were as follows:

		Faulty Operation	Normal Operation
6JU8	Pin No. 1	-14	-28
6JU8	Pin No. 3	+13	+28
6JU8	Pin No. 7	-12	-28
6JU8	Pin No. 9	+11	+28

This picture and voltage diagnosis points to a weak color burst condition. If the color picture looks strong, then the fault would be in the keyer stage. In this case a resistor in the keyer circuit had changed value and caused the weak burst. It is very important to always compare the voltages of the phase and color killer detectors.

The oscilloscope is most helpful for tracking down trouble in the color killer section. Scope waveforms throughout the chroma section and peak-to-peak voltage readings are "the in thing" for fast and accurate color circuit troubleshooting.

After it's certain the killer stage is at fault, voltage and resistance readings are used to isolate the defective part(s). Should the color chassis not be producing color, don't waste too much time checking the color bandpass amplifiers for faults as the killer stage may be defective and cause the color loss. With a "no color" symptom, always defeat the killer and ACC circuits to eliminate these stages as possible faults. It's best to check the 3.58 MHz color subcarrier oscillator, color killer, ACC system, chroma bandpass amplifiers, and color demodulators in that order for a rapid diagnosis. Off-value resistors, or open, leaky or shorted capacitors are the usual troublemakers that can cause color killer problems. And of course, defective tubes and any solid-state devices can and do cause trouble.

Some intermittent circuit troubles can be isolated by using a TV component failure formula. If the fault occurs in the first few minutes of set operation, this would indicate tube troubles, while from 10 to 15 minutes would mean a defective capacitor, and after 15 minutes or more could indicate resistors or anything else you might think. Of course, solid-state devices can go at any time. Then you can use a heat lamp, a can of coolant spray, VTVM and scope to track down the faulty components.

SOLID-STATE KILLER CIRCUIT OPERATION

The Sony color killer system, shown in Fig. 11-6, cuts off the second bandpass amplifier Q303 when black-and-white information is being received. To cause complete cutoff, the emitter of Q303 is returned to the fixed positive voltage at the junction of R321 and R323. The base voltage of Q303 is controlled by the conduction of color killer Q304.

Killer operation takes place as follows: When a color signal is received, the amplitude of the 3.58 MHz output signal of the CW driver increases, and a large negative voltage is developed at the anode of the combined ACC/killer detector D301. This negative voltage is applied to the base of NPN transistor Q304. The negative DC output of D301 acts to reduce conduction of Q304 during color reception and its collector voltage rises. A fraction of the collector voltage of Q304 is tapped off by R319 and R320 to form the base bias of Q303, the second bandpass amplifier. During black-and-white transmissions, the reverse occurs; Q304 conducts heavily and the base voltage for Q303 swings lower than the emitter voltage to effect cutoff.

The emitter of Q304 is returned to a variable positive voltage picked off at the arm of color killer control VR302. This control determines the amplitude of the 3.58 MHz signal that

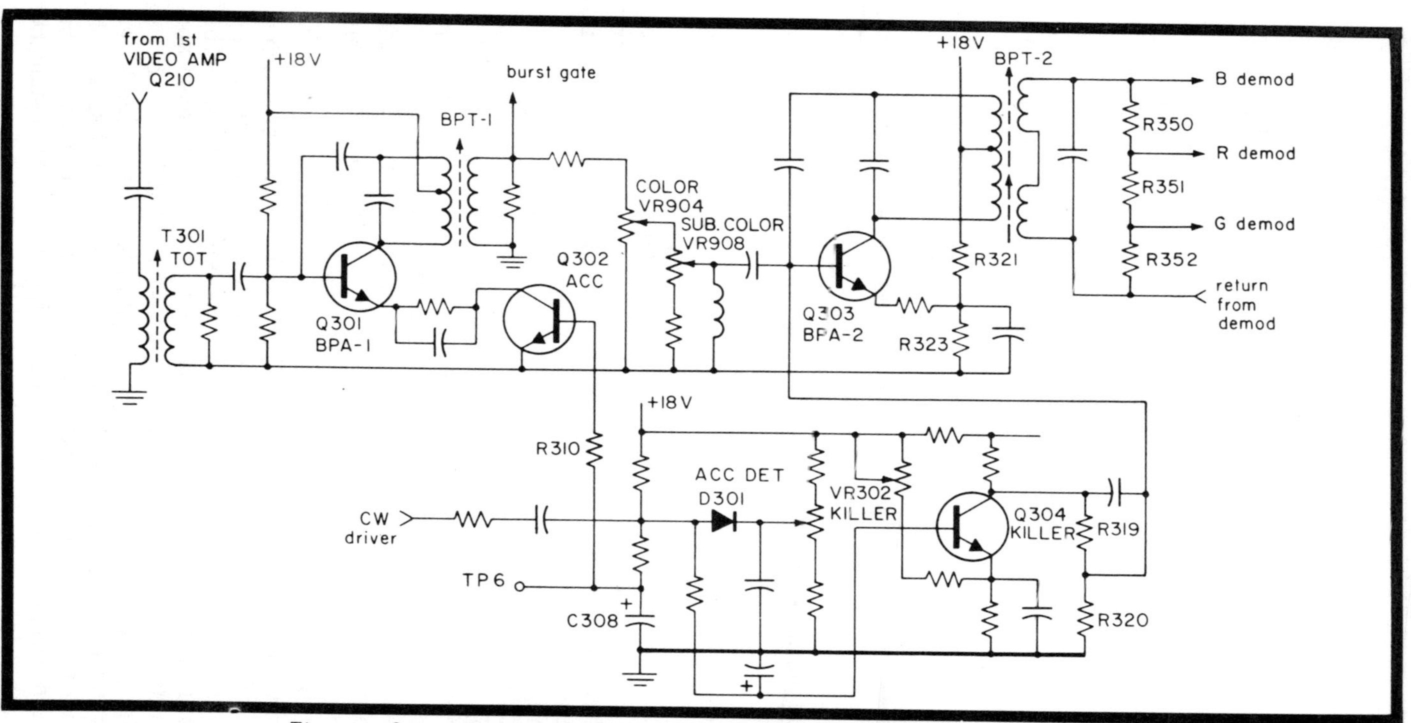

Fig. 11-6. Sony solid-state color killer, including ACC and bandpass circuits.

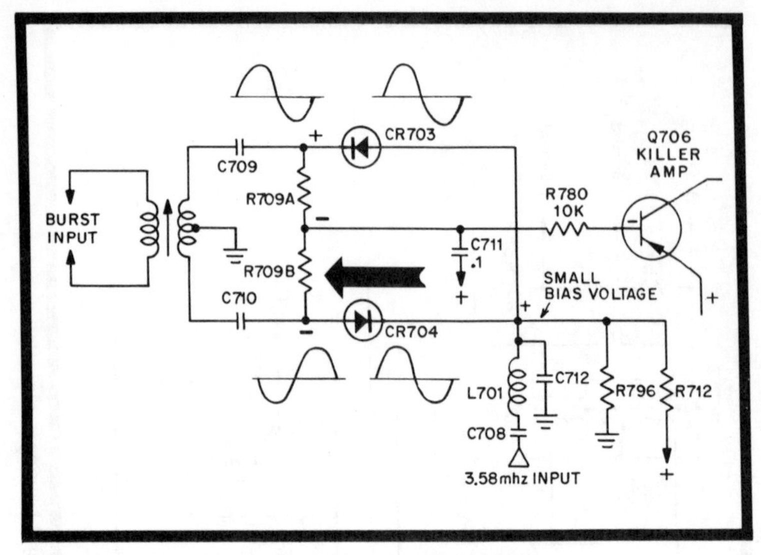

Fig. 11-7. ACC detector for CTC-40.

will permit Q304 to come into conduction. Thus, VR302 acts as the color-killer threshold control.

RCA SOLID-STATE KILLER SYSTEM

Depending on the type system used, color-killer action can be accomplished by cutting off any one of the chroma amplifiers, bandpass stages, or all of the color difference amplifiers. In the RCA CTC-40 chassis, color killer action is accomplished by cutting off the chroma bandpass stage.

The basic color killer sytem block diagram is illustrated in Fig. 11-8. Killer circuit control voltage is derived from the ACC detector shown in Fig. 11-7. The output of this detector is applied to the killer amplifier, which in turn controls the killer switch. The killer switch, as the name implies, "switches" the bandpass stage from a conducting to a nonconducting state when a black-and-white signal is being received.

A more detailed schematic of this color killer circuit is shown in Figs. 11-9 and 11-10. When a color program is being received (Fig. 11-9), the killer "switch" is biased into saturation by the conduction of the killer amplifier. The saturated killer switch effectively clamps both its base and emitter to its collector. Since the killer-switch emitter is direct-coupled to the bandpass amplifier base, the killer switch collector voltage (determined by divider network R740 and R741) essentially determines the forward bias of the bandpass amplifier. And as long as the killer switch conducts, the bandpass amplifier is forward biased or conducting.

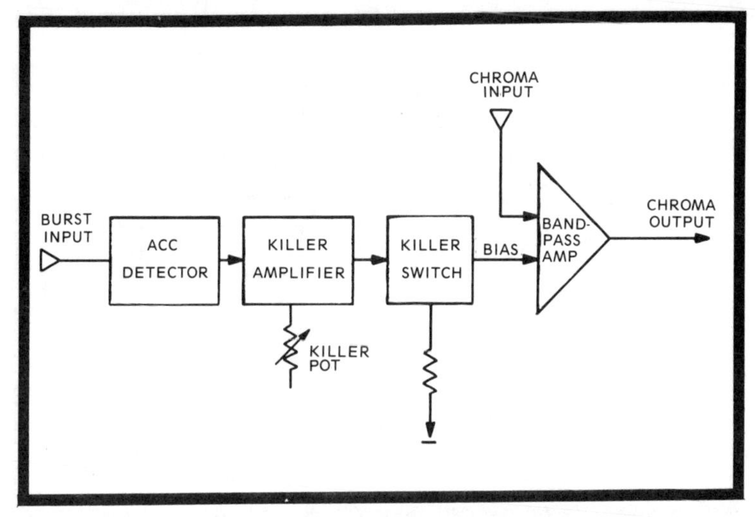

Fig. 11-8. RCA solid-state killer, CTC-40.

When a black-and-white transmission is being received, the absence of a burst signal (Fig. 11-10) causes the ACC detector to produce a positive output voltage. The positive ACC detector output cuts off the killer amplifier and thus stops the flow of forward-biasing current to the killer switch. The killer switch, being cut off, effectively "disconnects" the bandpass amplifier from its forward bias supply, which is the killer switch collector circuit.

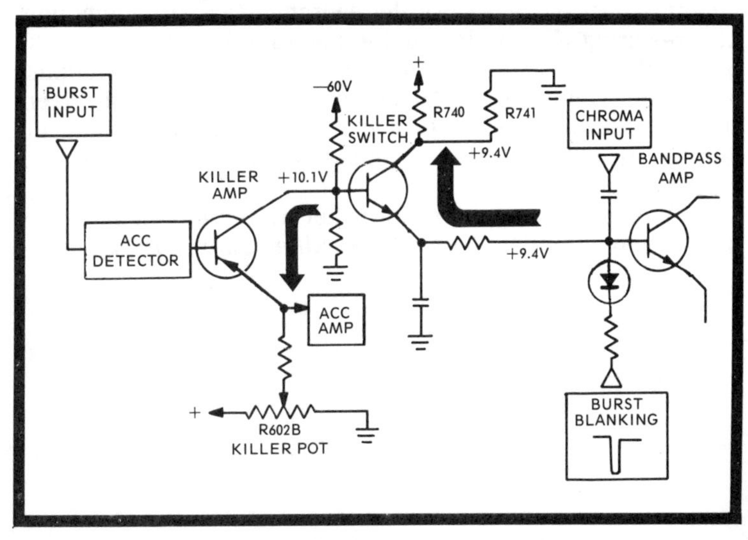

Fig. 11-9. CTC40 killer—color reception.

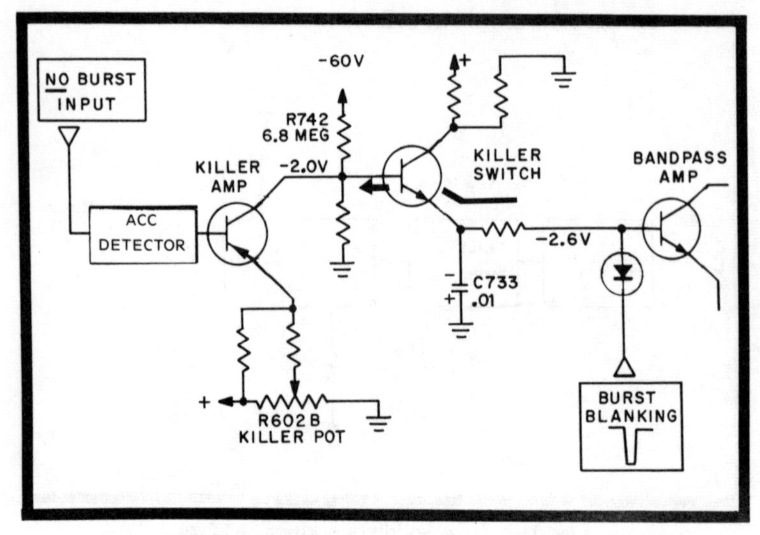

Fig. 11-10. CTC40 killer—B & W reception.

To assure complete bandpass amplifier cutoff, the base is held slightly negative from a voltage derived in the following manner. The loss of the killer amplifier conduction "unloads" a negative bias source coupled to the killer switch base from a -60-volt supply to the horizontal output circuit. This will result in a killer switch base bias of approximately -2.0 volts. Since both the killer switch and the bandpass amplifier are cut off, a negative potential resulting from the burst blanking pulses "charges up" capacitor C733. This negative voltage is sufficient to forward bias the killer switch base-emitter junction. The resulting action clamps the negative killer switch base potential to that potential found on the base of the bandpass amplifier. The base-emitter current flow which results from the killer switch being forward biased (under these conditions) is not sufficient to cause significant killer switch conduction.

The conduction point of the killer amplifier can be varied by altering its emitter bias with the killer control, R602B.

Chroma Demodulator and Amplifier Circuits

12

Color demodulators are phase-sensitive devices requiring two input signals for operation. When these two input signals are in phase, plate voltage will drop and as this signal reaches 180 degrees out of phase, the plate voltage rises. When the two signals are 90 degrees apart, plate voltage will be unchanged, thus, the demodulator produces maximum output when the phase difference between input signals is 0 or 180 degrees. The demodulator (also called a detector) practically ignores a signal whose phase is 90 or 270 degrees from the locally injected CW signal. However, at intermediate phase angles between 0 and 90 degrees, the demodulator output does have some intermediate value.

To recover the color difference signal, the 3.58 MHz CW subcarrier must be reinserted. This is done by feeding a 3.58 MHz signal from the local oscillator (see Fig. 12-1) to the cathodes of the X and Z demodulators. If the CW signals applied to a pair of demodulators are synchronized in phase with the CW fed to the modulators at the transmitter, the original color-difference signals (I and Q) can be detected. Color-difference signals R-Y, G-Y, and B-Y can be demodulated directly from the sideband signals by altering the phase of the CW signals applied to the demodulators.

The chroma amplifiers amplify the chroma signal and apply it to the chroma demodulators where the color difference signals R-Y, G-Y, and B-Y signals are demodulated, then applied to the red, green, and blue control grids of the color CRT.

Essentially, the chroma demodulators fall into three basic classifications. These are low level color demodulation, high level demodulation and the solid-state demodulators using transistors or ICs. Our first circuit analysis will be for the low level color demodulator system.

LOW LEVEL COLOR DEMODULATOR

Refer to Fig. 12-1. The color information from the bandpass amplifiers contains two chroma signals in quadrature. These two signals will each have to be recovered separately.

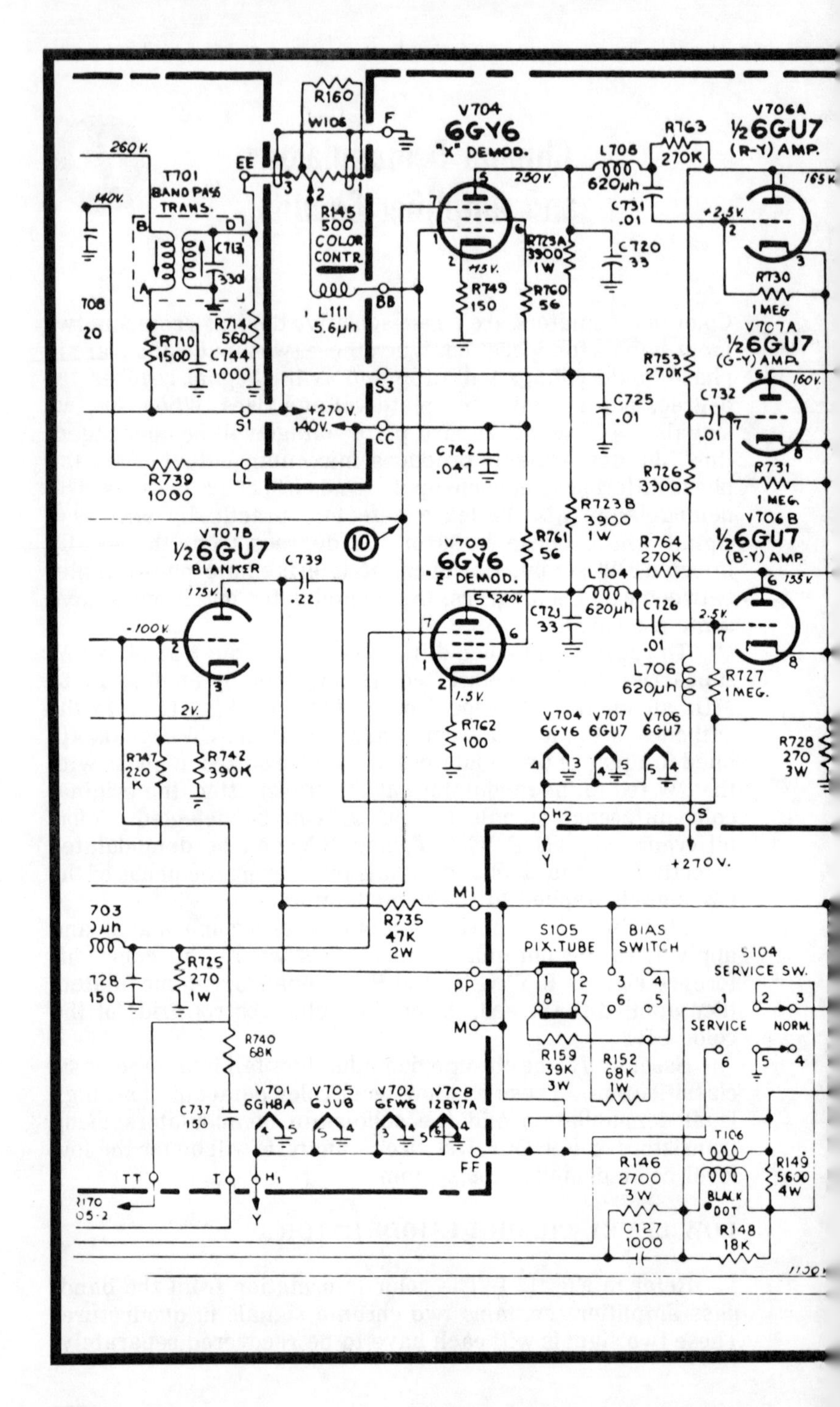
V704
6GY6
"X" DEMOD.
V706A
1/2 6GU7
(R-Y) AMP.
V707A
1/2 6GU7
(G-Y) AMP.
V706B
1/2 6GU7
(B-Y) AMP.
V709
6GY6
"Z" DEMOD.
V707B
1/2 6GU7
BLANKER
T701
BAND PASS
TRANS.
R145
500
COLOR
CONTR.
S105
PIX. TUBE
BIAS
SWITCH
S104
SERVICE SW.
SERVICE
NORM
+270V.

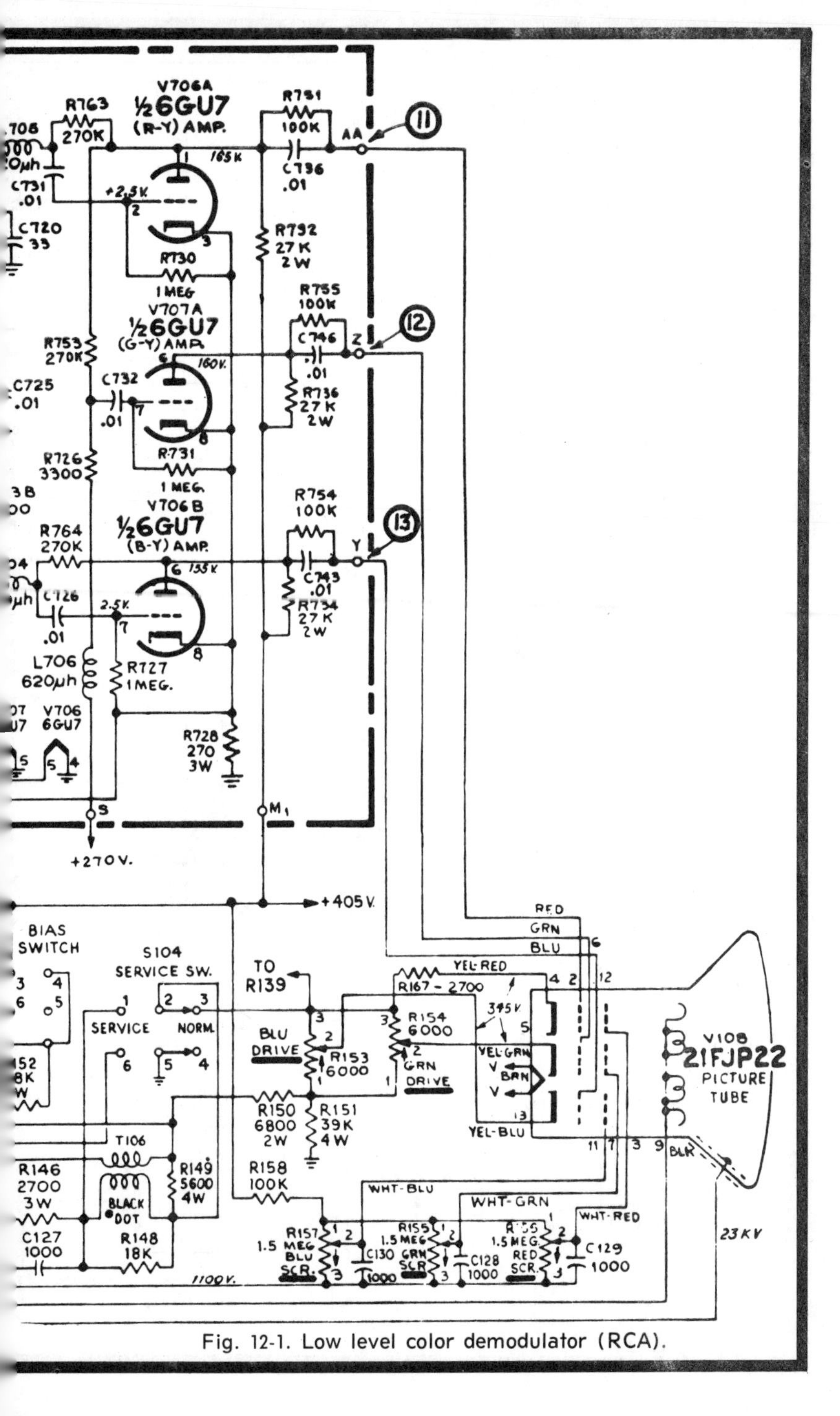

Fig. 12-1. Low level color demodulator (RCA).

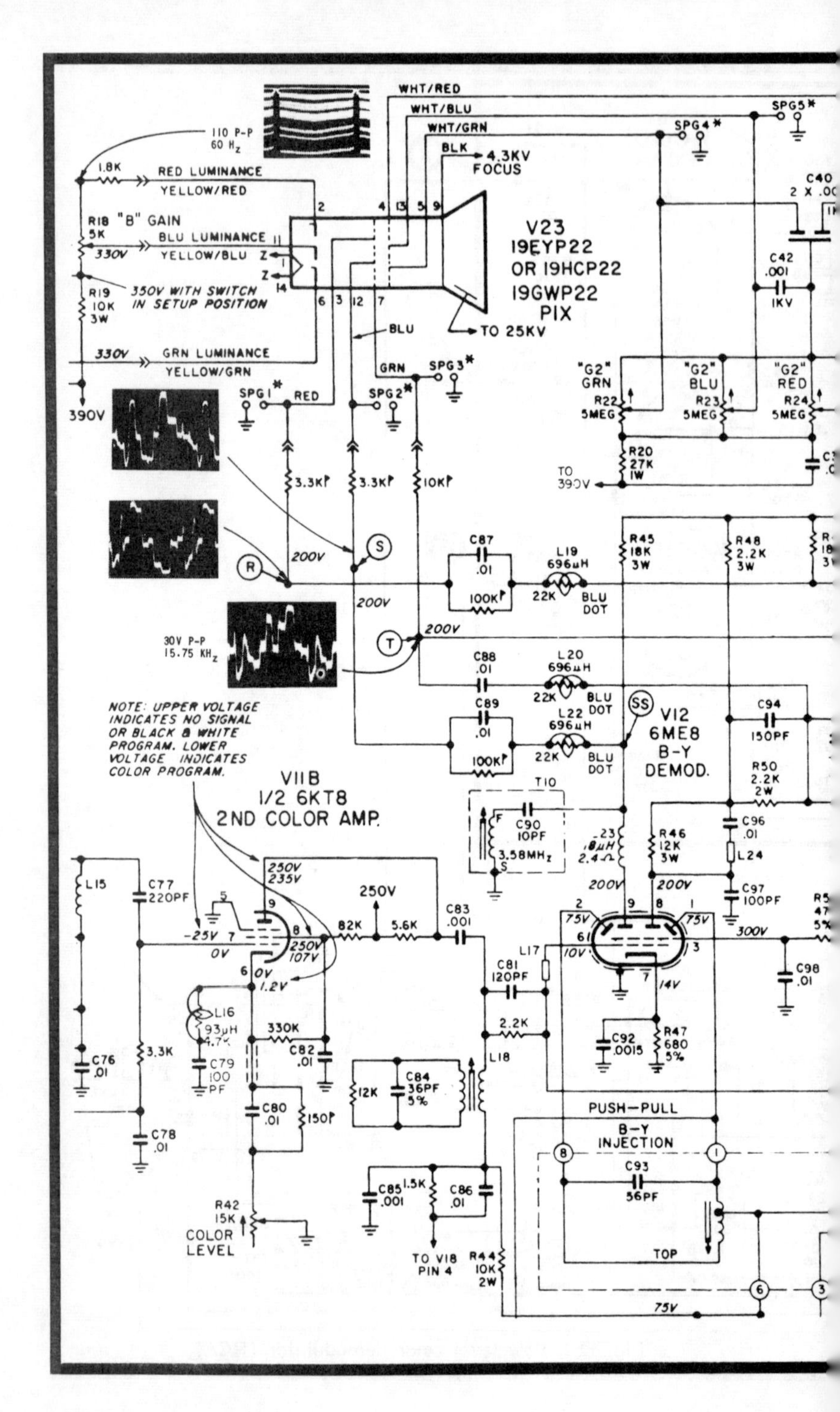

WHT/RED
WHT/BLU
WHT/GRN
BLK
4.3KV FOCUS
110 P-P 60 Hz
1.8K
RED LUMINANCE
YELLOW/RED
R18 "B" GAIN 5K
BLU LUMINANCE
YELLOW/BLU
330V
R19 10K 3W
350V WITH SWITCH IN SETUP POSITION
330V
GRN LUMINANCE
YELLOW/GRN
390V
V23
19EYP22
OR 19HCP22
19GWP22
PIX
TO 25KV
BLU
GRN
SPG1*
SPG2*
SPG3*
SPG4*
SPG5*
RED
C40
C42 .001 1KV
"G2" GRN R22 5MEG
"G2" BLU R23 5MEG
"G2" RED R24 5MEG
R20 27K 1W
TO 390V
3.3K
3.3K
10K
200V
200V
200V
C87 .01
100K
L19 696µH
22K
BLU DOT
R45 18K 3W
R48 2.2K 3W
30V P-P 15.75 KHz
C88 .01
L20 696µH
C89 .01
L22 696µH
V12 6ME8 B-Y DEMOD.
C94 150PF
R50 2.2K 2W
NOTE: UPPER VOLTAGE INDICATES NO SIGNAL OR BLACK & WHITE PROGRAM. LOWER VOLTAGE INDICATES COLOR PROGRAM.
V11B 1/2 6KT8 2ND COLOR AMP.
T10
C90 10PF
3.58MHz
.23 .8µH 2.4Ω
R46 12K 3W
C96 .01
L24
C97 100PF
250V 235V
250V
82K
5.6K
C83 .001
-25V
0V
250V 107V
0V
1.2V
75V
10V
14V
300V
C98 .01
L15
C77 220PF
L16 93µH 4.7K
C79 100 PF
330K
C82 .01
C81 120PF
L17
2.2K
C92 .0015
R47 680 5%
C76 .01
3.3K
C80 .01
150
C84 36PF 5%
12K
L18
C78 .01
PUSH-PULL B-Y INJECTION
C93 56PF
TOP
R42 15K COLOR LEVEL
C85 .001
1.5K
C86 .01
TO V18 PIN 4
R44 10K 2W

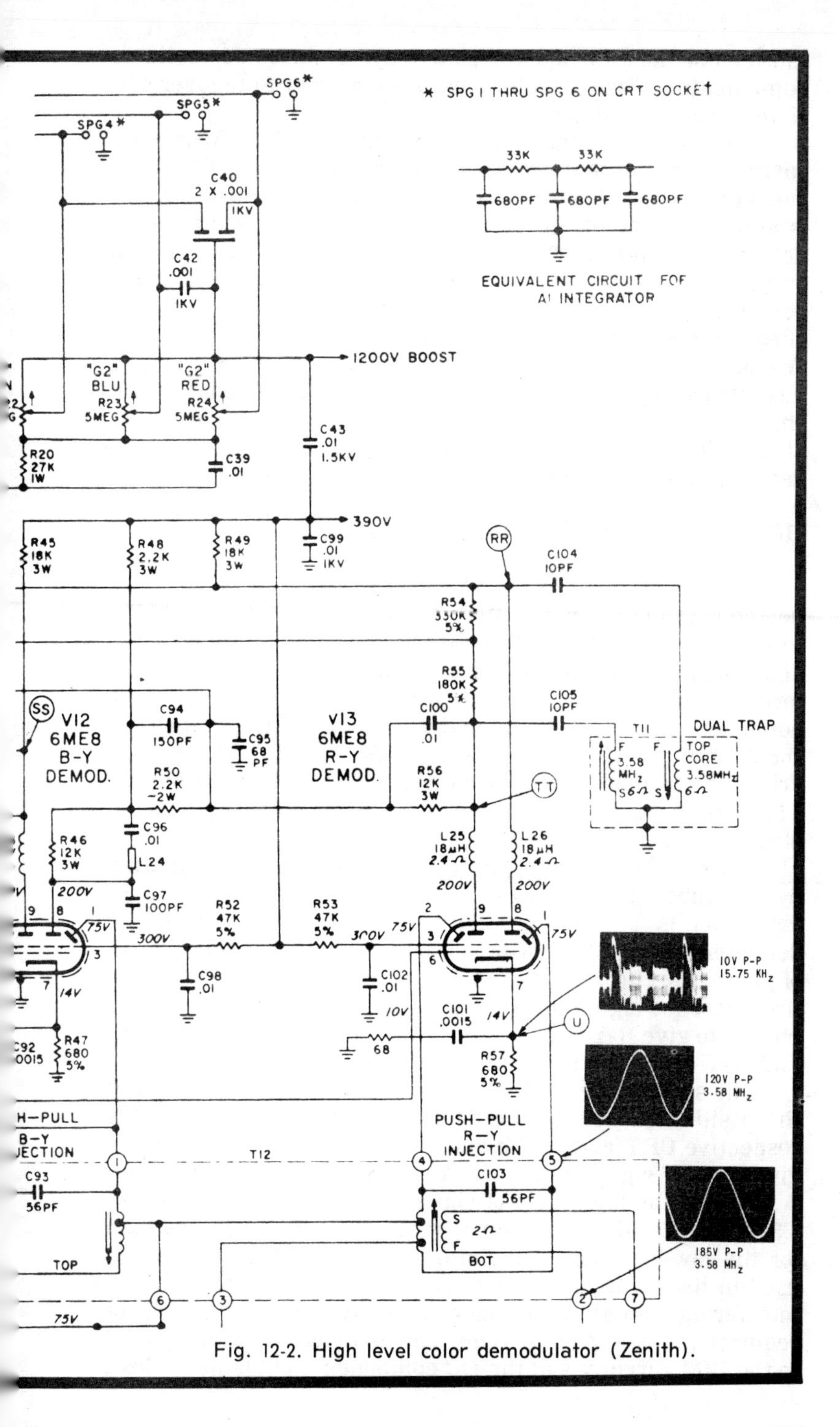

Fig. 12-2. High level color demodulator (Zenith).

This can be done by phase-sensitive synchronous detectors or demodulators. Two detectors are used in order to recover one of the two color signals.

The combined chrominance signal and 3.58 MHz subcarrier signal is applied to each detector. This 3.58 MHz subcarrier signal is regenerated in the color-sync circuit and is applied to one of the detectors directly, and to the other detector through a 90 degree phase-shifting network. The output from each detector depends upon the relative phase between the reinserted 3.58 MHz subcarrier and the chrominance signal. Since the reinserted subcarriers are 90 degrees apart, each detector will recover one of the two quadrature signals. Each detector will recover the quadrature chrominance signal that is either in phase or 180 degrees out of phase with the reinserted 3.58 MHz subcarrier applied to the detector, and will ignore the other quadrature signal.

HIGH LEVEL COLOR DEMODULATOR

The Zenith sheet beam demodulator is a high level color detector system. This system uses special 6JU8 or 6ME8 tubes. This tube has an indirectly heated cathode, control grid, and screen grid like an ordinary pentode. See Fig. 12-2. Beam-forming shields form the electron beam into a sheet. Two deflectors, similar to the deflection plates in a CRT, deflect the sheet beam to either of the two plates. These deflectors are driven in push-pull by the 3.58 MHz reference signal from the 3.58 MHz CW oscillator. The two plates of the 6ME8 receive the electron beam alternately.

The chroma signal is fed to the control grid, is amplified by the tube, and appears on the plate of the tube whose deflection is positive at that instant of time. Since the deflectors receive 3.58 MHz voltages that are 180 degrees out of phase, the chroma signal in the plates is tied to the phase of the reference signal. The phase of the reference 3.58 MHz is chosen to give R-Y and —(R-Y) from one demodulator (V13), and B-Y and —(B-Y) from the other (V12).

The scope patterns R, T, and S (note Fig. 12-2) show that the positive outputs of the demodulators are applied to the respective CRT grids. The negative outputs are matrixed to give the G-Y signal that is then applied to the green CRT grid.

Due to the high level of signal, the removal of unwanted 3.58 MHz after demodulation is more difficult to achieve than for the low-level demodulators. Tuned adjustable traps are used in the plate circuit of each demodulator to remove the interfering signal. Also, note that DC restoration is not required in this type of system. The detected signal is applied to the CRT directly and the DC component is thus never lost.

CHECKING THE COLOR DEMODULATORS

To checkout or test the color demodulators, the essential test instruments are a color-bar generator, oscilloscope, and VTVM.

To start this color test procedure, check the customer controls for correct adjustments and note any trouble symptoms. Examine the black-and-white picture (turn the color control down to zero if B + W transmission is unreliable) and if this appears normal you may assume the trouble is in the chroma section. Adjust the color killer control fully counterclockwise or defeat the color killer stage. Also, override the ACC action by a DC bias voltage or by grounding the ACC test point.

Connect the color-bar generator and look at the bar pattern across the screen. If the set is operating correctly, the color bars should appear as shown in the 10-bar pattern in Fig. 12-3. Of course, the tint or hue control must be adjusted for normal color reception. If no color is present on the screen, use the scope to signal-trace through the color circuits and look for the scope patterns like those shown in Fig. 12-2.

It should be noted (refer to circuit in Fig. 12-1) that the control grid of the demodulators is the last place for the

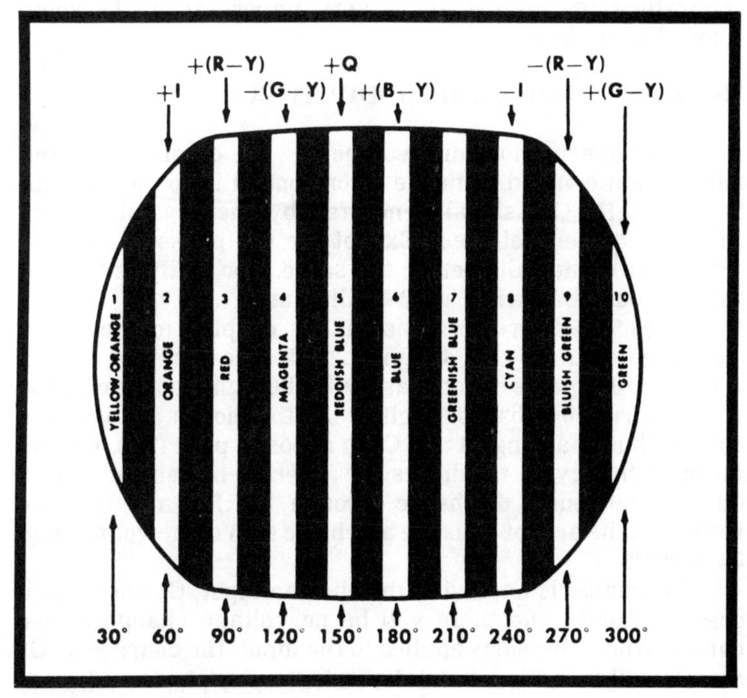

Fig. 12-3. Color-bar generator output on CRT (normal values).

chroma signals to appear. Beyond the control grid on the two demodulators, the signal follows two separate paths: The X signal is coupled to the R-Y amplifier and the Z signal to the B-Y amplifier. It is usually assumed that **one** fault has caused the color problem. Should there be no color at the CRT, **two** troubles would have to develop past the demodulators.

Some troubles will make the fleshtones appear too greenish or purple. In other cases, hues are missing and the condition cannot be corrected by the hue control. These symptoms indicate a fault in those circuits that produce or amplify the color-difference signals. This can be due to a phase error or the result of a missing or distorted color-difference signals. For some Zenith sets this would be the sheet beam demodulator, while in the RCA-type color system, the problem may be in the X or Y demodulators or the R-Y, B-Y, and G-Y amplifiers.

Missing color-difference signals produce a slightly different set of symptoms. Usually, the color picture appears to lack one of the primary colors, regardless of the hue control setting. This condition will be very obvious with the color-bar pattern displayed. For the Zenith demodulator use a scope to check the waveshapes at test points R, S, and T and compare them with those shown in Fig. 12-2. For correct color, the quadrature coil adjustment is very important on the sheet beam demodulator.

SOLID-STATE DIODE DEMODULATORS

In the synchronous diode detectors, the chroma sideband information coupled from the color control is combined with the 3.58 MHz CW signal generated by the crystal ringing circuit and then detected. Except for the phase of the CW signal, the diodes all operate the same. For analysis the one pair of diodes shown in Fig. 12-4 will be used.

The 3.58 MHz reference signal is coupled to the series connected diodes by L1, C1 and L2, C2. The signal at the Y2 anode is 180 degrees out of phase with the signal at the cathode of Y1. During one-half of each cycle, the diodes are forward biased, thus charging C1 and C2 to opposite polarities. During the other half cycle, the diodes are reversed biased and C1 and C2 simultaneously discharge through R1, R2, and R3. The voltage at the arm of R2 is the algebraic sum of the voltages on C1 and C2.

If no signal is applied to the chroma input, C1 and C2 will charge equally and there will be no voltage change at the output. When a signal is applied to the input, the charges on C1 and C2 will be unequal and dependent upon phase and amplitude of the input signal as compared to the reference signal

As the signal changes in phase or amplitude, the output voltage changes. This detected chroma signal is then applied to the grid of the color difference amplifier.

These demodulator circuits are simple and give very few troubles. The most common fault is either open or shorted diodes Y1 or Y2. Also leakage in capacitors C1 or C2 will occur. The balance control R2 may have to be adjusted. This is done during the color alignment procedure.

3.58 MHz CW REFERENCE SIGNAL

In order for any color demodulator to function, a 3.58 MHz reference signal must be applied to them. Do not overlook this 3.58 MHz CW signal while troubleshooting a "weak color" or "no color" symptom. The P-P voltage needed will depend on the type of demodulator used. The amount of P-P reference signal required is shown on many of the schematics. Use the oscilloscope to check for the presence of this CW signal and its P-P amplitude. If the amplitude of this 3.58 MHz CW signal is half what it should be, then you can expect the color saturation level to be down by about one-half of normal.

Demodulation errors will also occur when the phase of the 3.58 MHz CW signals applied to the demodulators is incorrect. A large phase error, one that cannot be corrected with the hue control, has this result. All colors and hues appear in the picture, but all colored objects are shown with incorrect hues. For example, fleshtones appear green, purple, blue, or some other unnatural hue.

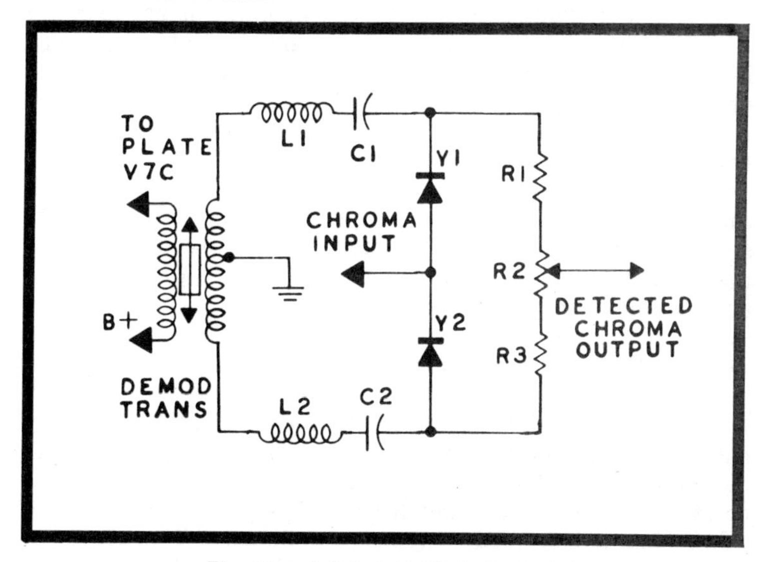

Fig. 12-4. Solid-state diode demods.

INTEGRATED CIRCUITS (CHIPS)

It seems that some color TV manufacturers will go directly from tubes to the "chips," sort of skip over the transistor era. For the color receiver this will create a cooler chassis, a more compact size, more reliability, and eventually a less expensive set. For the technician this means new techniques and troubleshooting methods will be called for. Also, many of the old standby methods can be brought back into play. One good "spin-off' from the "chip" is that the day of tube pullers and "diddle stick" twisters will be numbered.

Ways to Check the "Chip"

Transistors, and now ICs are more reliable than vacuum tubes. There are no cathodes to wear out, nor vacuum to lose, or any mechanical electrodes to vibrate or shake loose. However, integrated circuits can and do fail, even though advanced processing and better reliabliity checks are performed on these IC chips.

Obviously, ICs cannot be taken apart and repaired, so it is of no value to locate the internal defect. The technician's job is to positively determine that the chip is defective or if there is trouble in the outboard circuit component. And also to determine if a defective outboard component malfunction may have caused the IC to fail. So - the name of the game is - is it the "chip" or isn't it?

Because most components found in color circuits using ICs are in the "chip," fewer tests can be made or are found to make. Oh yes - voltage, resistance, and current checks are important, for example, but won't always be too efficient or practical. About the only way to check these little rascals is right in the operating color circuitry by feeding in a signal to the set from a color-bar generator. Then, with the oscilloscope, trace the input and output waveshapes through the various IC stages till an improper scope pattern is observed. This is a very fast and positive troubleshooting method, and just another form of the familiar signal injection and signal tracing used for many years.

Zenith's Color Demodulator "Chip"

Probably one of the most commonly used "chips" for color demodulation is Zenith's type 221-37 or 221-39 which has been in use for a few years now. This is, no doubt the "grand daddy" of all color demodulator "chips." This IC, called a "dual double-balanced synchronous detector," contains 21 transistors with lots of resistors plus other devices. Fairchild developed this chip for Zenith color receivers and designated it as the UA737. This is an integrated circuit providing a stable double-balanced demodulator and amplifier for the chroma signal. This "chip" is designed so that it will plug into a conventional 9 pin miniature tube socket and is keyed accordingly.

The schematic in Fig. 12-5 for a Zenith 14A9C29Z color chassis shows that the output of the second color amplifier chroma signal is coupled to this IC or color "chip" I.C. 201. These two chroma signals of opposite polarity - that scope out at 0.3 VP-P (see Fig. 12-6) - are coupled to terminals 2 and 3 of IC201. Color difference signals of —(G-Y), —(R-Y) and —(B-Y) appear at terminals number 7, 8, and 9 respectively. Shown in Fig. 12-7 is the familiar S shaped patterns that you have viewed on the scope for tube type demodulators and color-difference amplifiers that will also be seen at the "chip" pins 7, 8, and 9. These signals are fed to the control grids of a triple triode, color-difference amplifier 6MN8 tube at about 2 volts P-P. And as usual, the green signal will be a little less than the red or blue. A scope check at pins 4 and 5 will show a 1.1V P-P 3.58 MHz sine wave CW drive signal as shown in Fig. 12-8. This signal must be present for proper color demodulation. Now let your scope tell you if the IC is operating or if another stage is

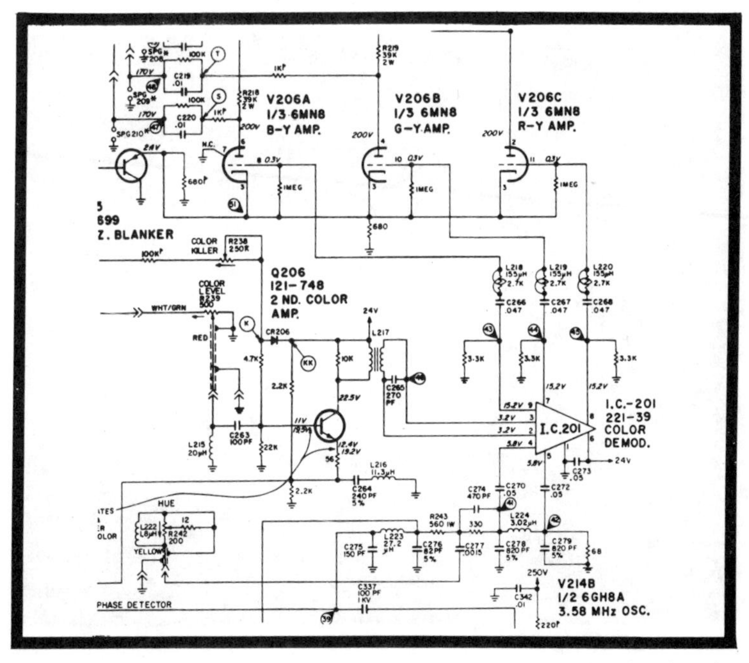

Fig. 12-5. Zenith 14A9C 29Z chassis schematic (partial) showing integrated circuit (IC chip) color demod.

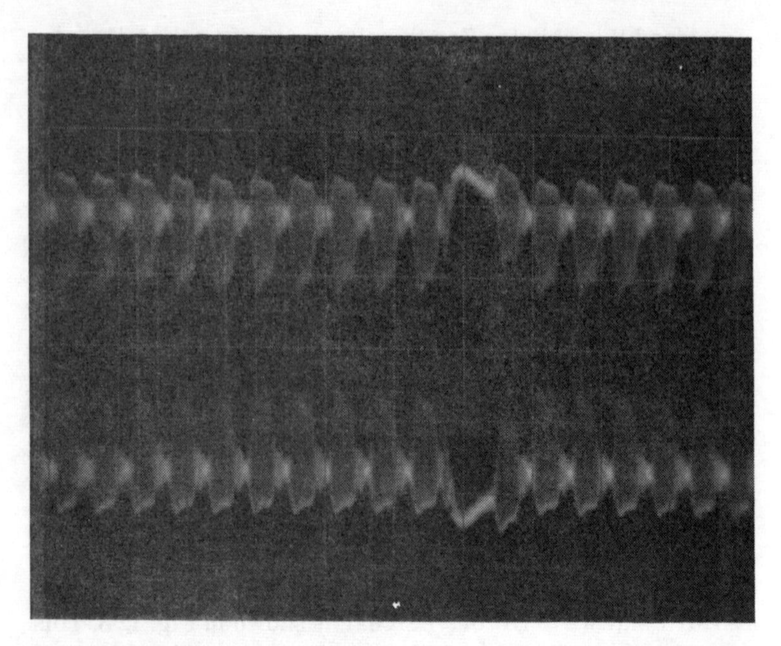

Fig. 12-6. Opposing polarity waveforms at pins 2 and 3 of I. C. 201 in Fig. 12-5. Amplitudes are 0.3V P-P. Scope is set at half the horizontal rate.

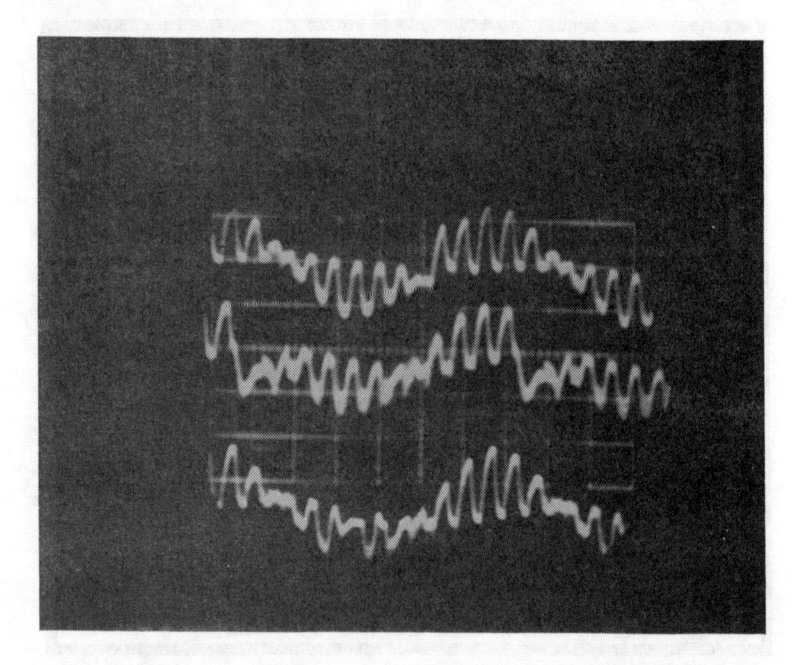

Fig. 12-7. Output waveforms from I.C. 201. Top trace is R-Y (red); middle is G-Y (green), bottom is B-Y (blue). Amplitudes approx 2V P-P.

defective. For a test, feed a color-bar pattern into the antenna and check for these three correct chroma output signals with your scope via a low-capacitance probe.

If we do not find a normal signal output from the "chip," the next step is to scope the input terminals (2 and 3) of I.C. 201. If we find this low signal (0.75V P-P) here, but have very low or no output, the third step is to measure DC voltages around the chip socket and check the current of the 24V DC supply voltage at pin 6. Caution: Do this before you pop in a new IC and maybe pop it out, too. Although the chroma output of the second color amplifier (at collector) is approximately 6V P-P, the step-down for impedance matching of L217 (2nd color amplifier transformer) to the "chip" input results in approximately 0.75V P-P input to terminals 2 and 3.

Voltage and Current Checks

All of the DC voltage readings can be taken from the chassis "top-side" as the leads for the 221-37 and 221-39 ICs are not enclosed. Also, all of the DC voltages are developed within the "chip" except for the +24-volt supply. All voltages at the "chip" pins will vary slightly as the color and hue controls are varied. Terminals 2 and 3 will read +3.4 volts, while pins 4 and 5 will check at about +6.2 volts. Pin 1 is grounded. If pins 2 and 3 do not have DC voltage agreement, the secondary winding of

Fig. 12-8. CW drive signal. Frequency is 3.58 MHz from local oscillator. Amplitude is 1.1 V P-P. Scope is set at half the horizontal rate. Top, pin 4 of I. C. 201; bottom, pin 5.

L217 is open. The color output pins 7, 8, and 9 should measure about +14.6 to 15.4 volts.

Should any of these voltages be off value, check all outboard components that are connected to the pins in question. One example would be a high or low voltage found at pins 7, 8 or 9. An off value resistor or leaky capacitor could cause this trouble. Also a defective chip will cause wrong voltages. One quick check is to pull the IC, which plugs into a standard 9-pin tube socket. Now, all pins should read zero volts if all outboard circuits are good, except pin 6 which is the +24 volt supply and it should go up to about +26.5 volts or so.

Of course, a check of the DC current drawn by the "chip" is a good test. The normal current for this IC is 21 ma give or take 1 or 2 mils. This will vary from IC to IC too. If you measure a lot more current drawn above normal, the "chip" is shored; and a lot less current indicates an open circuit somewhere in the "chip." When everything is "go" the chip will operate at a temperature that is warm to the touch.

ACTUAL CASE-BOOK "CHIP" TROUBLE SYMPTOMS

This trouble was found in a Zenith 14A9C29Z chassis. Note Fig. 12-5. The left half of the screen had a much greater bluish cast than the right side. This was very noticeable on black and white reception. The problem was found to be a shorted capacitor (C266 .05 uf) from terminal 9 of IC201 color demodulator to pin 8 of 6MN8 (B-Y amplifier) tube.

Also another odd trouble has cropped up with these ICs. The gray scale would seem to vary on this Zenith color set. And the colors did not look just right, but not often enough to draw a complaint. Now it was noticed that when the hue control was rotated - back-and-forth - the gray scale would go from a blue to a green tint. Most baffling to say the least. Yep—you guessed right, a defective "chip."

CHIP CIRCUITRY

Amplification through the IC is greater than 10 times, providing adequate signal amplitude (color-difference signal to the amplifier tube grids). The operation of the "chip" (similar to Zenith's switch-type tube operation) is such that the 2nd harmonic (7.2 MHz) or the 3.58 MHz must be trapped in the output circuit. (The 3.58 MHz fundamental is automatically canceled within the "chip.") This 7.2 MHz trapping is accomplished by coils L39, L45, and L51 in the output circuit. Note the "color diff. amps." in Fig. 12-9. A complete schematic of the UA737 color "chip" is shown in Fig. 12-10. Although a detailed description of this IC is unnecessary

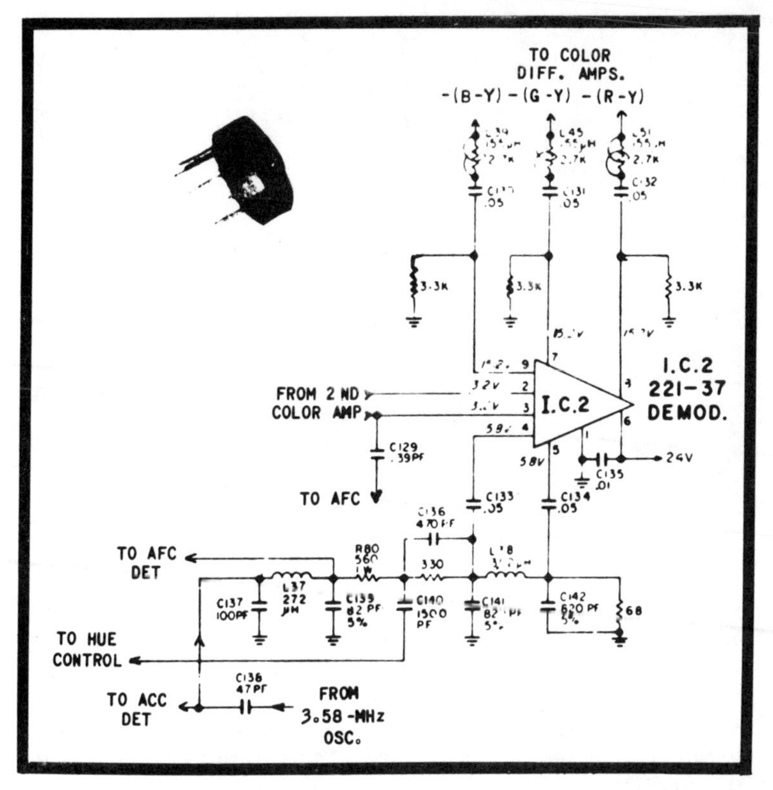

Fig. 12-9. Zenith 14Z8C50 partial schematic showing 7.2 MHz traps.

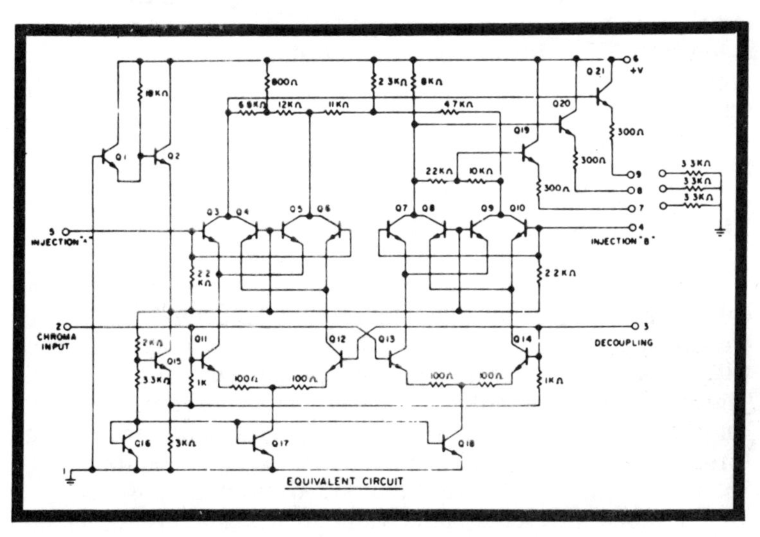

Fig. 12-10. Complete schematic of Fairchild UA 737 IC chip.

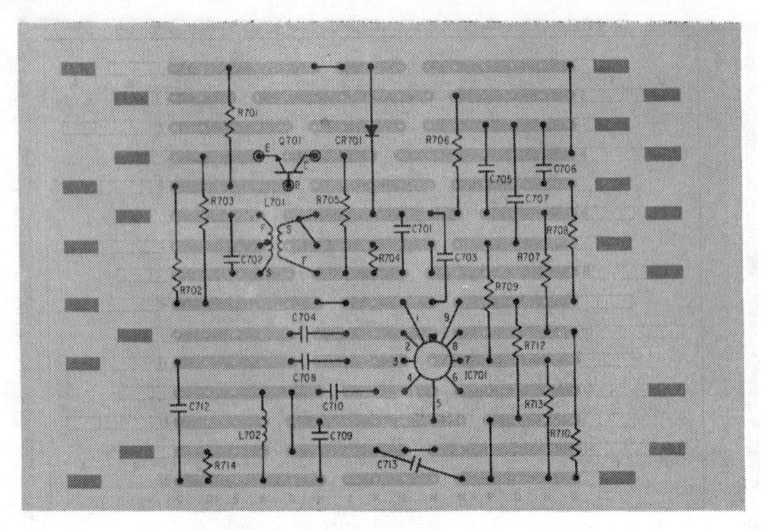

Fig. 12-11. Zenith plug-in dura-module.

for servicing (you sure can't get inside to fix it and you'd only want to know if it's working) it is interesting to see how the "chip" does function, so let's analyze it:

1. Chroma signals of opposite phase are coupled into the parallel connected inputs (bases) of the amplifiers, Q11, Q12, Q13, and Q14.

2. The color oscillator reference signals are coupled to the two sets of "switch" transistors Q3 through Q6, and Q7 through Q10.

3. The color difference output voltages from the matrix (color mixers) are coupled through emitter followers, Q19, Q20, and Q21.

4. The remaining transistors provide voltage division, regulation (Q1 acts as a Zener), and stability of operation.

THE PLUG-IN COLOR DURA-MODULE

For the past few years Zenith has been using a plug-in color dura-module unit that contains the IC demodulator, one transistor color amplifier, and other related components. The wiring diagram for this dura-module is shown in Fig. 12-11. The color signal enters the module at connection B10 and couples via C704 demodulator IC701. Following demodulation through the IC, color difference signals exit the module at U6, T5, and T7 and on to the transistor video output amplifiers.

The color difference signals and the Y signal are matrixed in these stages, and signals corresponding to R, G, and B

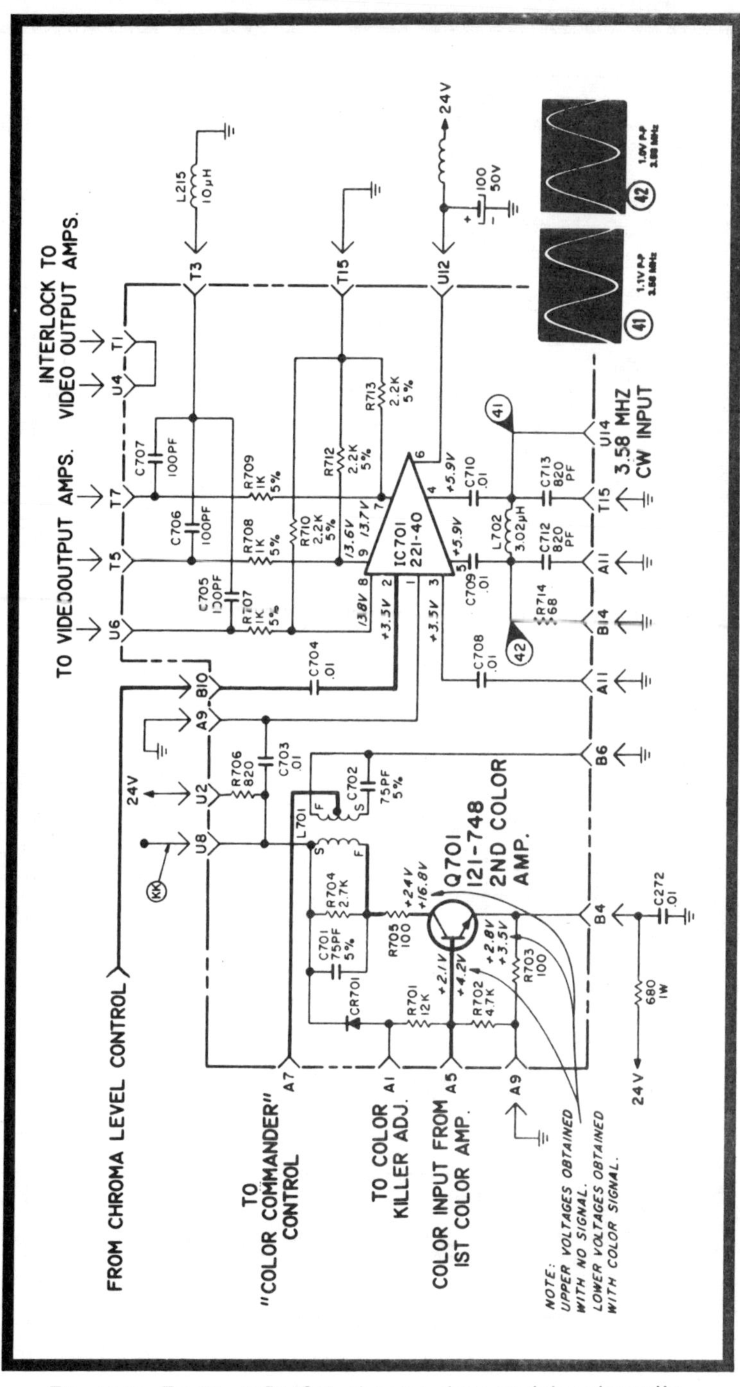

Fig. 12-12. Zenith 12B14C52, showing dura-module schematic.

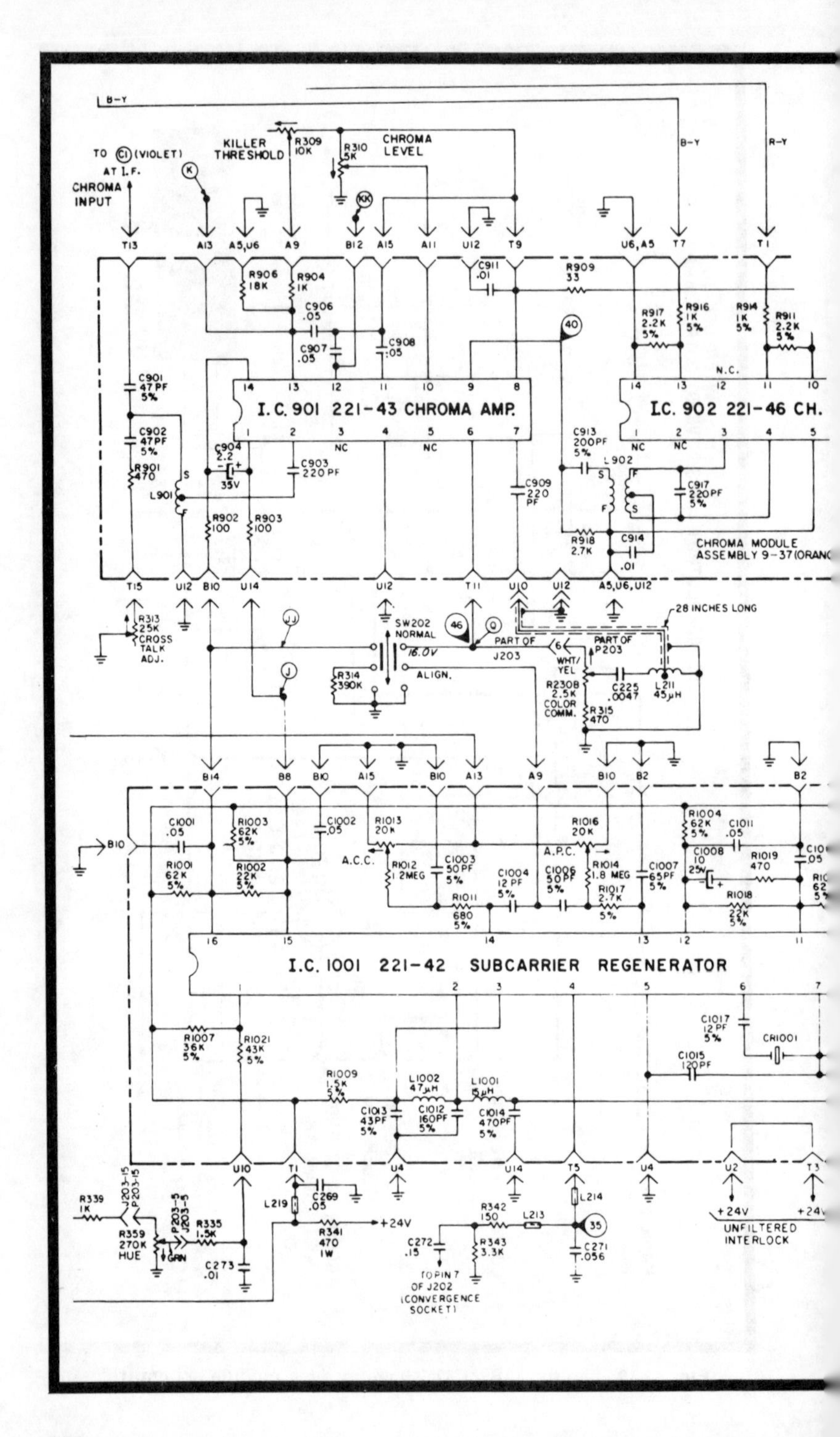

I.C. 901 221-43 CHROMA AMP.
I.C. 902 221-46 CH.
I.C. 1001 221-42 SUBCARRIER REGENERATOR
CHROMA INPUT
KILLER THRESHOLD
CHROMA LEVEL
CHROMA MODULE ASSEMBLY 9-37 (ORANG
28 INCHES LONG
SW202 NORMAL
ALIGN.
CROSS TALK ADJ.
COLOR COMM.
HUE
A.C.C.
A.P.C.
UNFILTERED INTERLOCK
TO PIN 7 OF J202 (CONVERGENCE SOCKET)

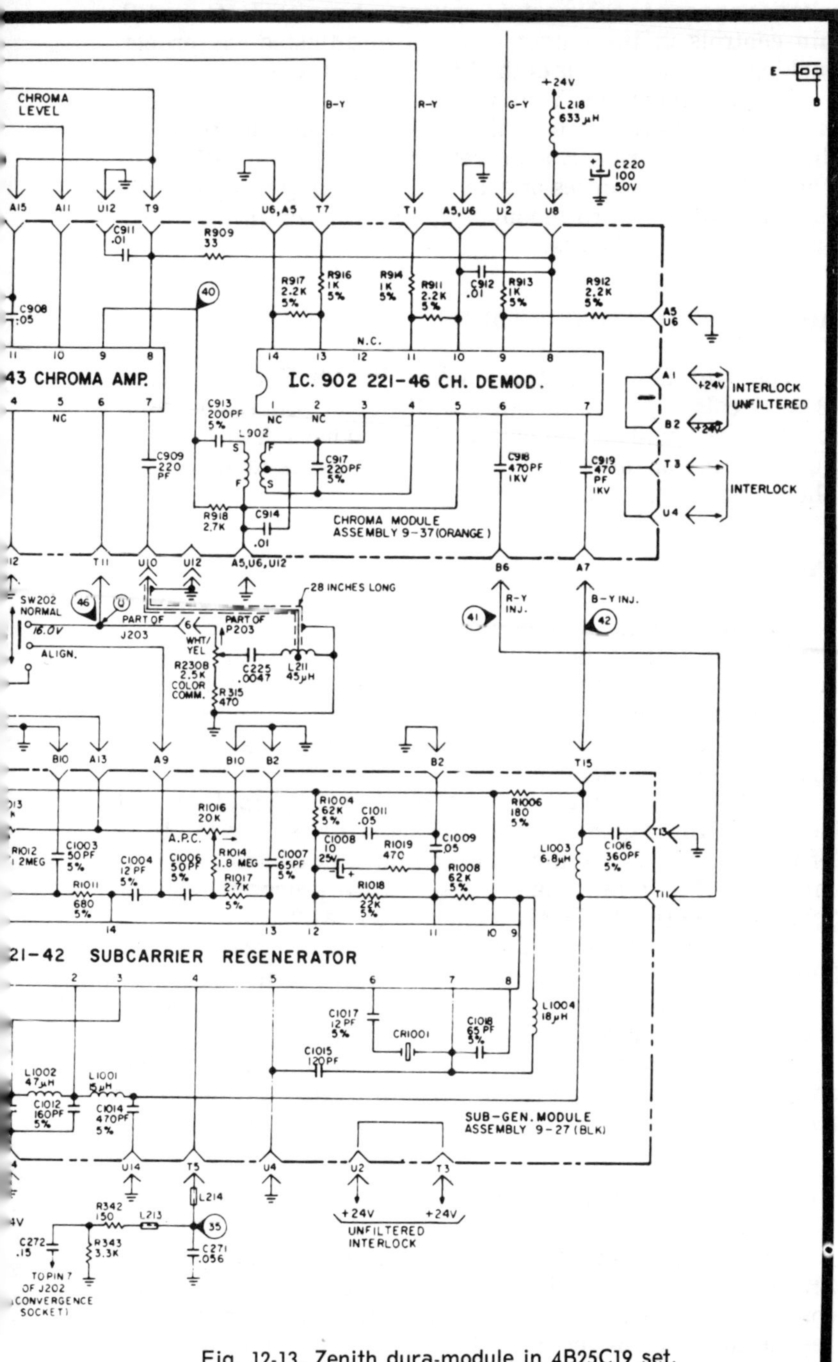

Fig. 12-13. Zenith dura-module in 4B25C19 set.

voltages appear in their output collector circuits. R, G, and B gain controls in the output circuit are adjusted for proper tracking of all three guns of the CRT (bright areas).

By premixing the Y and color signals, the final video amplifiers can process both the video and color information. Thus, the CRT grids can be returned to a fixed DC level, reducing the possibilities of gray-scale tracking troubles. The schematic of this dura-module system is shown in Fig. 12-12 for the 12B14C52 chassis.

MORE ZENITH COLOR "CHIPS"

You will now find three color ICs in Zenith's latest 4B25C19 color TV chassis. Most of the color processing components are found on two plug-in dura-modules. Two "chips" plug into one of the modules while only one is found on the other. Note Fig. 12-13.

Chroma Amplifier I.C.901

Chroma information enters the circuit at T13 (from test point C1) and the top left of Fig. 12-13, and couples to the input matching network and cross talk potentiometer R313. The cross talk adjustment is made while viewing color bars on the screen and adjusting for best sharpness of both leading edges and sides of the color bars.

I.C. 901 consists of a number of differential amplifiers that increase chroma gain. The chroma signal exits from the IC at terminal 6 and goes to the color commander control, then reenters the IC at terminal 7. After more processing in the "chip" the chroma signal exits at terminal 9.

The chroma level, and killer threshold controls, are DC bias adjustments for the circuitry within the "chip." Shorting across test points K and KK defeats the color killer function and "opens" the color channel for continuous color operation.

The coupling transformer at terminal 9 of the chroma amplifier I.C. 901 couples the chroma signal (push-pull) to terminals 3 and 4 of the chroma demodulator "chip".

Chroma Demodulator I.C.902

The chroma demodulator IC is essentially like the type 221-39 "chip" previously discussed except for its physical construction which, like the chroma amplifier and subcarrier regenerator IC, is contained in an in-line multiterminal package which plugs into its associated socket.

Subcarrier Regenerator I.C. 1001

As you may have already guessed, this "chip" regenerates the 3.58 MHz CW signal. Terminal 6 is the oscillator input while the oscillator output (3.58 MHz CW) appears across terminals 7 and 8. The subcarrier output signal appears across terminals 2 and 3 and is coupled to a phase delay network. The hue control functions as a DC bias adjustment and controls the phase of the subcarrier output signal which is processed within the IC. The hue range is approximately plus and minus 45 degrees.

Chroma information enters the subcarrier regenerator IC across terminals 13 and 14. The color burst is processed and then develops the ACC and APC (automatic phase control) voltages. ACC voltage, developed within the IC, appears across terminals 15 and 16 (test points J and JJ). Normal ACC voltage is very low, approximately 0.035 volt.

RCA TWO-CHIP COLOR SYSTEM

These two color "chips" are employed in RCA's new "Argosy" portable color TV which makes use of the CTC-49 chassis. Each IC is mounted in its own modular unit which is then plugged into the main chassis. See Fig. 12-14. The chroma module, called MAC, processes chroma frequencies between 3.08 MHz and 4.08 MHz. All active devices are in one "chip" that contains a chroma-bandpass amplifier; burst and blanking functions are also accomplished within this IC. The MAE module performs the demodulation and tint control

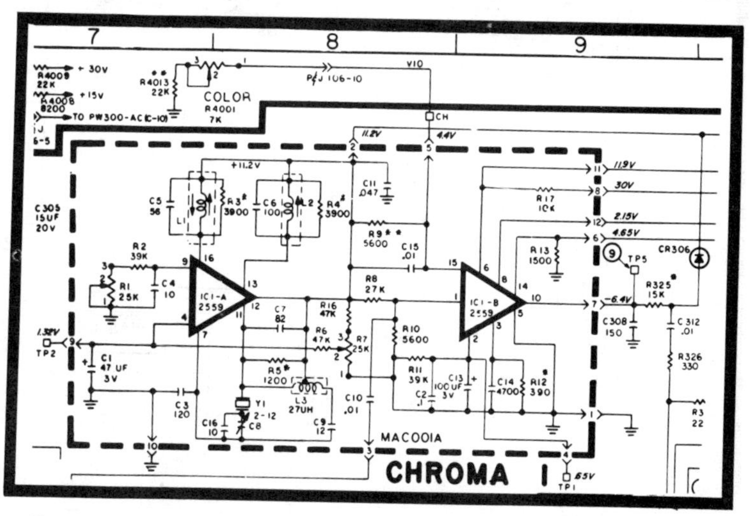

Fig. 12-14. RCA MAC module, CTC-49, part of 2-chip chroma system.

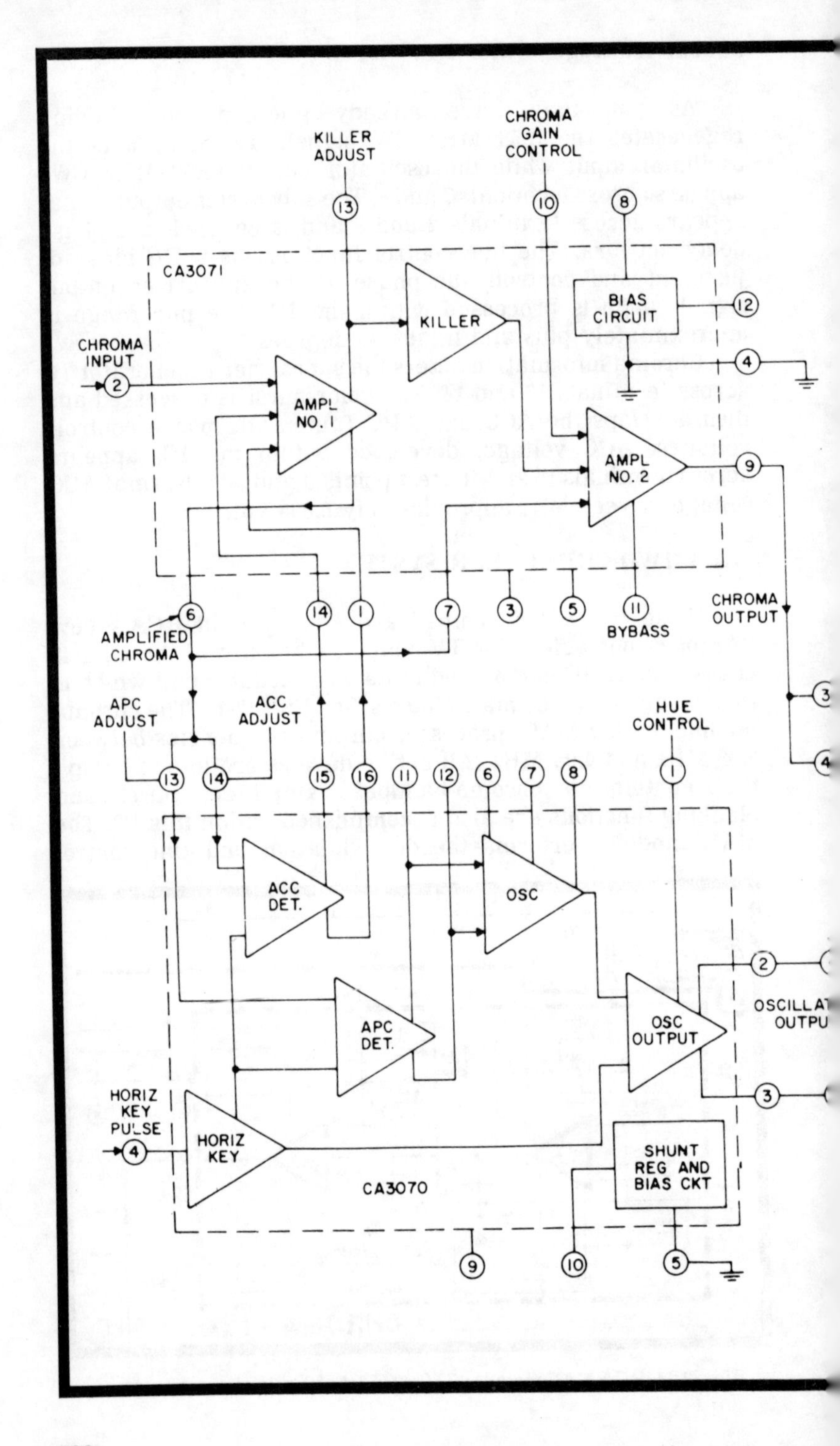
KILLER
ADJUST
CHROMA
GAIN
CONTROL
13
10
8
CA3071
BIAS
CIRCUIT
12
KILLER
CHROMA
INPUT
2
4
AMPL
NO. 1
AMPL
NO. 2
9
6
14
1
7
3
5
11
BYBASS
CHROMA
OUTPUT
AMPLIFIED
CHROMA
APC
ADJUST
ACC
ADJUST
HUE
CONTROL
13
14
15
16
11
12
6
7
8
1
ACC
DET.
OSC
APC
DET.
OSC
OUTPUT
2
3
HORIZ
KEY
PULSE
4
HORIZ
KEY
SHUNT
REG AND
BIAS CKT
CA3070
9
10
5

The RCA CA3070, CA3071, and CA3072 are monolithic silicon integrated circuits that constitute a complete chroma system for color television receivers. The CA3070 is a complete subcarrier regeneration system featuring a new concept of phase control applied to the oscillator circuit. The CA3071 is a chroma amplifier system and the CA3072 performs the demodulation function.

The CA3070 utilizes the 16-lead plastic dual-in-line package; the CA3071 and CA3072 are supplied 14-lead plastic dual-in-line packages.

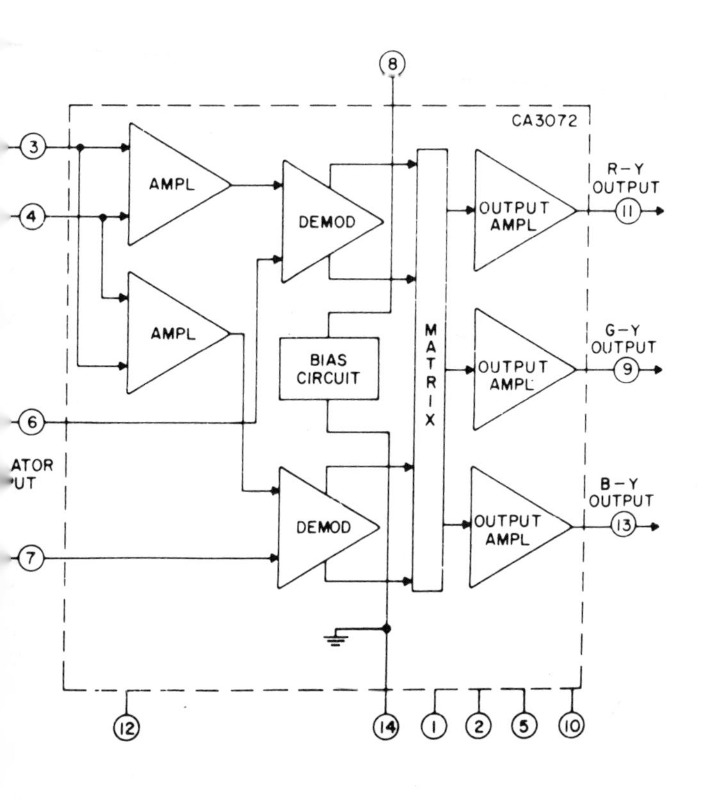

Fig. 12-15. RCA 3-chip chroma system.

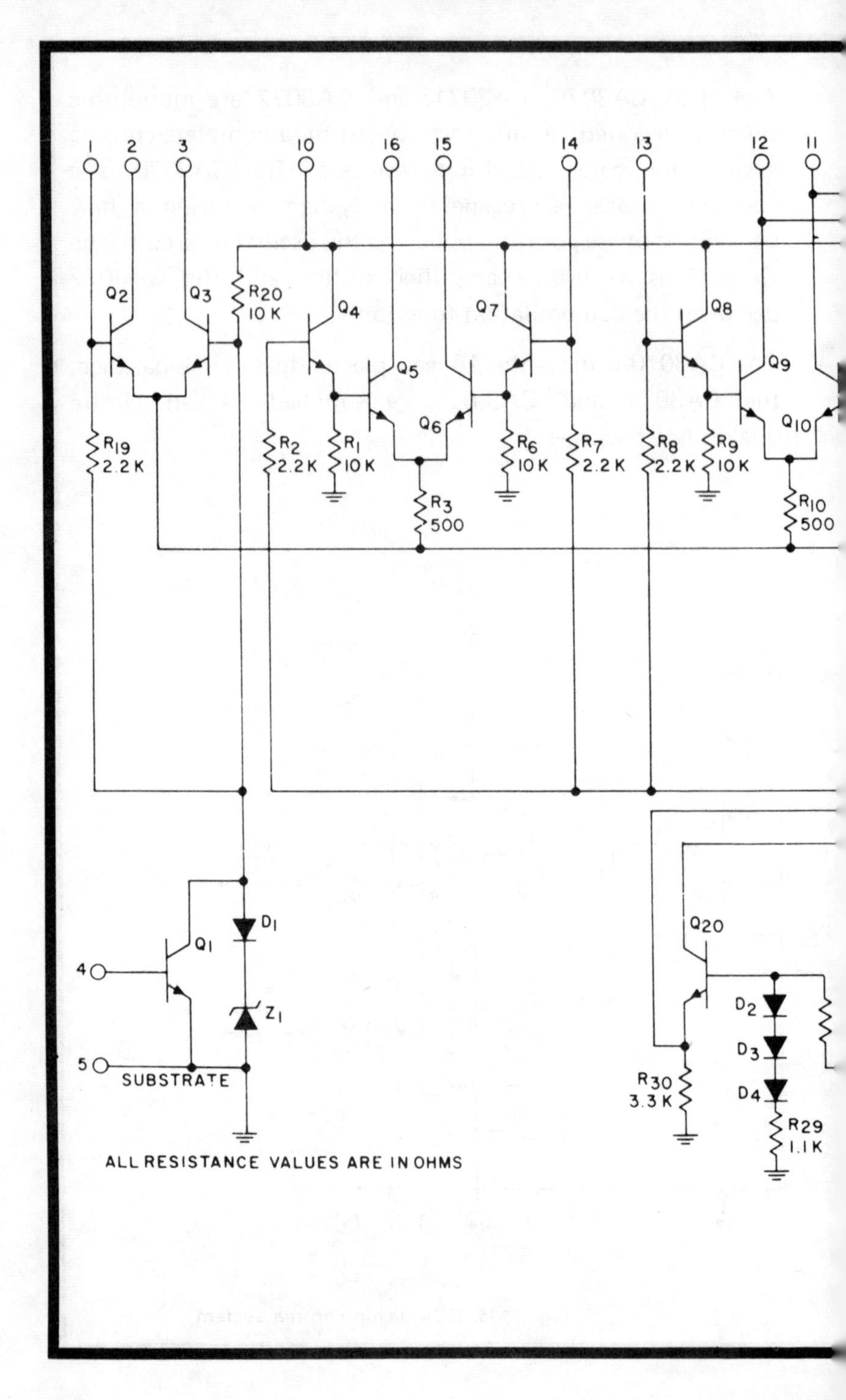
1
2
3
10
16
15
14
13
12
11
Q2
Q3
R20
10 K
Q4
Q5
Q6
Q7
Q8
Q9
Q10
R19
2.2 K
R2
2.2 K
R1
10 K
R3
500
R6
10 K
R7
2.2 K
R8
2.2 K
R9
10 K
R10
500
Q1
D1
Z1
4
5
SUBSTRATE
Q20
D2
D3
D4
R30
3.3 K
R29
1.1 K
ALL RESISTANCE VALUES ARE IN OHMS

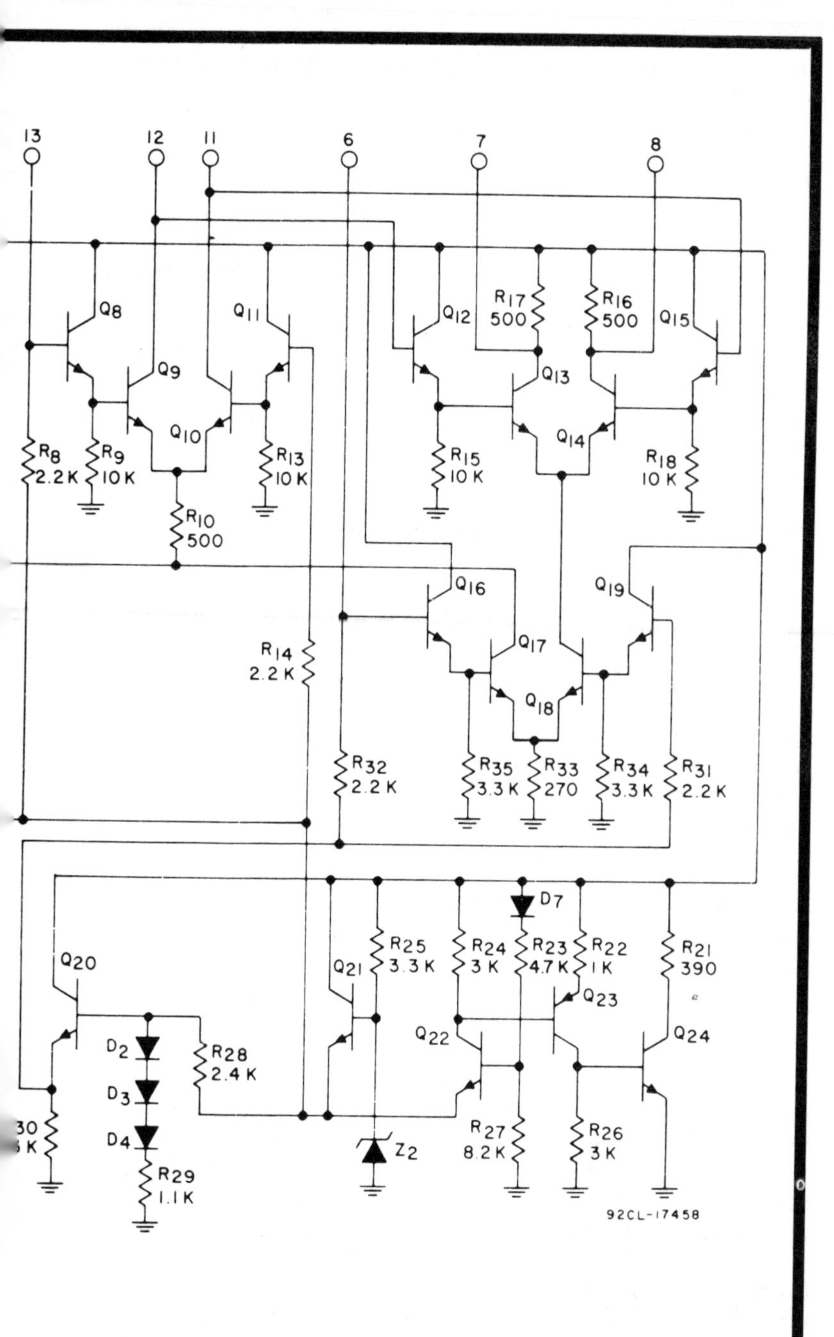

Fig. 12-16. RCA CA3070 chroma signal processor chip.

functions. RCA calls these two "chips" the IC-A and B2559; they are monolithic silicon integrated circuits that constitute a complete chroma for color TV receivers. Each IC device utilizes a 16-lead quad-in-line plastic package.

RCA THREE-CHIP COLOR SYSTEM

These three "chips", which RCA has designated CA3070, CA3071, and CA3072 have just gone into production and will no doubt soon be seen in the deluxe RCA solid-state color chassis. These are monolithic silicon ICs that make a complete chroma system for a color TV receiver. The CA3070 utilizes the 16-lead plastic dual-in-line package, while the CA3072 uses the 14-lead plastic dual-in-line packages. The simplified block diagram for this TV chroma system is shown in Fig. 12-15. Now for a look at how they operate.

CA3070 Chroma Signal Processor "Chip"

This is a complete subcarrier regeneration system with automatic phase control applied to the oscillator. An amplified chroma signal from the chroma amplifier IC is applied to terminals 13 and 14, which are the automatic phase control (APC) and the automatic chroma control (ACC) inputs. APC and ACC detection is keyed by the horizontal pulse which also prevents amplified oscillator output during the burst interval. Note the schematic diagram of this chip in Fig. 12-16.

The ACC system uses a synchronous detector to develop a correction voltage at the differential output, terminals 15 and 16. This control signal is fed to the input terminals, 1 and 14, of the chroma amplifier "chip" CA3071. The APC system also uses a synchronous detector. The APC error voltage is internally coupled to the 3.58 MHz oscillator for balance, and the phase of the signal at terminal 13 is in quadrature with the oscillator.

To accomplish phasing requirements, an RC phase shift network is used between the chroma input and terminals 13 and 14. The feedback loop of the oscillator is from terminals 7 and 8, then back to pin 6. The same oscillator signal is available at pins 7 and 8, but the DC output of the APC detector controls the relative signal levels at terminals 7 and 8, because the output at terminal 8 is shifted in phase as compared to the output at terminal 7, which is fed directly to the crystal circuit. Control of the relative amplitudes at terminals 7 and 8 alters the phase in the feedback loop, thereby changing the frequency of the crystal oscillator. Balance adjustments of DC offsets are provided to establish an initial no-signal offset control in the ACC output, and a no-signal, on-frequency ad-

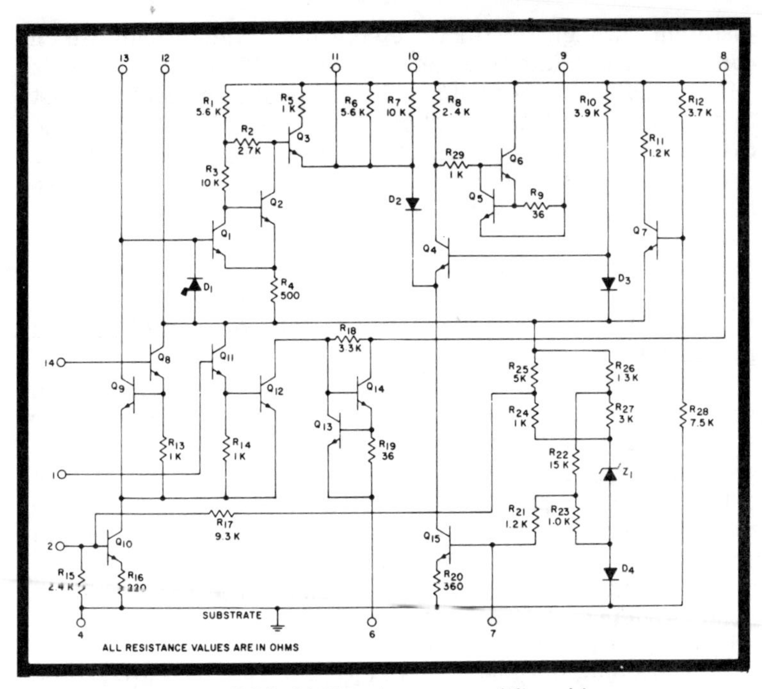

Fig. 12-17. RCA CA3071 chroma amplifier chip.

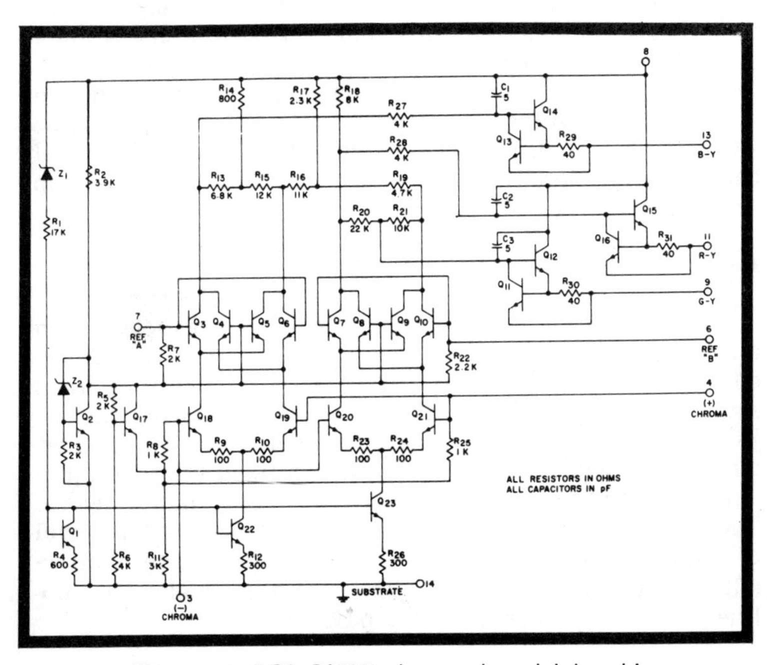

Fig. 12-18. RCA CA3072 chroma demodulator chip.

justment through the APC detector-amplifier circuit which controls the oscillator frequency. The oscillator output stage is differentially controlled at pins 2 and 3 by the hue control input to pin number 1. The hue phase shift is accomplished by the external R, L, and C components that couple the oscillator output to the demodulator input terminals . This "chip" also has a shunt regulator to establish a 12-volt DC supply.

CA3071 Chroma Amplifier "Chip"

The CA3071 is a combined two-stage chroma amplifier and functional control circuit. Refer to Fig. 12-17.

The input signal is received from the video amplifier and goes to pin 2 of the input amplifier stage. The first amplifier stage is part of the ACC system and is controlled by differential adjustment from the ACC input, pins 1 and 14. The output of the 1st amplifier goes to terminal 6 and this signal may be applied to the ACC detection system of the chroma signal processor "chip." The output of terminal 6 is also applied to terminal 7 which is the input to the 2nd amplifier stage. Another output of the 1st amplifier also goes to the killer adjustment circuit found at pin 13.

The DC voltage level at pin 13 rises as the ACC differential voltage decreases with a reduction in the burst amplitude. At a preset condition determined by the killer control adjustment, the killer circuit is activated and causes the 2nd chroma amplifier stage to be cut off. The 2nd chroma amplifier stage is also gain controllled by the amount of DC voltage applied to pin 10. The output for the 2nd chroma amplifier stage will be found at terminal 9. This output circuit provides a differential chroma drive signal for the demodulator circuit coming up next.

CA3072 Chroma Demodulator "Chip"

The CA3072 has two sets of synchronous detectors with matrix circuits to produce the R-Y, G-Y, and B-Y color-difference output signals. Note Fig. 12-18. The chroma input signal is applied to terminal 3 and 4 while the oscillator injection signal is brought into pins number 6 and 7. The color-difference signals, after matrix, have a fixed relationship of amplitude and phase nominally equal to DC voltage levels. The outputs of this "chip" are suitable for driving high-level color-difference or R, G, B output amplifiers. The emitter-follower output stages used to drive the high-level color amplifiers also have short-circuit protection.

Servicing the Sound Detector and Audio Circuits

13

In this chapter we will look at the audio circuits that develop the sound at the speaker in TV receivers. These modern sound systems include gated-beam quadrature detector, ratio detector, and dual-control quadrature detector. All TV sets now utilize the intercarrier sound system. We'll review the latest sound systems that incorporate the IC "chips" and then some sound alignment "tips" plus some information on how to cure some of those hum and buzz troubles that bug us all from time to time.

SOUND SYSTEM REVIEW

The television audio signal is transmitted on a separate FM frequency sound carrier. This sound frequency, recovered in the TV receiver by heterodyning, appears at the sound takeoff diode detector as a 4.5 MHz signal. Color and B & W TV sound systems are somewhat similar (Fig. 13-1). However, the color chassis uses a separate diode detector for the sound and another for video signals. One of the reasons for this is the beat frequency between the sound and color signal carriers. Also, the IF signal is trapped out of the last video IF stage to eliminate this 920 kHz beat between the 4.5 MHz sound IF carrier and 3.58 MHz color subcarrier, which would then occur in the output of the video detector. Fig. 13-2 illustrates a complete sound section of a color TV receiver.

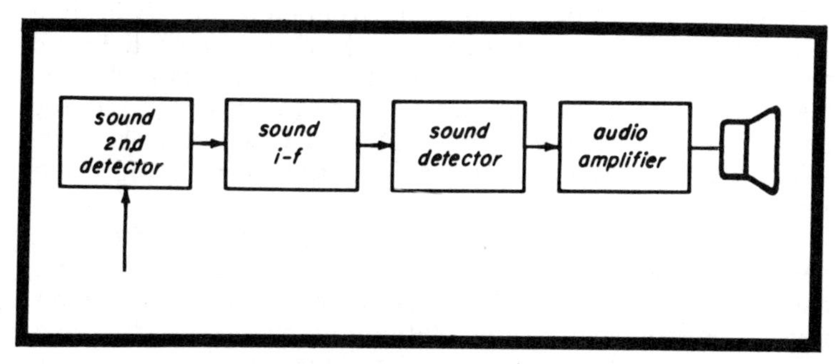

Fig. 13-1. Block diagram of TV sound system.

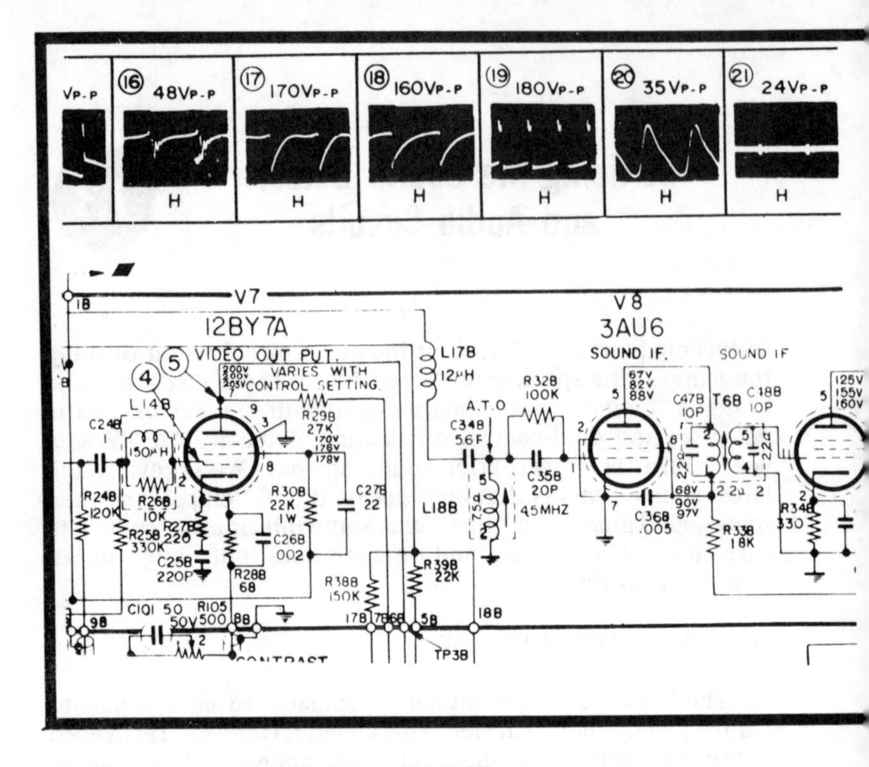

GATED-BEAM DETECTOR SYSTEM

The gated-beam detector is a very simple FM detector circuit and alignment is very easy. The RCA and Zenith versions of this detector circuit are about the same; therefore, the early Zenith design will serve us with an adequate description. This is a very successful multielement vacuum-tube detector and its solid-state counterpart will undoubtedly be used in the near future.

The limiting and demodulation characteristics are determined by the design and construction of the 6BN6 tube (Fig. 13-3). All of the newer sets use the 6Z10 type pentode and beam power tube and it's still the same special construction as the 6BN6. Since no discriminator transformer is required, the circuit is no more complex than a standard AM detector. Thus, alignment is simplified by the fact that no detector transformer is required.

The 6BN6 tube is of a very unconventional structural design, as shown in the sketch. The rectangular cathode, enclosed on three sides by a grounded focusing electrode, faces a positive accelerator plate on its open side. Electrons emerging from the cathode are shaped into a compact, sheet-

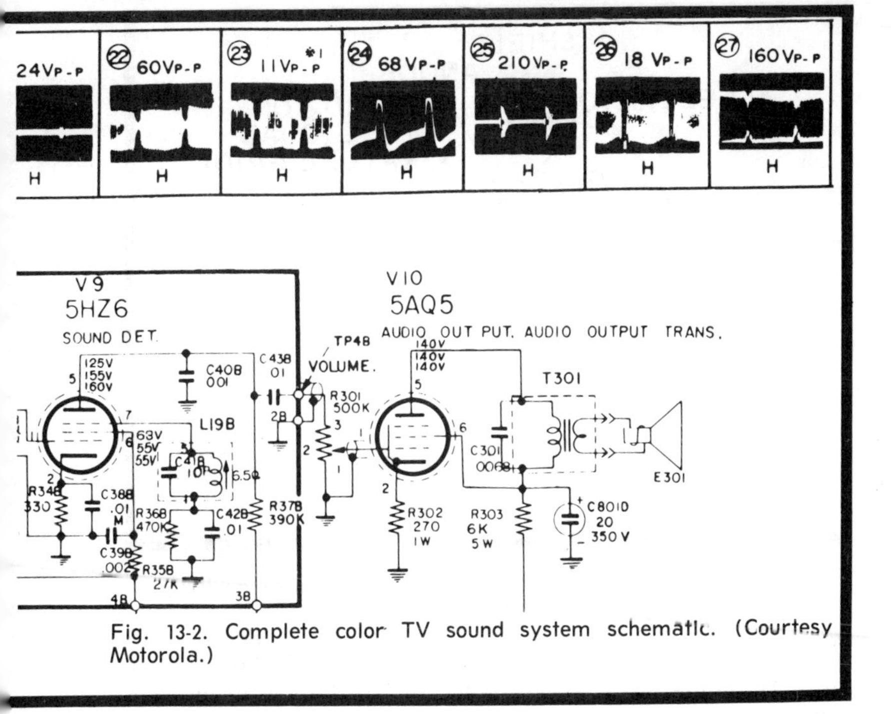

Fig. 13-2. Complete color TV sound system schematic. (Courtesy Motorola.)

like beam which is projected against the limiter grid through the narrow slot in the accelerator. The limiter grid acts like a gate: with a zero or a positive potential applied, it passes the beam, but with a negative potential of a few volts or more on the grid, the electron beam is blocked. Thus, the beam is deflected like a stream of water which falls back on the accelerator.

When a sine-wave signal of 1 volt RMS or more is applied to the limiter grid, the beam is alternately passed and blocked, so that the beam current behind the input grid has the form of a square wave. As long as the input signal remains over 1 volt RMS, variations in its amplitude do not affect the intensity of this square wave. Hence, limiting takes place right at the first grid.

In normal operation, the limiter grid is biased 2 or 3 volts negative with respect to the cathode. Input signals may easily be 5 to 10 volts, so that the limiter grid is very positive during part of each cycle. After leaving the limiter grid, the beam passes through the second accelerator (a somewhat wider slot with a screen) and is projected against the quadrature grid. The quadrature grid is tuned to the signal frequency and driven by space-charge coupling. The anode is placed behind the quadrature grid and a positive potential is applied to it.

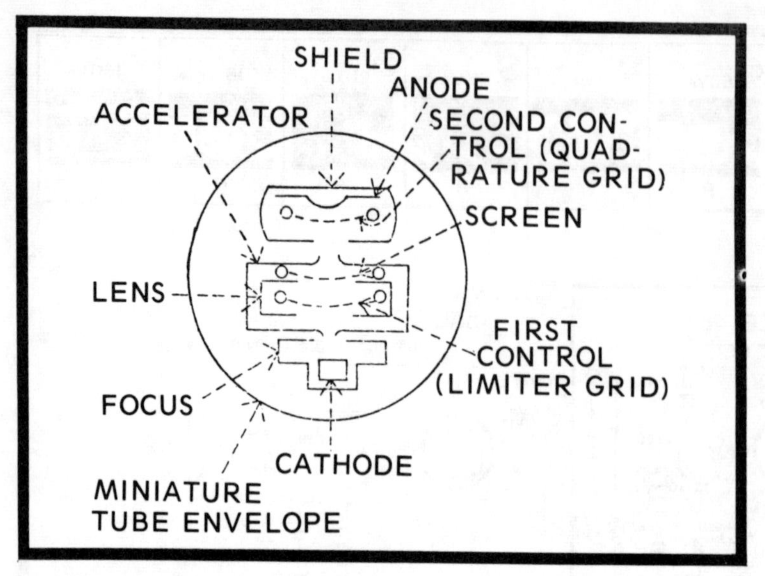

Fig. 13-3. 6BN6 gated-beam tube detector. (Courtesy Zenith.)

The circuit drawing in Fig. 13-4 shows the gated beam limiter-detector. A 4.5 MHz IF signal between 1 and about 50 volts is applied to the limiter grid and for half of each cycle the electron beam is passed and projected upon the quadrature grid. The 4.5 MHz IF signal found at the limiter grid should produce about 5 volts P-P across this tuned circuit.

The quadrature grid clips the leading portion from each half-cycle pulse and passes on to the anode periodic pulses of current about one-quarter cycle in length. Modulation of the signal frequency affects the phase displacement between the half-cycle electron stream and the voltage produced on the quadrature grid, causing the same variations in length of the anode current pulses. The plate load resistor is bypassed for IF, but the DC voltage drop across this resistor is proportional to the pulse length and is, therefore, a direct function of the original modulation signal.

The 500 ohm variable resistor (in the cathode circuit of the 6BN6) is generally known as the buzz control since it sets the operating point of this tube so that the 60 Hz vertical pulse will not appear in the plate circuit and in the audio output. This buzz control adjusts the bias and therefore the limiting level of the control grid. Ordinarily, the buzz and quadrature coil adjustments are made for the cleanest sound and minimum buzz as heard at the loudspeaker.

After adjustment of either the quadrature coil or buzz control, the other should be touched up slightly. If buzz still persists and cannot be eliminated with either of these ad-

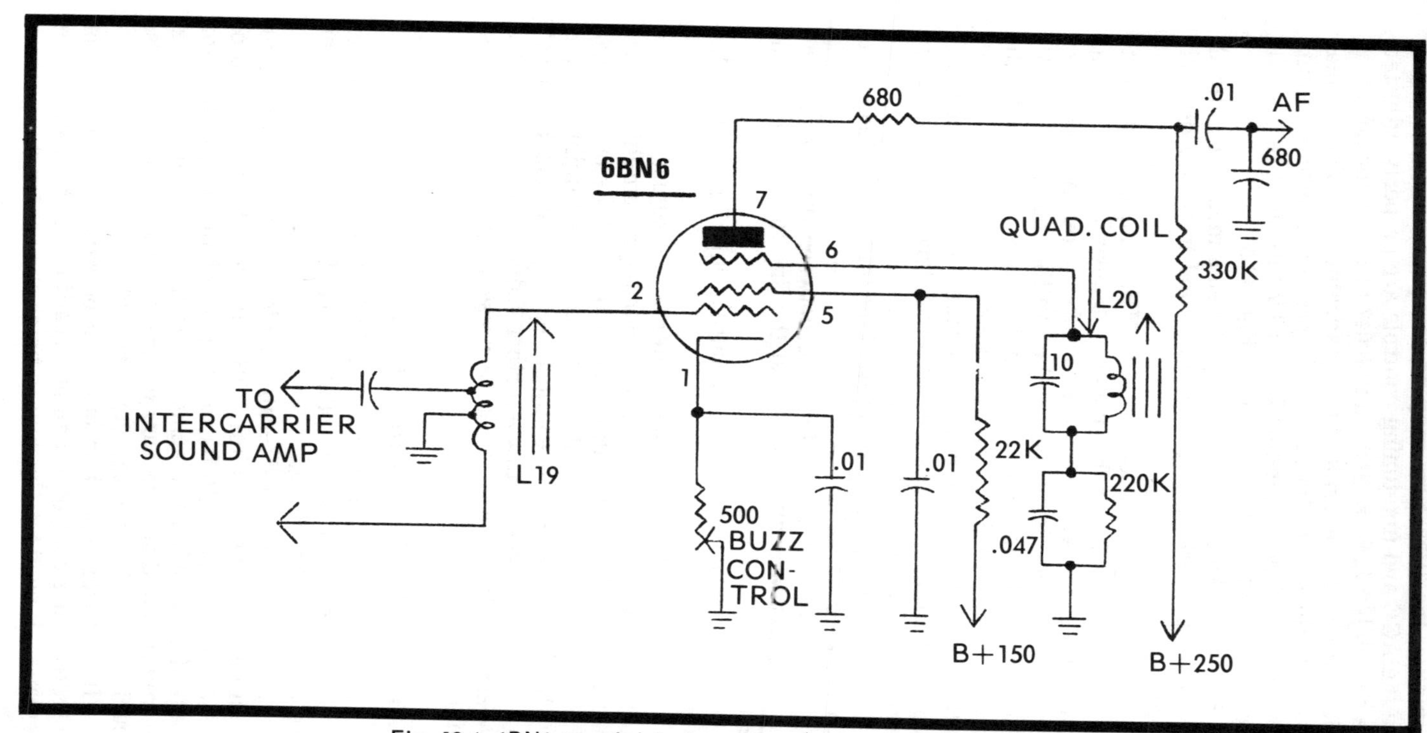

Fig. 13-4. 6BN6 sound detector schematic. (Courtesy Zenith.)

justments, the 6BN6 tube may be defective. However, make sure the AGC and fine-tuning controls are properly adjusted. The paralleled 220K resistor and 0.047 uf capacitor circuit in series with the ground side of the quadrature coil places a bias voltage on the quadrature grid to improve the noise immunity of this circuit. The late model Zenith TV sets now use the 6Z10 tube and sound circuits shown in Fig. 13-5.

Before making the sound adjustments make sure the IF stages are properly aligned and all sound traps are properly tuned.

Proper alignment of the 4.5 MHz intercarrier sound channel can be made by using a transmitted TV signal. However, the signal to the receiver antenna terminals must be kept at a reduced level, below the limiting point of the 6Z10 sound discriminator. This level can be easily identified by the "hiss" which can be heard along with the sound. Various methods may be used to reduce the signal level into the set. A step attenuator at the tuner antenna terminals is recommended for most satisfactory results.

The sound alignment for most of the late model Zenith sets is performed as follows:

1. Connect the step attenuator between the antenna and the TV antenna terminals. Adjust the fine tuning control for best picture and sound.

2. Adjust the step attenuator until a "hiss" is heard in the sound.

3. Referring to Fig. 13-5, alternately adjust the sound limiter plate transformer (T13), sound-sync amplifier plate transformer (T6), quadrature coil (L27); and the buzz control (R60) adjustments for best quality sound and minimum buzz.

Note: Any one of these adjustments may cause the "hiss" to disappear, necessitating a further reduction of the received TV signal. This "hiss" must be present for accurate sound alignment.

GATED-BEAM SOUND DETECTOR TROUBLES

Let's now go into some actual gated-beam detector troubles and see what the solutions are. Refer to Fig. 13-5 for the following discussion.

A rare case of considerable audio buzz was traced to an open neutralizing capacitor (C98) that goes from the grid of the sound IF stage (V8B) to the grid of the sound detector stage of V14A. After replacing C98, retune sound limiter transformer T13 and set the buzz control for minimum buzz. Also touch up coil L27 for the clearest audio.

Intermittent sound or buzz can be caused by a dirty or defective buzz control. Clean and readjust the buzz control or replace it if need be. Use a good tuner cleaner for this control.

Some of the Zenith sets using the 6Z10 may develop an intermittent buzz in the sound and the audio level will go up and down. One particular chassis had to operate an hour or two before the trouble would appear. After a long hunt the defect was tracked down to some cold solder joints at the terminals of coils T6 (sound takeoff) or T13 (intercarrier sound transformer). To repair, remove the shields from coils T6 or T13 and resolder all connections. If this does not cure the problem, the coils will have to be replaced.

Another set with this same chassis had a loud buzz in the sound. By adjusting the buzz control it was noted that neither the buzz nor audio level would vary. This condition was caused by a burned out 750 ohm buzz control and a faulty (shorted) 6Z11 tube.

Other problems that have occurred with the gated-beam detector are no sound or, again, buzz and a distorted audio output. One fast way to isolate trouble in these type detectors is to perform all alignment adjustments. Should one of the controls or transformers not act properly, this is a good clue as to which circuit component is defective.

When quadrature coil L27 opens, no audio output will be received. Should L27 short or develop leakage the sound may be distorted or a buzz will be heard in the sound output. When you adjust L27 in the alignment procedure and no effect is noted it's very likely the quadrature coil is defective or may have some cold solder joints. Resolder or replace this coil.

The solder joint where quadrature coil L27 connects to chassis ground may be faulty and this will cause some very odd sound troubles. This poor ground solder joint has been known to cause motorboating, squeals, howls and the sound level to cut in and out at a very intermittent rate. Resolder this one and any other ground connections in the sound system to be sure.

To improve the sound quality for some of the early model Zenith color sets using the 6Z10 tube, add a 0.01 uf capacitor from cathode (pin 8) to ground. (This parallels C91.) After the capacitor is installed be sure to retune the quadrature coil and readjust the buzz control. Later production chassis already have this capacitor installed.

Motorboating Audio

When this set was turned on the sound and picture began motorboating which indicated some stage was oscillating. This did not seem like oscillation in the tuner or IF stages, so the audio section was checked out. The circuit, see Fig. 13-6, uses a 6Z10 tube and some type of component failure had to be causing a feedback problem. So various components were

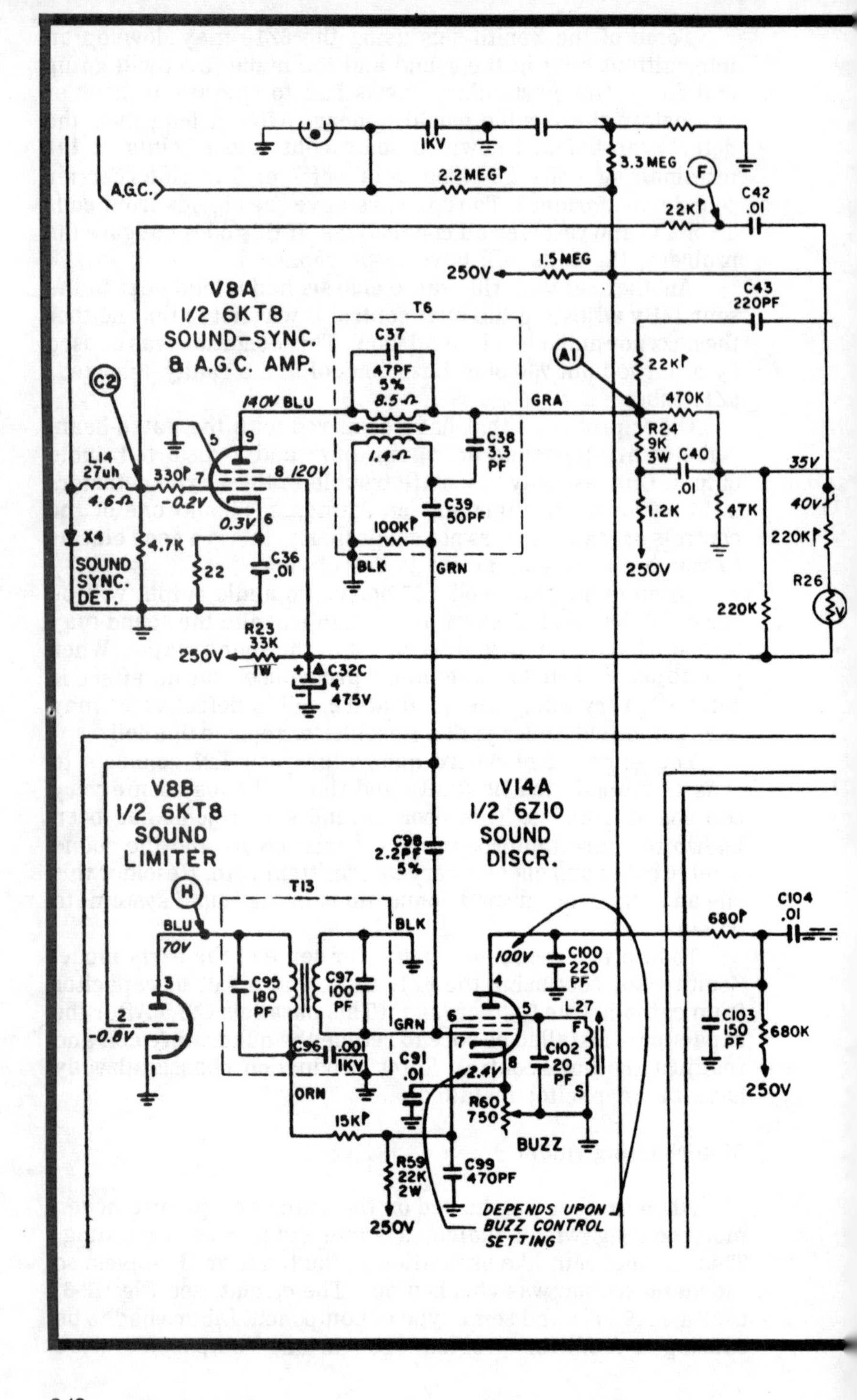

A.G.C.
1KV
2.2 MEG
3.3 MEG
F
22K
C42
.01
1.5 MEG
250V
C43
220PF
V8A
1/2 6KT8
SOUND-SYNC.
& A.G.C. AMP.
T6
C37
47PF
5%
8.5-Ω
1.4-Ω
C38
3.3
PF
C39
50PF
100K
BLK
GRN
GRA
A1
22K
470K
R24
9K
3W
C40
.01
1.2K
47K
250V
35V
40V
220K
R26
220K
C2
140V BLU
L14
27uh
4.6-Ω
330
-0.2V
120V
0.3V
4.7K
22
C36
.01
X4
SOUND
SYNC.
DET.
R23
33K
1W
250V
C32C
4
475V
V8B
1/2 6KT8
SOUND
LIMITER
V14A
1/2 6Z10
SOUND
DISCR.
C98
2.2PF
5%
H
T13
BLU
70V
BLK
C95
180
PF
C97
100
PF
GRN
-0.8V
C96
.001
1KV
C91
.01
ORN
15K
R59
22K
2W
250V
C99
470PF
100V
C100
220
PF
L27
F
S
C102
20
PF
2.4V
R60
750
BUZZ
680
C104
.01
C103
150
PF
680K
250V
DEPENDS UPON
BUZZ CONTROL
SETTING

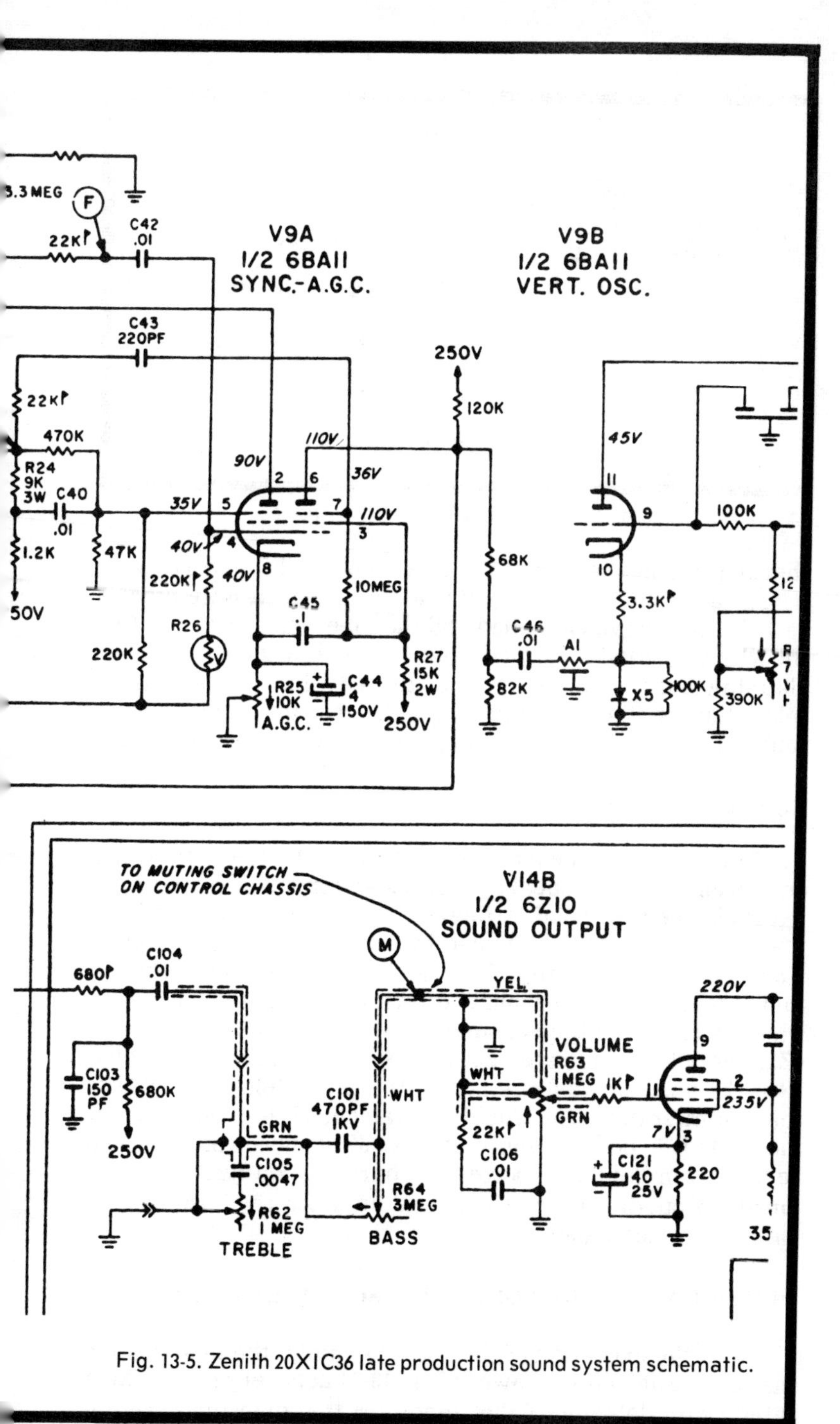

Fig. 13-5. Zenith 20X1C36 late production sound system schematic.

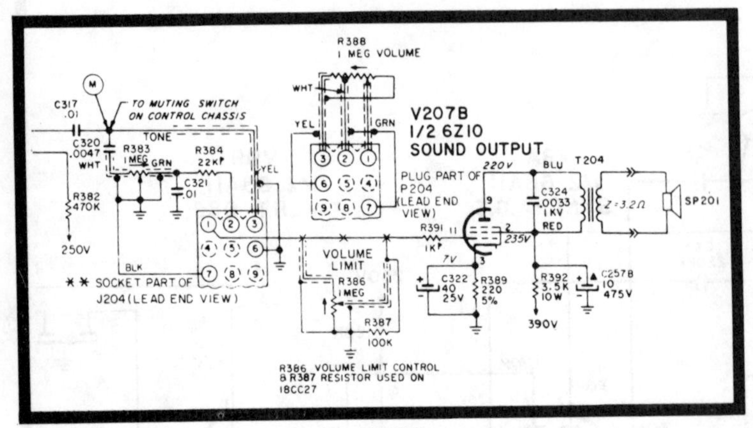

Fig. 13-6. Zenith 18CC29 sound output.

checked out in the audio output circuit. A burned resistor, R391, was found in the control grid circuit of the 6Z10 tube. A new 1K resistor was installed and the tube was checked to find out why the resistor burned. The tube had a high resistance plate-to-control-grid short and turned this stage into a relaxation oscillator circuit. When the resistor and tube were replaced, the motorboating stopped and audio was restored.

No Sound

This set had a fine picture but the complaint was no sound on this chassis. The speaker was tested for an open voice coil but it checked out good. A finger check at the control grid of the audio output tube indicated by a loud buzz that this stage was operating correctly. The next step was some voltage measurements in the sound discriminator stage of V207A which is shown in the schematic of Fig. 13-7. At pin 6 of the 6Z10 the VTVM indicated zero volts. This should be 75 volts. Also the voltage was missing at the plate (pin 3) of V203B sound limiter tube. The trouble was an open R244 resistor from the 250V B+ power supply. A new 6Z10 tube was installed because it might have shorted and burned open R244. Now we tuned the quadrature coil L226 and adjusted the buzz control for a good clean sound.

DUAL-CONTROL QUADRATURE SOUND SYSTEM

With the exception of the way it's made, the dual-control quadrature detector (shown in Fig. 13-8) acts very much like a gated-beam detector. Tubes made for this detector purpose, such as the 6T10, 6DT6 and 12AE10, are specially constructed sharp-cutoff pentodes used in circuit arrangements quite like that of a 6BN6. An important function of this tube is its ability to cut off plate current sharply with both the control grid and the suppressor grid. Note the similarity of the circuit arrangements in Figs. 13-5 and 13-8.

The buzz control is not used in the 12AE10 circuit, because the bias requirements are not so critical as for the 6BN6 tube. The 680 ohm linearity resistor in the plate circuit of the 6BN6 is not used, and the biasing network for the quadrature grid circuit must be used with the 12AE10. The internal construction of these two pentodes is quite different, although the schematics for them appear to be similar. The 12AE10 has the usual pentode construction while the 6BN6 is as in Fig. 13-3.

For high-level signals, the control grid of the 12AE10 draws current. This flow of grid current reduces the tuned circuit Q and causes limiting action. Otherwise, the control and quadrature grids have the same action as the corresponding grids in the 6BN6.

On low-level signals, the circuits act like a locked-in oscillator. When no signal appears, there is no control-grid current and thus no loading of the tuned circuit. The feedback circuit from the quadrature grid to the control grid is used to sustain oscillation at 4.5 MHz. A weak signal which causes no control-grid current will lock this oscillator frequency in step with the frequency of the incoming signal. This oscillator signal is varied about 4.5 MHz by the incoming signal and will

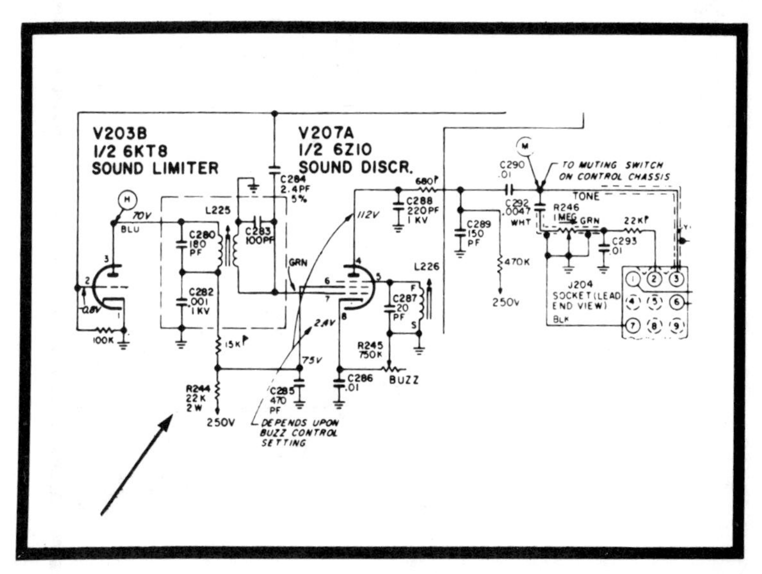

Fig. 13-7. Zenith 14A10C17 sound limiter and sound descriminator circuit.

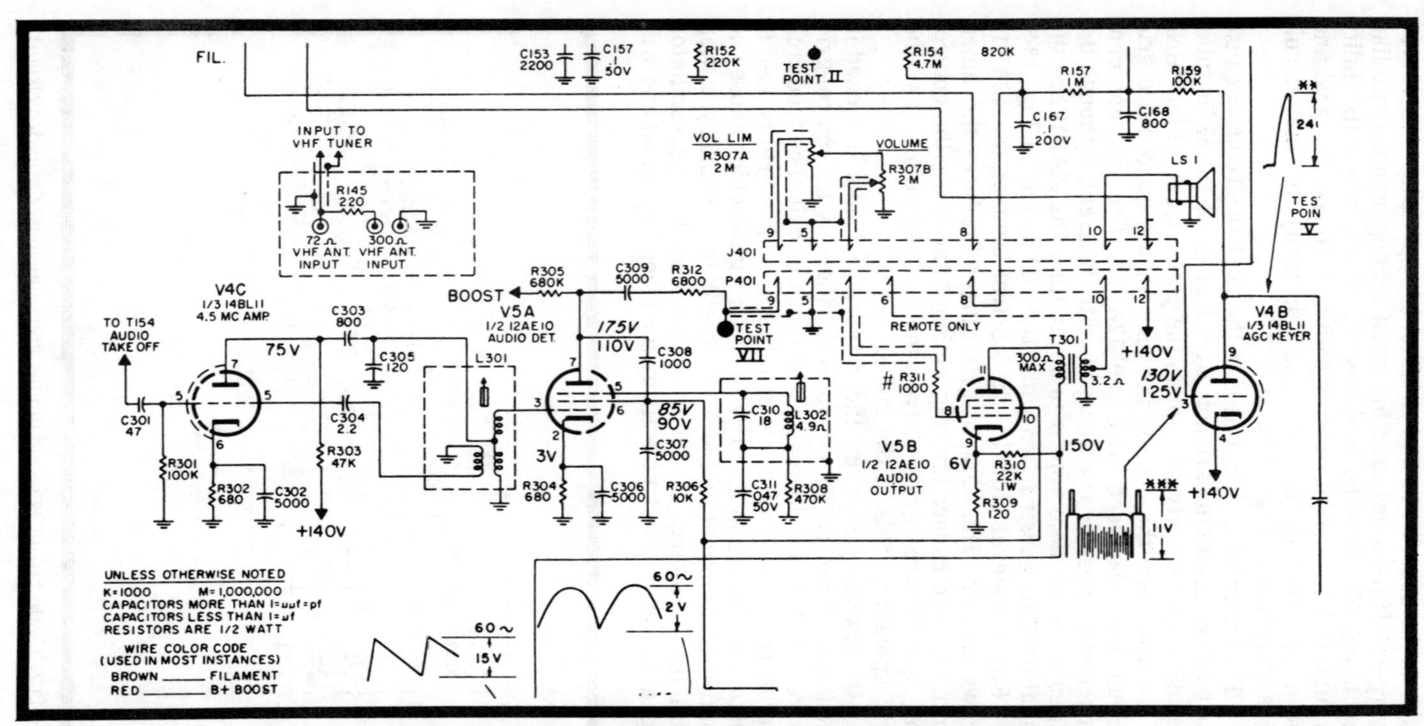

Fig. 13-8. Dual-control quadrature detector. (Courtesy General Electric.)

produce a high-level FM signal having the same response as an incoming low-level signal. This signal is then demodulated by the 12AE10. Use the same alignment procedures as with the 6BN6 tube circuit detector.

Like the 6BN6, the 12AE10 circuit acts as a limiter, detector, and audio voltage amplifier.

SOLID-STATE SOUND SYSTEMS

Many of the later model TV receivers are using transistors and ICs in the sound section. For this reason, a large part of this chapter will be devoted to these solid-state devices.

Zenith Sound Module

The Zenith IC sound module shown in Fig. 13-9 is contained in a shielded metal can that is mounted to the chassis like an IF transformer. It is soldered to the chassis to insure a good ground connection. This assures complete containment (shielding) of the 4.5 MHz sound IF signals. Should this 4.5 MHz RF be permitted to "seep out," then excessive snow may appear in the picture for weak RF signals. Circuitry within the module has been simplified by use of a plug-in IC chip.

Composite video information appearing at test point C4 is coupled through capacitor C602 to the primary winding of sound takeoff transformer T601. This network also serves to filter out low frequency components present in the video. The secondary winding of T601 is connected to pins 4 and 6 of the IC chip. Pin 6 is bypassed for AC by C605. The IC consists of a three stage limiter and a balanced product detector. T602,

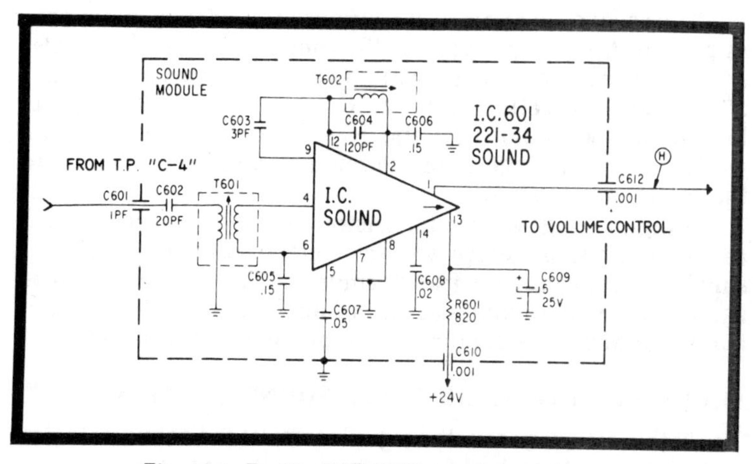

Fig. 13-9. Zenith 22AB55 IC sound module.

connected to pins 2 and 12 of the IC, serves as the quadrature coil. Tune it for minimum buzz while trying to attain maximum audio output level.

Detected audio appears at pin 1 (test point H) and is coupled to the audio output section. Capacitor C608 connected between pin 14 and ground provides the proper audio de-emphasis. Approximately 12V of B+ is fed to pin 13. Pins 7 and 8 are grounded. The other voltages around the "chip" are the results of internal voltage drops. Metering these voltages is a convenient way to check out the IC's performance. Circuitry within the IC is shown in Fig. 13-10 comprising a total of 19 transistors, 18 resistors, and 6 diodes.

Motorola Sound Module

The diagram for one of Motorola's color TV IC sound systems is shown in Fig. 13-11. The three sections, shown by the triangles, include the amplifier/limiter, FM detector, and audio preamplifier.

Because it is desirable to keep 4.5 MHz out of the video, the 41.25 audio carrier is taken out by traps ahead of the video detector. The 45.75 video and 41.25 audio carriers are picked off the 3rd IF by a coupling capacitor. Audio IF of 4.5 MHz is produced by mixing the video carrier with the frequency modulated audio carrier in the detector transistor located on the "B" panel (not shown here). The output of this mixer is tuned to 4.5 MHz by a coil, capacity tapped to minimize loading and provide coupling to the IF amplifier in the IC.

A new and unique phase detector in the IC provides automatic recovery of the audio and at the same time cancellation of the 4.5 MHz carrier. A 90 degree phase shift coil is needed, and this is L1, externally connected at pins 8 and 13 of the IC.

The recovered audio at the output of the detector appears at pin 2 of the IC, and goes to the tone and volume controls. Audio leaves the board at terminal 4D and returns at terminal 5D. The audio is returned to the IC at pin 9 for low-level amplification. The audio from the IC is direct-coupled from an emitter follower stage at pin 10 for impedance matching into the audio output stage.

A high voltage silicon transistor (not shown here) is used as an audio output device with the collector tied to the 95V DC supply. A conventional transformer-speaker arrangement is used to develop an audio output of about 3 watts. The output transistor and transformer are not mounted on this panel.

SERVICING THE SOLID-STATE SOUND SECTION

In this sound system the signal starts at the sound detector amplifier (Q103) in the IF module unit. (Refer to the complete

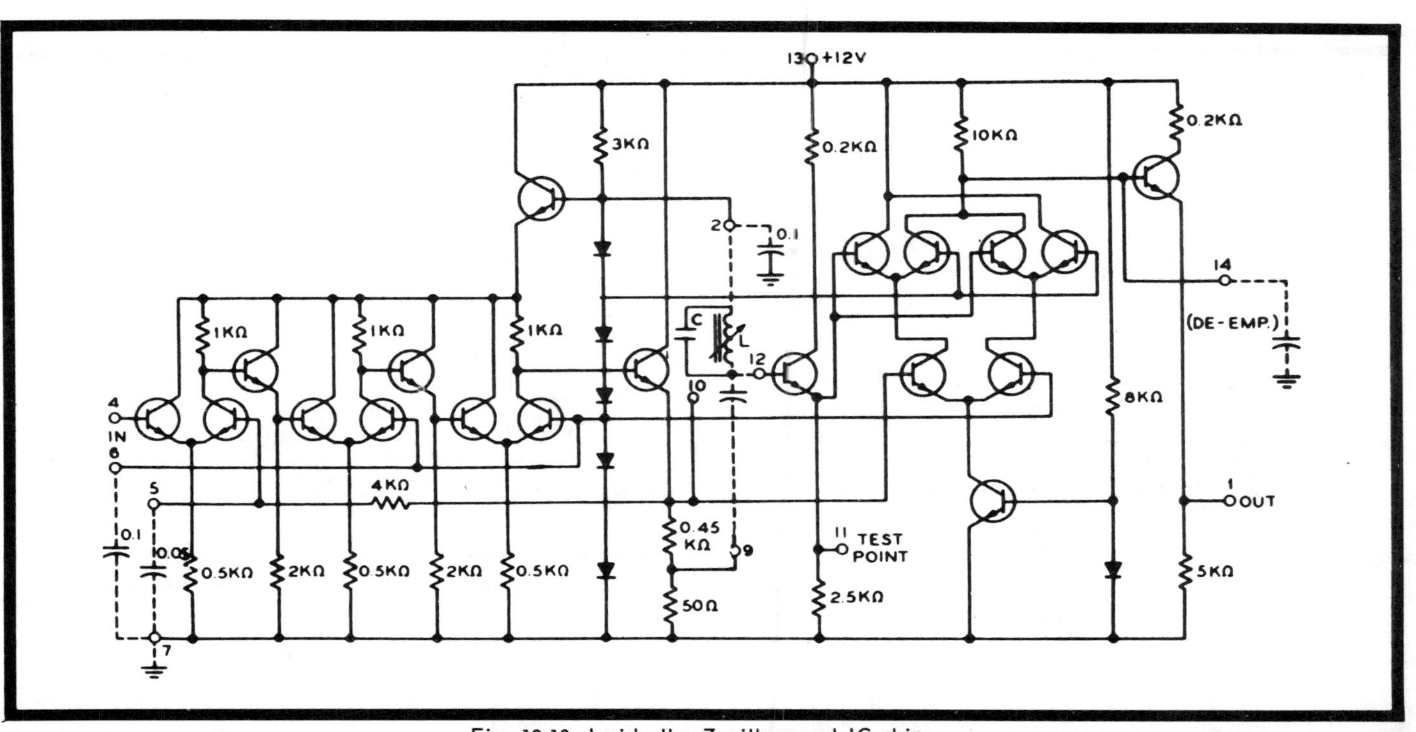

Fig. 13-10. Inside the Zenith sound IC chip.

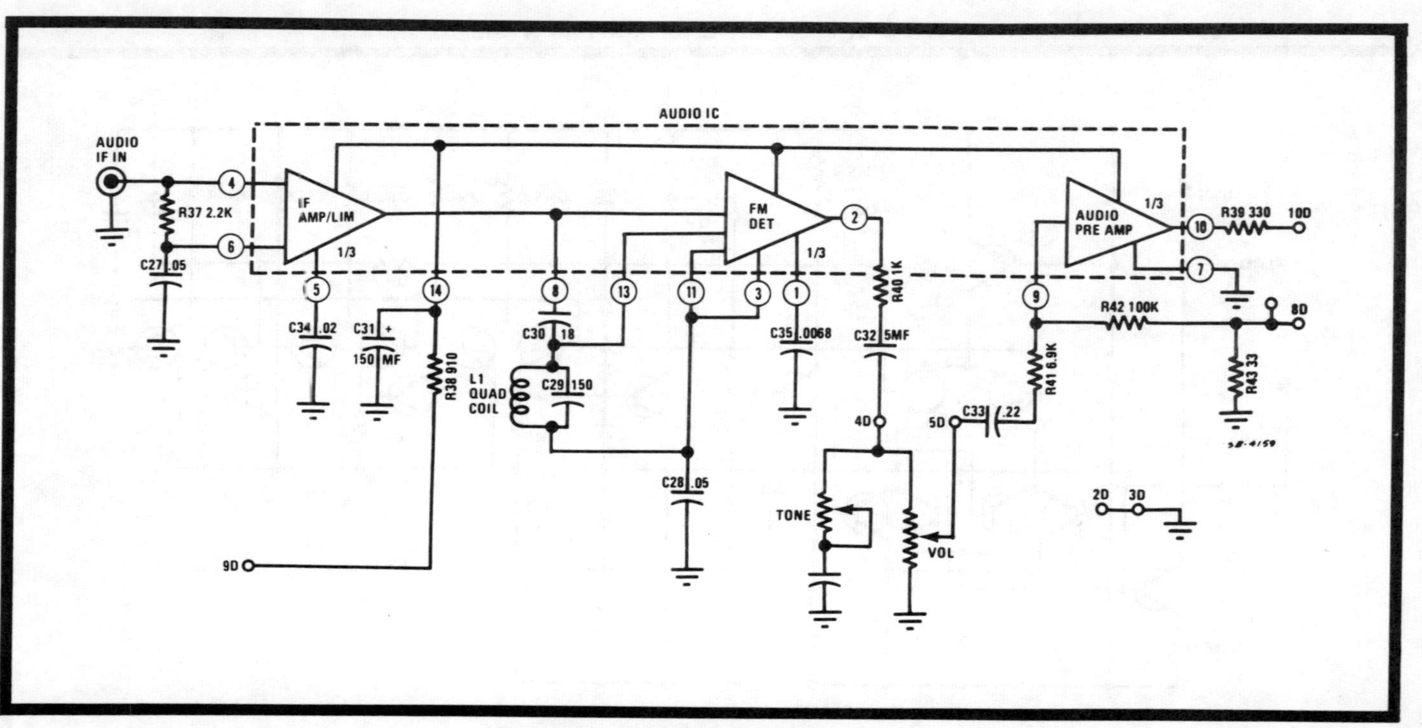

Fig. 13-11. Motorola IC sound module.

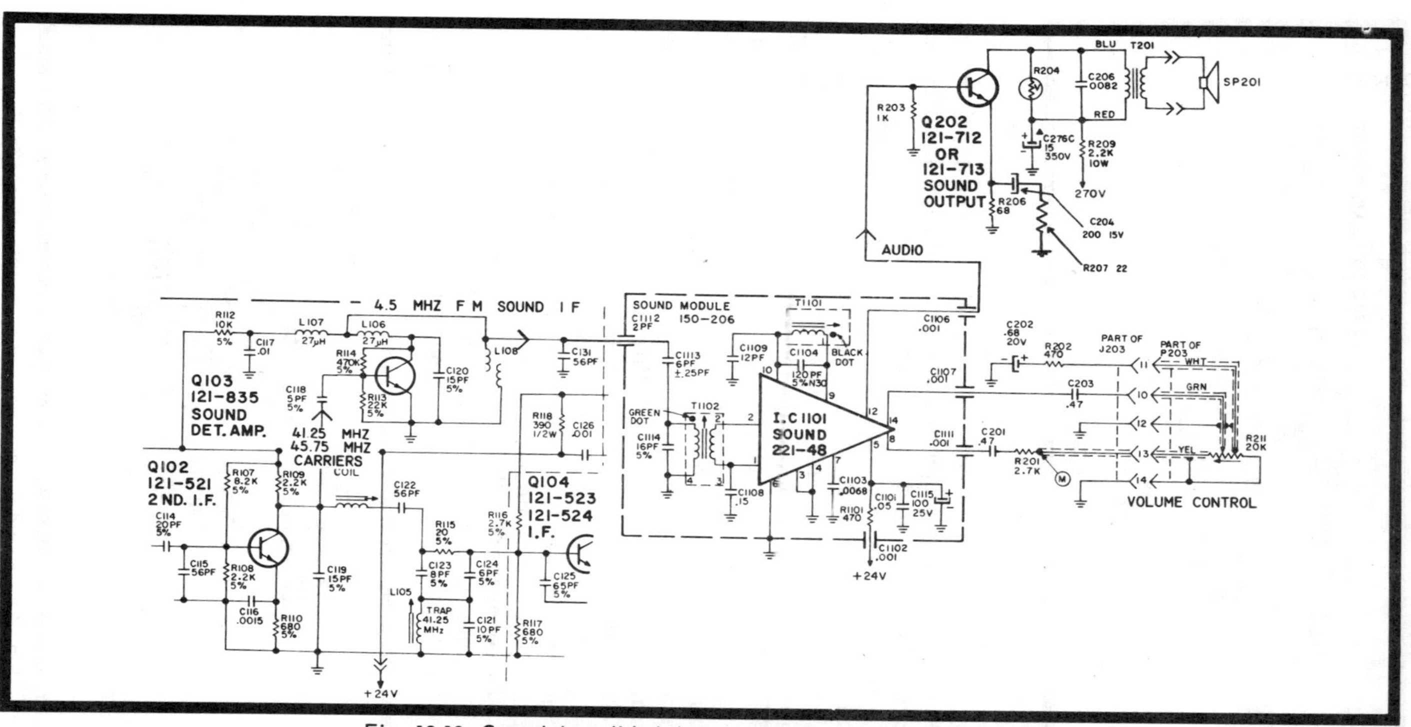

Fig. 13-12. Complete solid-state sound system. (Courtesy Zenith.)

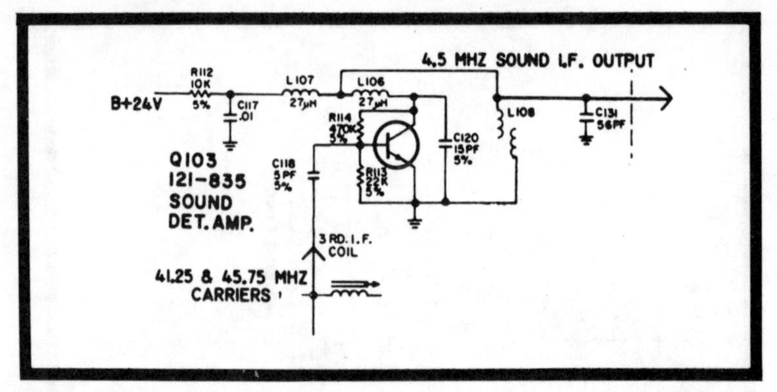

Fig. 13-13. Sound detector amplifier details. (Courtesy Zenith.)

sound IF and audio system in Fig. 13-12.) The 121-835 transistor is biased as a mixer and produces the 4.5 MHz difference between the IF sound and picture carriers.

This sound detector amplifier (located in the IF module and shown in Fig. 13-13) can be easily checked out.

There will be a slight forward bias on the base (checked with a VTVM) and a low collector voltage (through R112, 10K resistor), which will indicate that it is drawing current. Now remove the lead between the video IF and sound modules. With an oscilloscope, check at the output of Q103 for sync information. However, this is only a test, since 4.5 MHz information is all that can pass through C1113 (6 pf) to the 4.5 MHz transformer (T1102) on sound module 150-205.

A normal sound module can be checked with a variable frequency 4.5 MHz generator that is unmodulated. Refer to Fig. 13-14 for this module circuit.

Disconnect the yellow lead (audio output from the chip to the volume control) and connect a VTVM to this lead. Use a low DC range on the VTVM. Disconnect the white lead from the

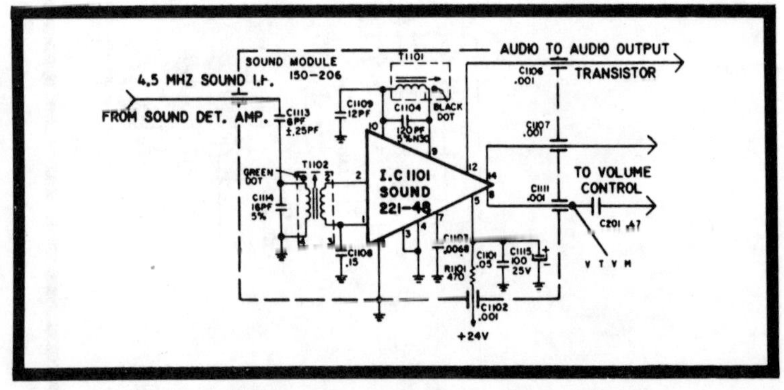

Fig. 13-14. Sound detector, 1st audio amplifier chip. (Zenith.)

sound detector amplifier and connect this lead to the 4.5 MHz generator. Now apply 24V B+ to the sound module. As you slowly vary the carrier on either side of 4.5 MHz, this will produce a varying DC swing at the sound output as indicated on the VTVM. For a particular frequency, varying the input carrier amplitude will vary the output DC level up to a point above which an increased input will not affect output.

The two tests above confirm both FM detector and limiting actions. If performance cannot be achieved, check adjustments of coils T1101 and T1102 as well as the volume control and components.

The sound output stage (Q202) is a normal transistor, forward biased through the chip. This audio output stage is shown in Fig. 13-15. Its collector voltage is obtained from the 270V B+ line through dropping resistor R209 and the primary winding of output transformer T201. Distorted, weak or no audio, yet with normal transistor collector voltage, and apparently normal operating sound module, may be caused by failure of either VDR R204 or capacitor C206 across the primary winding.

Some failures of Q202 have been due to a loose mounting screw, thus causing a high heat build-up and the transistor to be ruined. This is on the newer, flat case transistor "type 77" with a mounting hole in the center and three leads coming out one end. When a replacement is installed, apply lots of silicone grease where it mounts onto the heat sink. Make sure the mounting screw is good and tight in order to have maximum heat transfer. This should help to prevent future transistor failures.

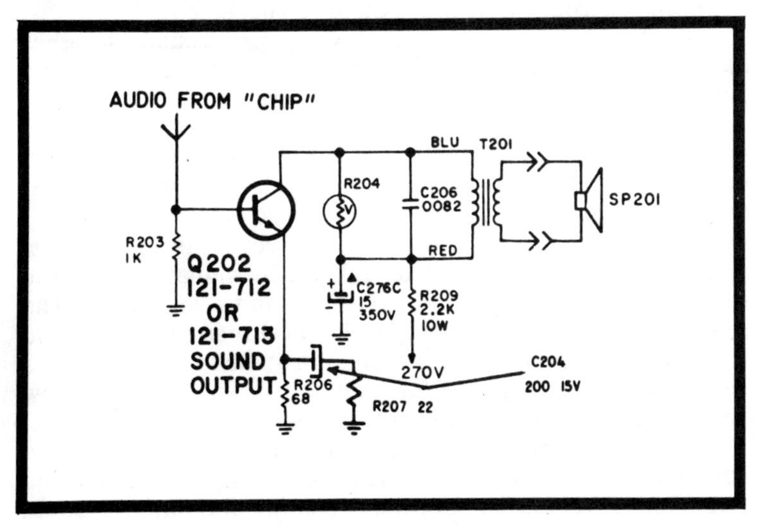

Fig. 13-15. Transistor audio output (Zenith).

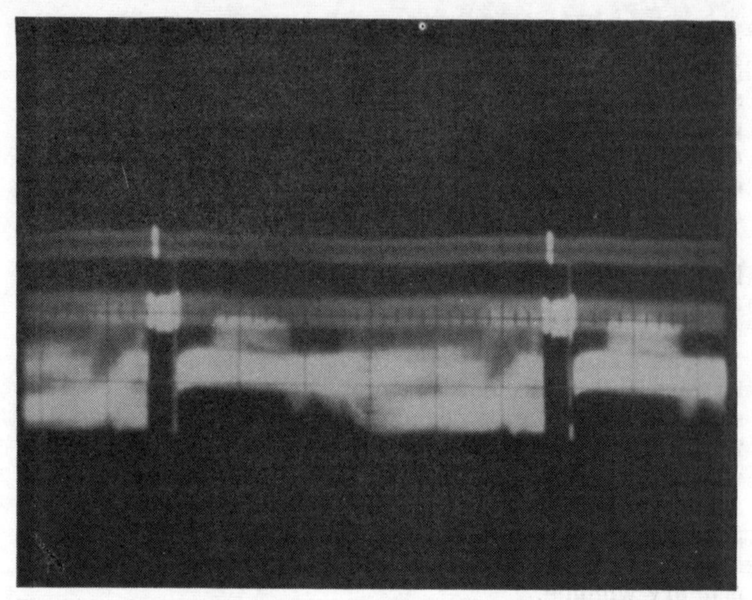

Fig. 13-16. Sync pulses in the sound IF. These demodulated 60 Hz sync pulses can have either negative or positive polarity. (Scope set at 60 Hz rate.)

SOME "SOUND" TROUBLESHOOTING TIPS

In most receivers audio problems will be located in the sound circuits; of course, there will always be some exceptions. Some of the exceptions will include buzz or hum in the sound. Also, should both the picture and sound be affected, a common circuit fault must then be suspected.

Some sound circuit problems will cause troubles in other stages of the TV chassis, such as sound bars in the picture and radiation interference from the sound stages.

Other sound troubles are distorted sound, weak sound, no sound, buzz, motorboating, and microphonics. To help locate these troubles, measure voltages and resistance with a VTVM and make a visual inspection of the sound section. An overall alignment and response check will also reveal these troubles. A quick way to isolate sound faults is by signal tracing the sound section with an audio generator and oscilloscope. The scope or an audio amplifier type signal tracer can be used to trace the TV station signal through the audio system. This is a very good method to use for tracking down distortion, low volume level, and any intermittent condition.

When a scope is employed for signal tracing, use a demodulator probe to follow the sound IF signal through the FM sound IF amplifier. Set the scope for maximum vertical gain and touch the probe, in turn, to the input and output of the sound IF amplifier stages. You will be looking for the presence of demodulated (60 Hz) sync pulses and this will indicate active circuit operation up to the point where the probe is connected. Note scope trace in Fig. 13-16. These can be either positive or negative sync pulses. Looking at the schematic in Fig. 13-17, these signal pulses should be found at pin 5 and 7 of V4C and at pin 3, control grid, of V5A, while the set is tuned to an active station.

The audio signal may be checked at the output of the FM detector and audio amplifiers by using the scope and a direct probe. The amplitude of the audio signal on the scope will increase as you move the probe toward the speaker. Referring to Fig. 13-17, an audio signal will be found on pin 7 of V5A, at the volume control (which is a good test point), pins 8 and 11 of V5B, and at the speaker voice coil. The amplitude of the signal will also depend upon the setting of the volume control as you move past that point. An audio signal tracer can also be used in the same way.

Signal injection can also be used to pinpoint trouble in the sound section. To find the faulty stage, inject a 600 Hz audio signal to point "B" of the sound circuit in Fig 13-18, which is the audio output of the ratio detector. If the tone is heard in the speaker, the audio amplifier system is working properly. Next, use a 4.5 MHz signal with 60 Hz FM modulation and feed this composite signal into the base of the first sound IF, Q206. (This type of signal is available from an RF sweep generator.) If no sound is heard, then apply this signal to the base of Q207 (second sound IF stage); and if the 600 Hz sound is now heard, this isolates the fault to the first sound IF stage. A VTVM is now used to find the faulty component. In one faulty set, voltage checks showed zero volts on the base and emitter

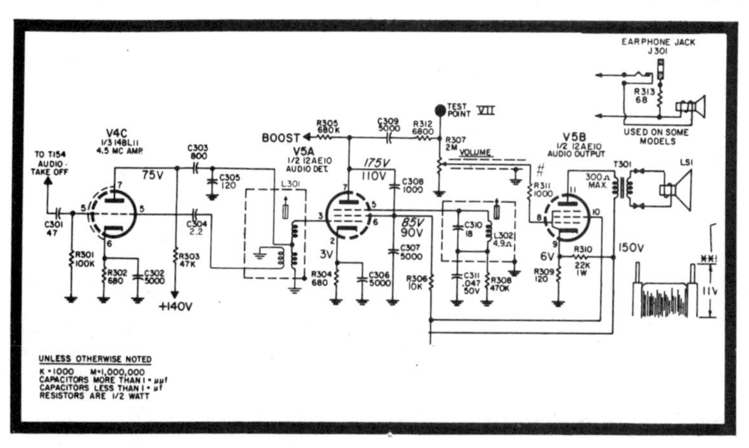

Fig. 13-17. Tube-type audio circuits. (G.E.)

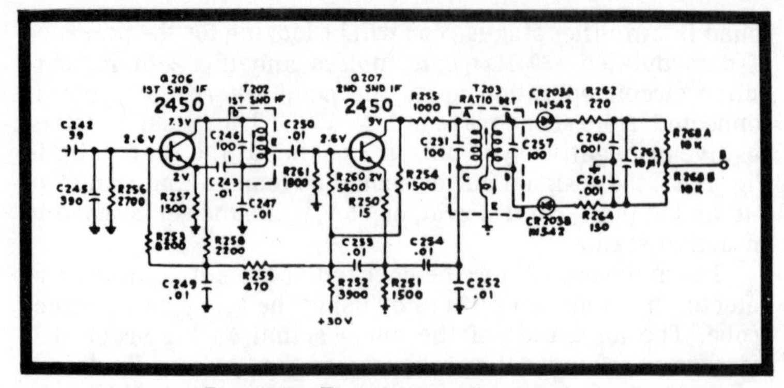

Fig. 13-18. Transistor audio circuits.

while the collector voltage was higher than normal. The base circuit was suspected and an ohmmeter check showed C245 shorted.

By following the signal path through a particular circuit, each stage can be evaluated from the standpoint of input signal versus output signal.

In the process of signal injection and signal tracing, the technician can make use of this information to isolate troubles in tube or solid-state TV receivers first to a particular stage, then by voltage and resistance measurements to a specific component.

The foldout section that follows contains schematics of representative TVs to illustrate various means used by manufacturers for signal amplification, takeoffs, etc. These, in concert with the specific Zenith Troubleshooting flow charts in the foldout section, will aid you in directing your efforts toward logical analysis of faults and their isolation.

The following foldout section consists of pages 257 through 288.

Low Voltage Power Supply Functions and Checks

Because of its very natural requirements, the power supply (high or low voltage) is the very heart for all types of electronic equipment. The power supply furnishes the proper DC voltage to the various circuits and AC voltage to the tube filaments or only to the CRT in the case of an all-solid-state TV chassis.

Many of the sets use a power transformer to produce the desired voltages but some are directly fed from the AC power line and are not isolated. Most of the TV sets now use silicon diode rectifiers in a bridge circuit configuration and we very seldom see the old 5U4 rectifier tubes any more.

MODERN TV POWER SUPPLY CIRCUITS

Before going into actual power supply problems let's look at a few of the modern power supply systems and see how they function. A good understanding of circuit operation will help you find faults more efficiently.

Basic Half-Wave Power Supply

A simple half-wave power supply (Fig. 15-1) is a basic circuit found in many black-and-white TV receivers. This is a two-section capacitive-input filter circuit that may use a power transformer or be direct AC line operated.

Another type is the inductive-input filter system (Fig. 14-2) that has a series inductor next in line right after the diode rectifier. An inductive-input filter provides a lower output voltage but offers better B+ voltage regulation than a capacitive-input filter. The capacitive-input filter also produces a much higher peak current through the rectifiers.

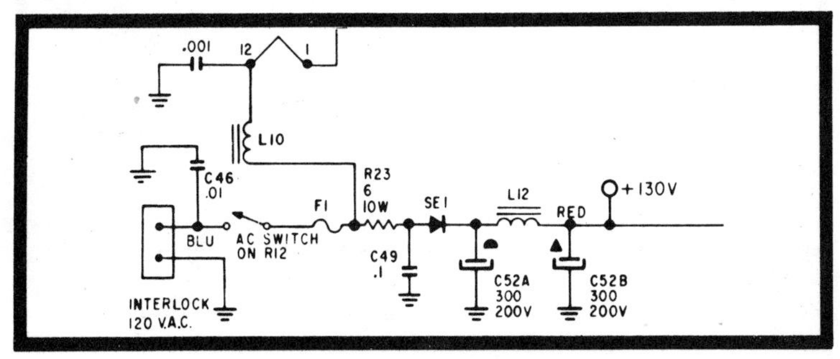

Fig. 15-1. Basic half-wave power supply.

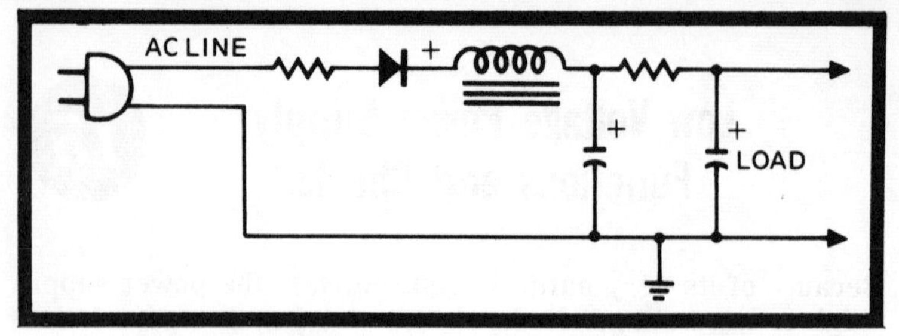

Fig. 15-2. Half-wave power supply with choke input.

Half-Wave Doubler Power Supply

The basic half-wave doubler power supply (Fig. 15-3) was used in many of the middle-aged color TV receivers. The filter capacitors and the choke, or resistor preceding each capacitor, are considered as primary filter circuits. Capacitor C1 refines the pulsating DC (at the diode rectifier output) into a steady DC voltage with some ripple. C2 (with a choke or resistor preceding it) reduces the ripple further, producing a smoother DC voltage with a small semi-sine-wave ripple. Also, C2 bypasses stray signals induced into the B+ line along the way, thus preventing undesired inter-stage coupling. Ultimately C3 and C4 smooth the DC further, while providing increased decoupling between stages and a ground return for all signals on the B+ line.

Full-Wave Doubler

A full-wave voltage doubler circuit is shown in Fig. 15-4; the DC voltage produced will be approximately two times the RMS value of the supplied AC voltage.

The oscilloscope waveforms shown are for a properly operating doubler rectifier circuit. Waveform W1 is a 60 Hz sine wave ripple equal to 10 percent of the DC voltage produced. In a full-wave circuit W2 has a sawtooth frequency

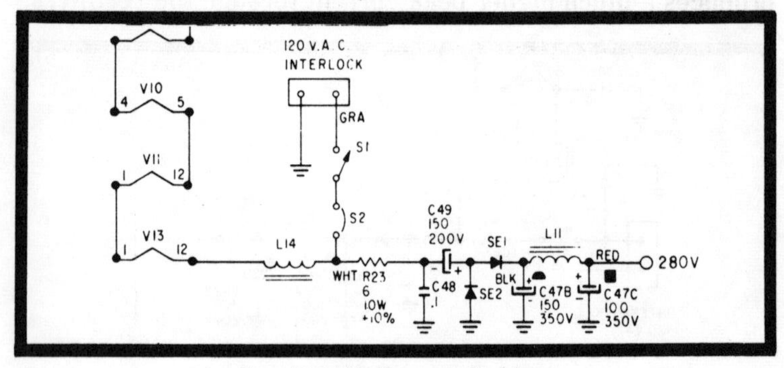

Fig. 15-3. Half-wave doubler.

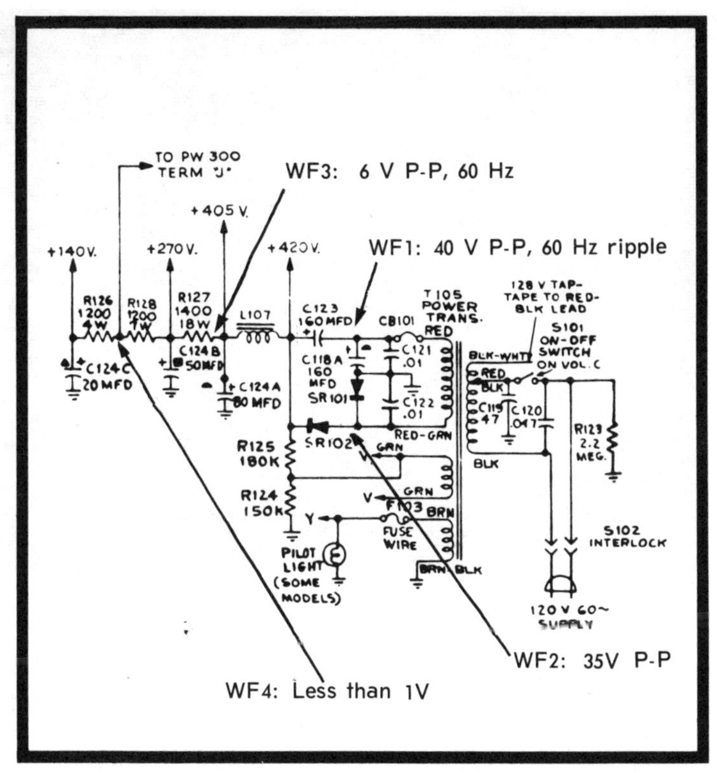

Fig. 15-4A. Full-wave doubler.

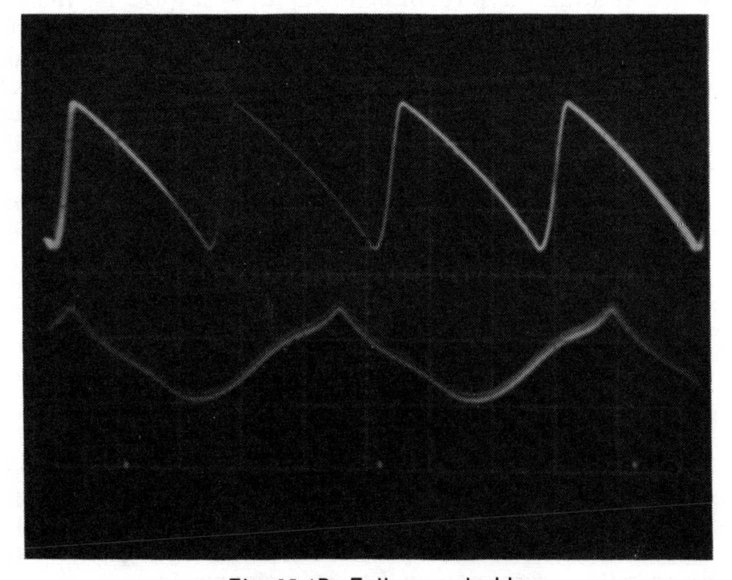

Fig. 15-4B. Full-wave doubler.

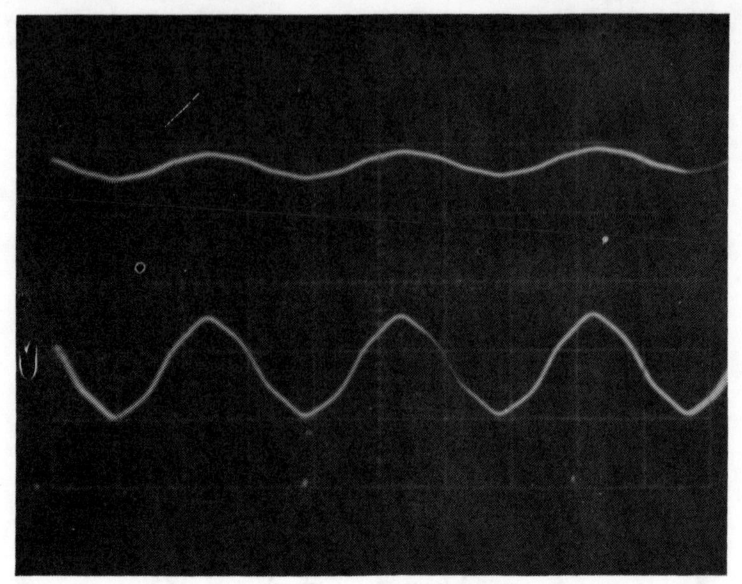

Fig. 15-4C. Full-wave doubler.

of 120 Hz. W3 will have about 6 volts P-P and W4 should be less than 1 volt and have only a slight ripple.

Motorola's solid-state TS-915/919 power supply circuit is shown in Fig. 15-5 with some scope waveforms. The four silicon diodes in a full-wave bridge supply high current which is well filtered by an input choke. Two diodes of the bridge serve as a full-wave rectifier to supply +35V DC for most of the receiver primary signal circuits. A half-wave rectifier develops +255V DC for the video output amplifiers. Various other voltages are divided off the main bridge supply for specific circuit needs. Automatic degaussing is featured with a thermal relay to bypass the degaussing coil when the receiver is operating. Filaments to the picture tube are always on to increase the life of the CRT and have fast CRT warmup. A quick-on defeat switch is provided to turn off the CRT filaments if so desired. This chassis is protected against a high current surge by a conventional thermal cutout system. The total power input for this chassis varies from 125 to 200 watts, depending upon CRT brilliance.

Wattmeter Testing

While we're thinking of TV receiver power consumption, the AC wattmeter can be a valuable piece of test gear. Simply plug the set into the wattmeter and monitor the AC line input current drain. Then compare the normal wattage with that given in the service data to the wattmeter reading.

Bridge Rectifier System

The following color TV bridge rectifier will develop the B+ voltages from three separate rectifier circuits to produce 400, 285, and 180 volts.

This GE "KE" power supply is balanced for current through the degaussing coil. The power transformer has a secondary winding split at the center into two equal windings. A tap on each winding provides a source for full-wave rectification of the +285-volt supply. The degaussing circuit is connected between the center taps of the two secondary windings.

Refer to the simplified schematic in Fig. 15-6 of the balanced power supply as we dig into this system.

The three DC power sources are all developed from the full-wave rectifier systems. With the degaussing circuit connected at the center of the complete system at the junction of the two secondary windings, all of the current from both the negative and positive half cycles flows through the degaussing coil when AC power is first applied.

Two 160 ohm resistors R154 and R160 are connected in series across degaussing coil L111 to connect the two secondary windings together and provide a connection for CR108 at the center of the secondary winding without creating a short circuit across the coil.

When the set is first turned on, thermistor RG102 is cold and at a high resistance, thus most of the current flows through the low resistance of the degaussing coil. The small current through RG102 causes it to heat, and having a negative coefficient, its resistance becomes lower and the current increases through RG102. Then the resistance of RG102 becomes so low that the small current flowing in L111 is switched off by VDR101. The low resistance of RG102 is now effectively a short circuit, connecting the two secondary windings together, and it also connects R154 and R160 in parallel at 80 ohms.

The power transformer has been designed to accommodate three separate rectifier circuits to produce 400 volts, 285 volts and 180 volts DC, while lower voltages are derived from taps in the filter networks. The 285 volts is obtained by full-wave rectification. This system works in a different manner than a conventional center-tapped system because the return circuit is through the bridge rectifier diodes. CR110 functions from the potential existing between terminals 2 and 4 while CR109 functions from the potential between terminals 1 and 3.

It may look as if both CR109 and CR110 might conduct at the same time since both diodes are included in the circuit for each half-cycle of rectification. This happens only for an in-

NOTE: THE PEAK-TO-PEAK COLOR SIGNAL AT EMITTER OF Q-6B CAN VARY AND IS DEPENDENT ON THE AGC SETTING OF A GIVEN RECEIVER. THE 1 VOLT PEAK-TO-PEAK READING IS GIVEN AS A REFERENCE. THE PEAK-TO-PEAK WAVEFORM MEASUREMENTS SHOWN ON THE SCHEMATIC ARE RELATED DIRECTLY TO THE 1 VOLT P-P MEASURED AT EMITTER OF Q-6B. AMPLITUDES OF ALL SIGNAL WAVEFORMS TAKEN MUST BE RELATED DIRECTLY TO THE AMPLITUDE OF PEAK-TO-PEAK SIGNAL MEASURED AT THE EMITTER OF Q-6B.

TO EXTEND THE RELIABILITY OF THE HORIZONTAL OUTPUT TRANSISTORS, THE HORIZONTAL RETRACE TIME HAS BEEN INCREASED. HENCE, THE HORIZONTAL BLANKING TIME IS ALSO INCREASED. THIS RESULTS IN PARTIAL BLANKING OF VIDEO INFORMATION ON THE EXTREME LEFT OF PICTURE.

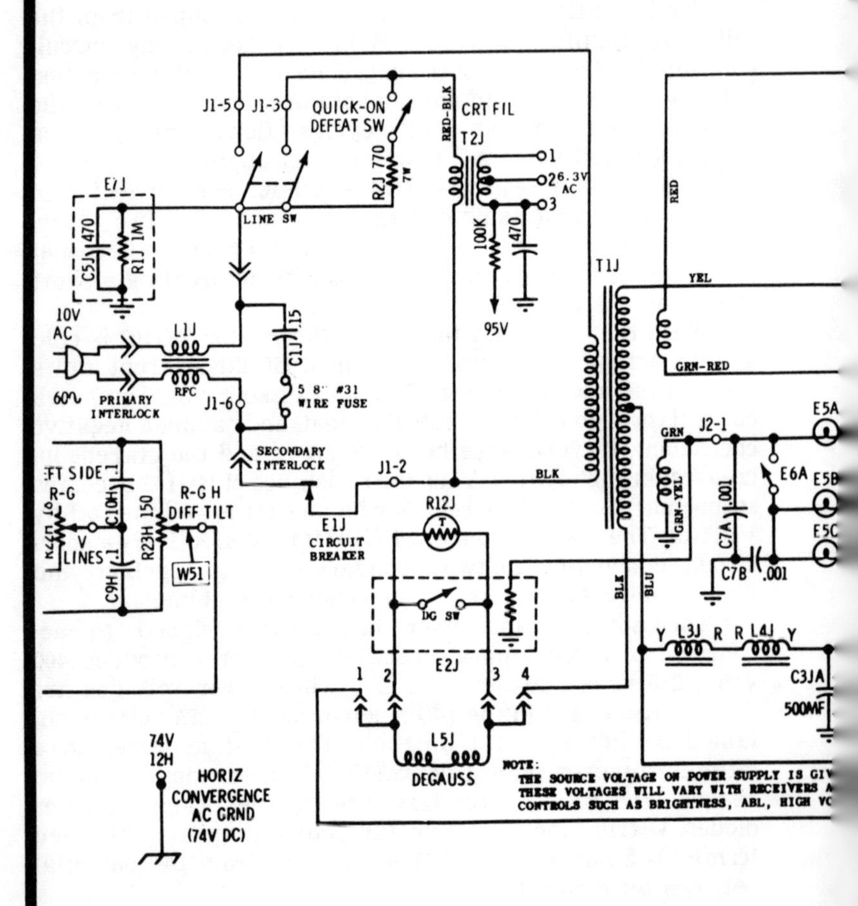

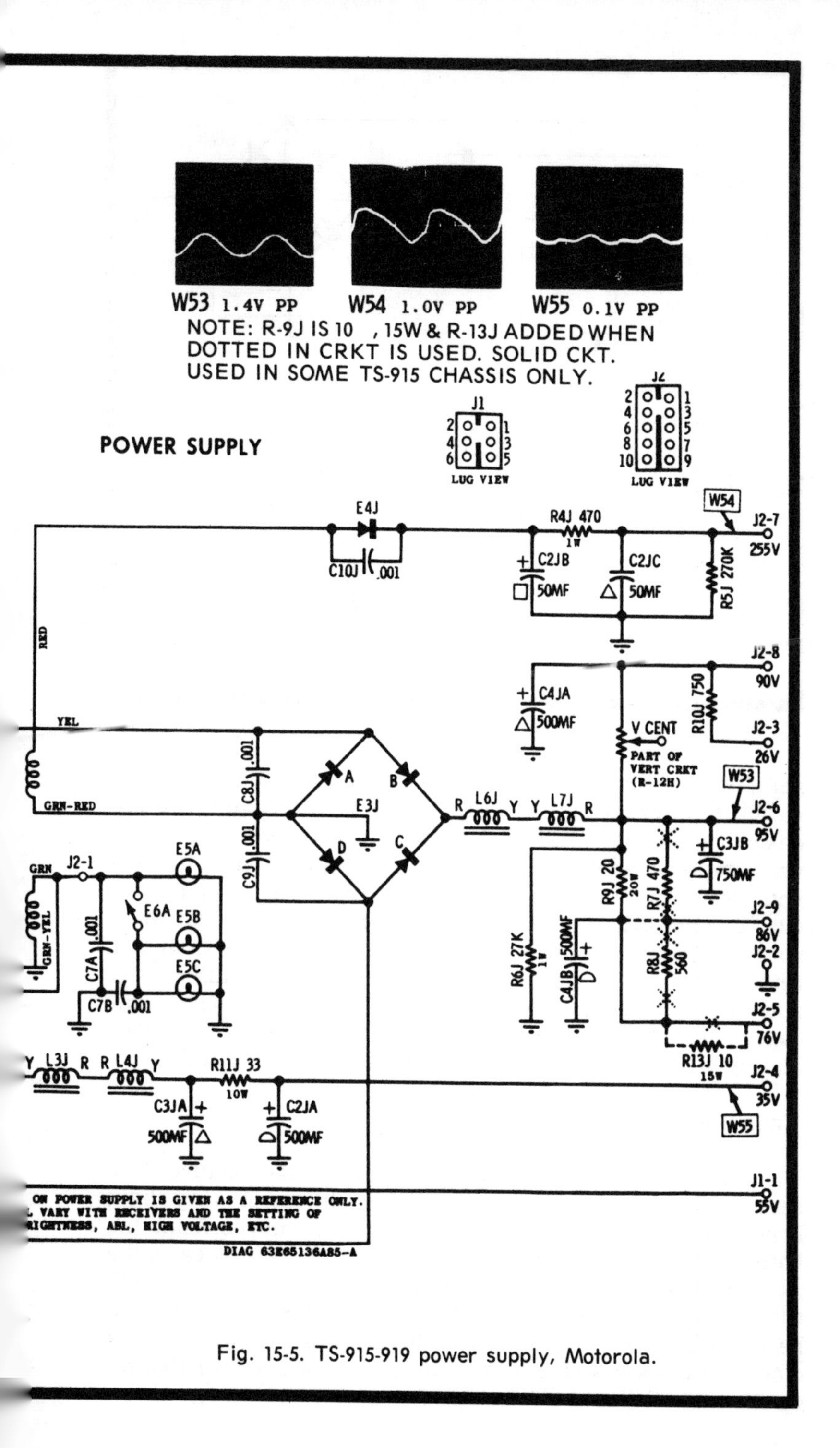

Fig. 15-5. TS-915-919 power supply, Motorola.

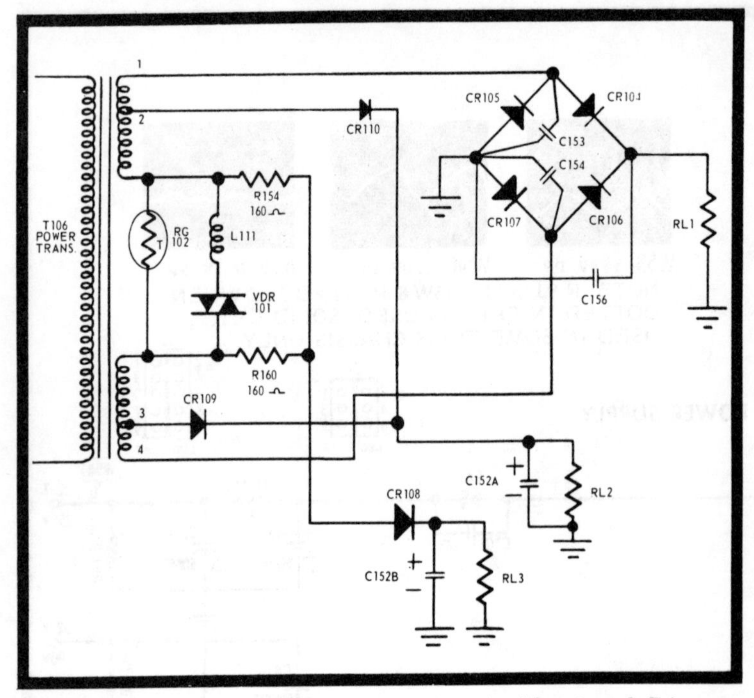

Fig. 15-6. Late "KE" balanced power supply. (Courtesy G.E.)

stant when the power supply is first turned on and no charge is on C152A. For example, when CR110 is functioning as just described, the potential between terminals 3 and 4 on CR109 is very small as compared with the potential on CR110. Initially both diodes do conduct at the same time with no charge on C152A. As C152A becomes charged, however, it reaches a potential which exceeds that between terminals 3 and 4, resulting in CR109 being cut off. C152A attains its full charge and DR110 continues to rectify one half-cycle. The same conditions, of course, apply to the rectification of the other half-cycle when the potential is between terminals 1 and 3.

Hum Bars

A light or dark horizontal bar across the picture tube screen is an indication of a hum condition. A very light horizontal bar (about 2 inches wide) that moves from the bottom to top of the screen may come from interference on the AC line or diode rectifier RF radiation. This is caused by RF radiation action from within the diodes themselves. This can be eliminated by connecting 0.001 uf capacitors across the diodes. A weak hum bar across the picture was eliminated by connecting C153 and C154—note power supply in Fig. 15-6—

across the CR105 and CR107 diodes. Capacitor C156 protects the bridge rectifier diodes from transients due to any arc that might occur within the high voltage circuits.

THE ALL ELECTRONIC POWER SUPPLY SYSTEM

The conventional power transformer shown in Fig. 15-7 operates at the line frequency and considerable iron core is needed to transfer the 60 Hz field from primary to secondary. If the frequency is increased, less iron is required to have the same primary to secondary power transfer. This is done in the Motorola electronic power supply by use of the horizontal scan frequency generated from a separate oscillator, to operate an electronic switch transistor. As shown in Fig. 15-8, a DC voltage is supplied to the transformer primary and through the electronic switch to ground. The switch is turned off and on at the horizontal scan rate and develops an interrupted DC in the primary winding. This moving field (15,750 Hz) transforms the same required power to the secondary as the conventional transformer but with a much smaller and lighter transformer.

The Switch Control Stages

The three control stages that are needed to operate the switch at the required 15,750 Hz rate are shown in Fig. 15-9.

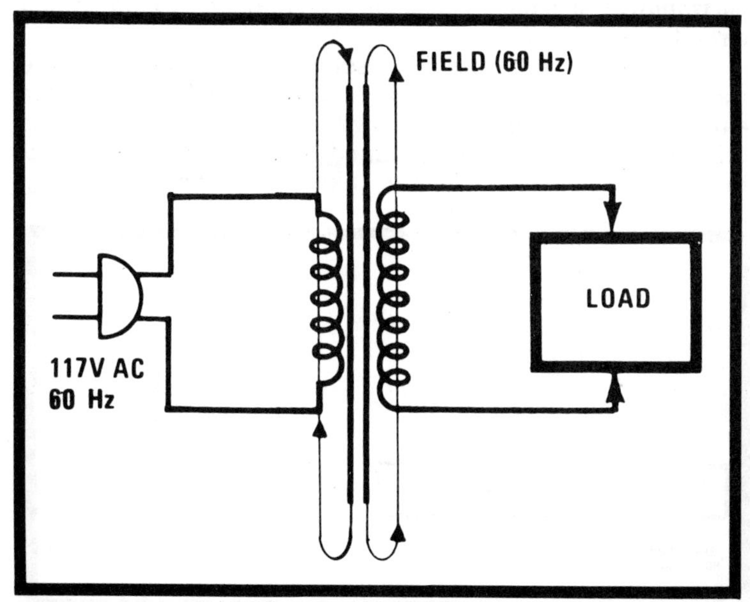

Fig. 15-7. Conventional power transformer for 60 Hz. (Courtesy Motorola.)

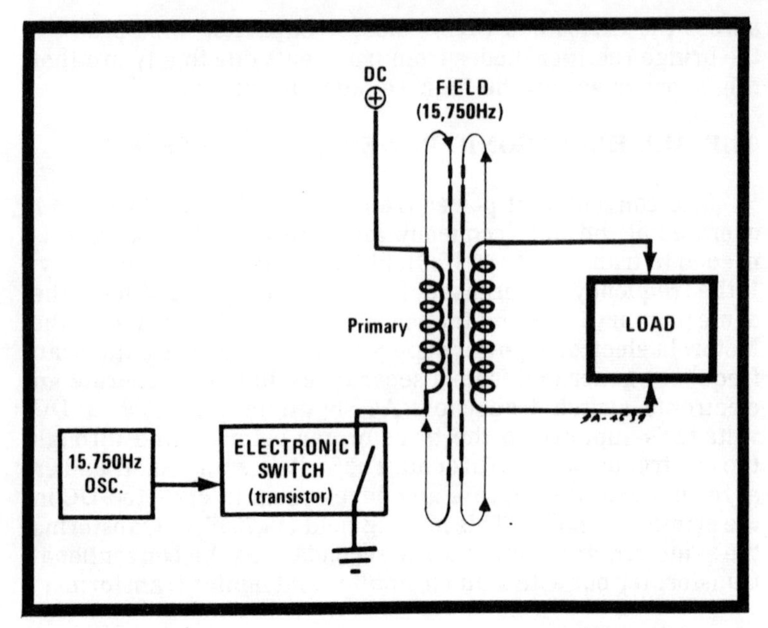

Fig. 15-8. Electronic power supply basic circuit (Motorola).

The same period of time between pulses is shown by the shaded areas.

For needed stability, the oscillator is locked to the horizontal sync in the TV by a pulse from the horizontal sweep circuit.

The shaper converts the oscillator signal to a square wave. The driver amplifies this square wave to a proper amplitude for turning on the switch transistor as the CRT brightness is increased, the demand for more power rises, and the load changes. The power taken from the low voltage

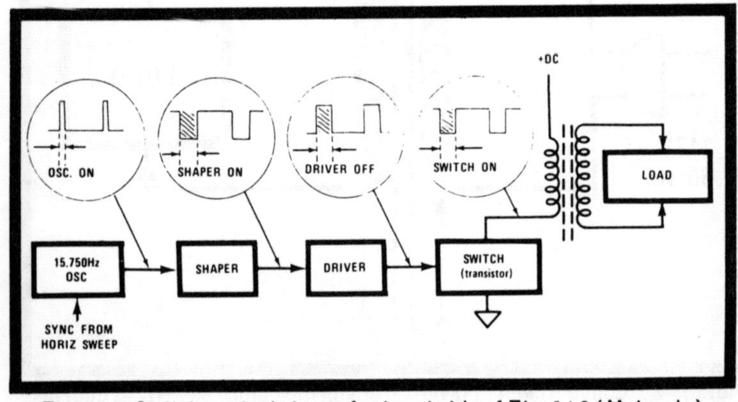

Fig. 15-9. Switch control stages for input side of Fig. 14-8 (Motorola).

supply secondary must be furnished by the primary. This causes the primary power to follow the secondary power demands.

In a conventional transformer, the primary AC voltage remains relatively constant over a wide range of secondary power demands. Increases in secondary load (power demand) is then reflected as increases in primary current.

With the electronic supply (refer to Fig. 15-10), we see that the DC voltage is applied to the primary. However, as mentioned earlier, the DC is interrupted by the switch to produce pulsating DC for the transformer action.

When the secondary demands more power, the required increase in primary current is accomplished by closing the switch for a longer time. This allows the primary current to increase further to fulfill the secondary loading. To do this, the square wave switching signal is altered by applying a control voltage to the shaper stage.

Adjustment of the square wave is accomplished by a separate secondary winding. As the load increases, all secondary voltages will decrease. The decrease in voltage is sensed by the regulator which in turn acts as the shaper to alter the square wave. The square wave switching signal now turns on the switch transistor longer (which is referred to as the duty cycle). Now the frequency of the square wave does not change, but remains locked in at the horizontal scan rate. Instead the **width** increases and turns on the switch longer and allows the primary current to rise further.

The waveforms in Fig. 15-11 show that for an increase in power demand, the regulator will cause the shaper to conduct longer, while the driver conducts less and the switch conducts

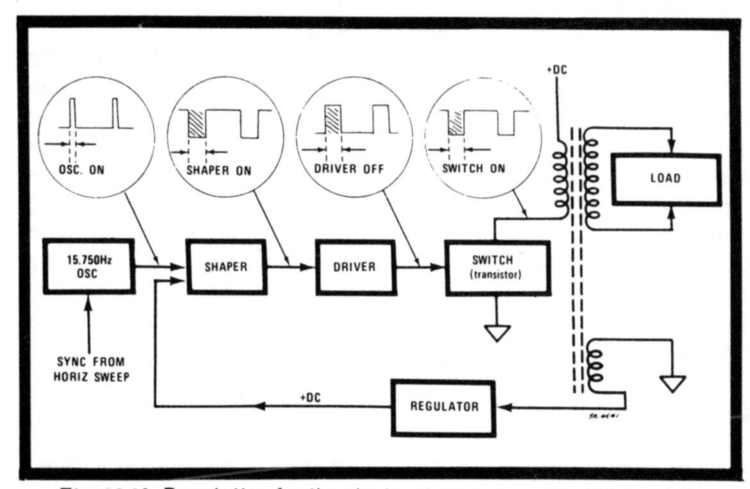

Fig. 15-10. Regulation for the electronic power supply (Motorola).

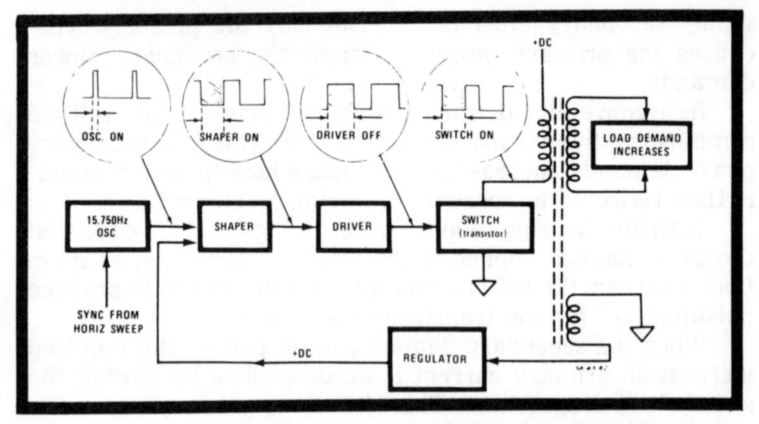

Fig. 15-11. Regulator action for increased load (Motorola).

longer. This increase in power demand is satisfied by increasing the width of the turn-on signal for the switch.

The opposite occurs when the load decreases as the secondary voltage now rises. This rise is sensed by the regulator, and the regulator in turn acts on the shaper to alter the switching square wave. The shaded portion of each waveform (in Fig. 15-11) reduces in width and the duty cycle of the switch reduces, reducing the primary current, in turn reducing the secondary power. The reduced load has now been compensated for.

Load Regulation and Voltage Changes

Another advantage of turning the switch transistor on for longer or shorter periods to provide load compensation is the added benefit of better voltage regulation.

Since the regulator senses secondary voltage changes, the net result is that a rather constant secondary voltage is maintained. The same voltage compensation occurs when the line voltage changes as when the secondary voltage changes with the load. If the line voltage rises, the secondary voltages rise. However, the regulator senses the voltage increase and reduces the "on" time (width) of the square wave at the shaper switch. With the switch on for less time, the primary current is reduced. This reduces the field and lowers the induced secondary voltage. This has now compensated for the rise in AC line voltage.

Circuit Overload Protection

The block diagram in Fig. 15-12 shows the overload shutoff stage. An SCR (silicon controlled rectifier) is used. The

SCR samples the switch transistor emitter resistor voltage. The SCR is designed to "fire" (turn on) when an overload occurs that causes the switch emitter voltage to reach a predetermined level. Once the SCR fires, the switch signal is removed and the switch shuts off to protect the supply from damage.

SCRs will remain on until all power is removed and, in this case, this means turning the set off with the on-off switch. The SCR will then be non-conductive and allow the supply to operate when turned back on **if the overload has been cleared.**

Note in Fig. 15-2 that the DC voltage applied to the transformer primary is developed by a voltage doubler circuit. The doubler rectifier diodes are a plug-in unit for easy replacement.

The oscillator, shaper, driver, overload shutoff, switch, and regulator circuits are all contained on a plug-in panel as is the lightweight, powered-iron-core power transformer.

Trouble Symptoms and Adjustments

Looking at the electronic supply schematic in Fig.15-3, you will note that there are two adjustments on this supply panel. In most cases they will not require field adjustments. The regulator control adjust "pot" R19 sets the proper output voltage. The coil T2 adjusts the horizontal sync level to the oscillator stage.

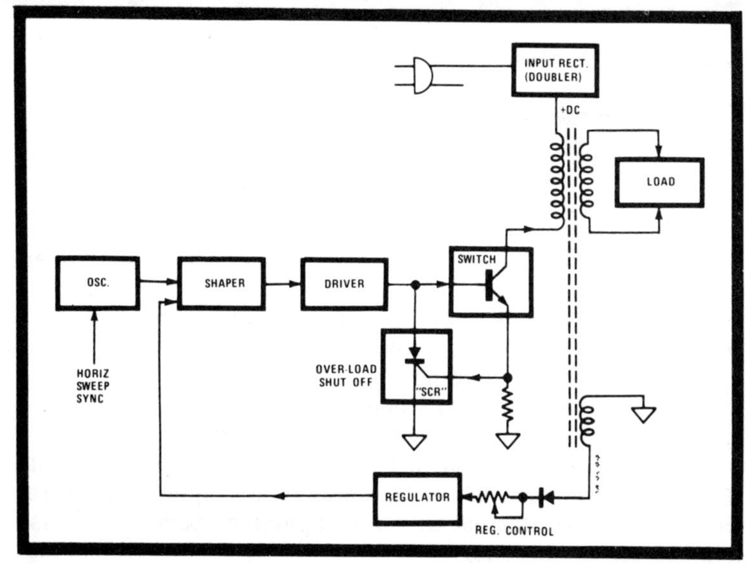

Fig. 15-12. Complete block diagram, including overload shutoff (Motorola).

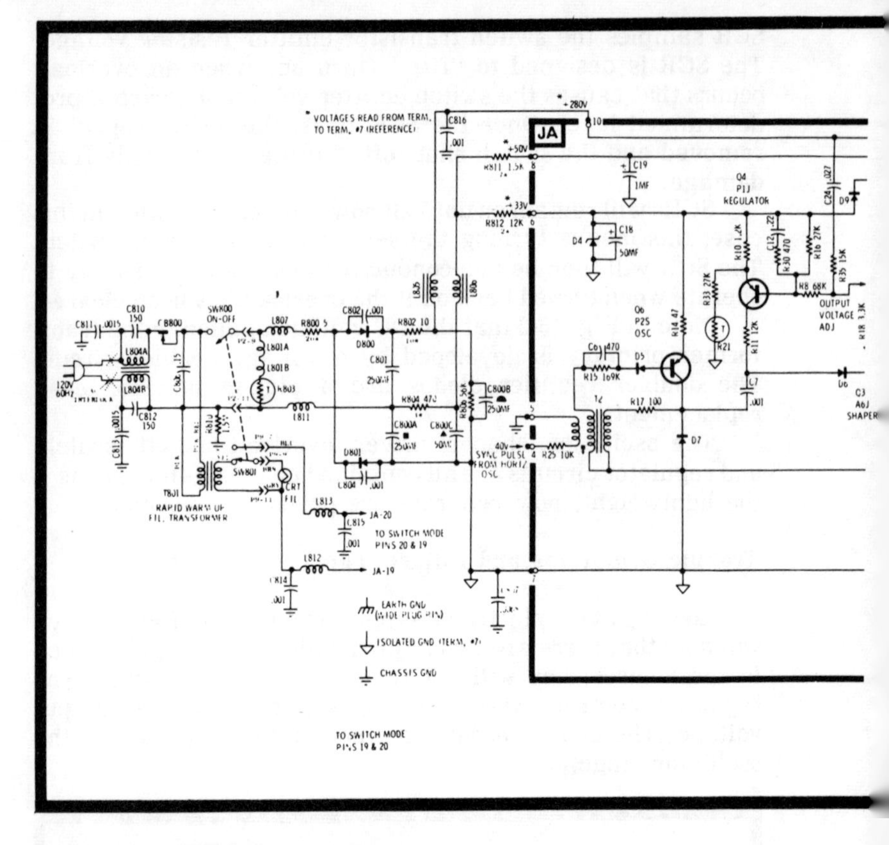

Symptoms, Analysis, and Cure

A defective electronic power supply designated as the "JA" panel may cause the following problems:

A completely dead set.
Circuit breaker opens up.
No raster on the screen.
Excessive screen brightness that cannot be controlled.
No vertical sweep.

A defective B+ rectifier doubler (plug-in package) may cause the following problems:

A completely dead set.
Circuit breaker kicks open.
A small raster or no brightness at all.

Should the 70-volt line of this power supply (refer to Fig. 15-3) only measure 45 volts, suspect an open C808 filter capacitor. When this 500 uf filter capacitor opens it will also trip the circuit breaker.

Should the set come on with a horizontal line across the screen or if the picture comes on momentarily, look for a

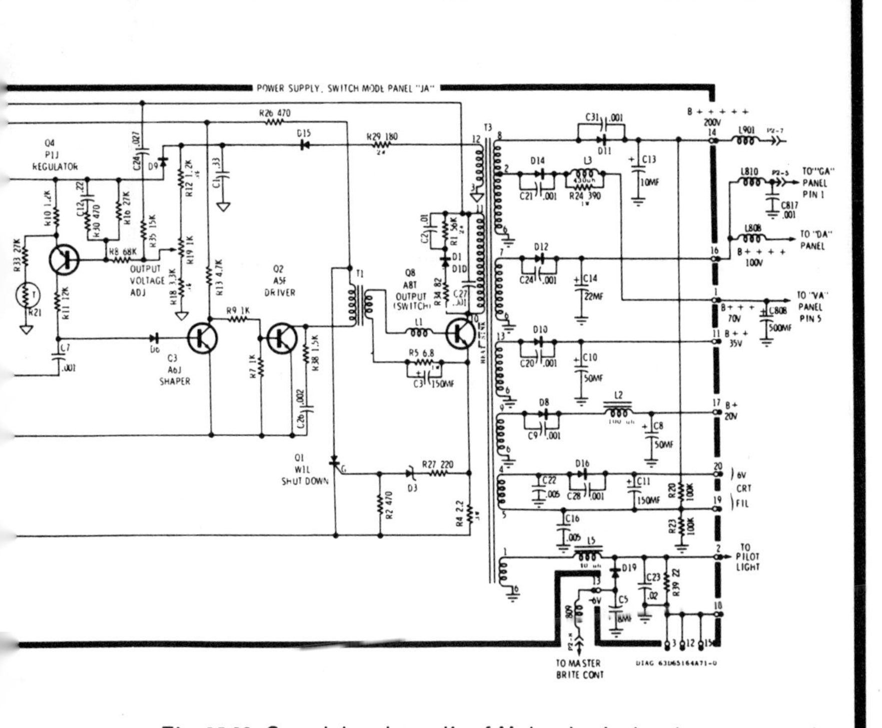

Fig. 15-13. Complete schematic of Motorola electronic power supply.

defective "VA" panel. This panel, when defective, will also cause the 70 volts to be shorted.

If the horizontal sync is lost to the power supply oscillator Q6, the vertical edges of the picture lines will have a small sawtooth look. Use your scope to check at pin 4 of the "JA" plug-in panel to see if the horizontal sync pulse is present.

Basically two plug-in units are needed for "in-home" service. Most power supply troubles can be corrected with a service call by changing either the snap-in power supply panel or plug-in B+ rectifier card.

As these color receivers go out of warranty you may want to repair the power supply panel instead of exchanging it with a new replacement unit.

For the first initial check look at the panel for any cracks or cold solder connections.

With an oscilloscope (refer to Fig. 15-13) check through the transistors from the oscillator Q6 to the switch Q8 for a proper square-wave signal at the horizontal frequency. Refer back to Figs. 15-10 and 15-11 for these square-wave scope patterns.

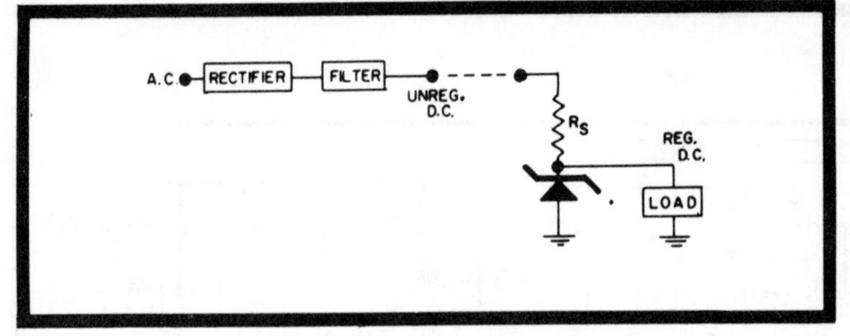

Fig. 15-14. Shunt zener diode regulator (Zenith).

If the square-wave signals are not correct, make all routine voltage and resistance checks in the faulty section. Then use a transistor checker to test out all of the solid-state devices.

Now check transformers T1, T2, and T3 with an ohmmeter for any open windings. After these checks and appropriate repairs have been made, the electronic power supply should now be okay.

Regulated B+ Power Supplies

The most commonly used power supply is one that is unregulated, which means that as line voltage and temperature vary, the rectified and filtered output will also vary. This type of power supply is only used when the load is capable of maintaining proper operation within limited AC line voltage changes.

Because transistors and ICs are sensitive to overvoltage, regulated supplies are used to prevent these changes. With some of the sophisticated circuitry now encountered, the degree of change in the source voltage becomes more critical. In some of the new all solid-state color TV chassis, a high degree of regulation is necessary for stable operation. Thus, a regulated power supply is very necessary. The simplest type of regulation that can be used is zener regulation.

Semiconductor diodes which maintain a fixed voltage across the junction when the "zener" or reverse-voltage breakdown point is exceeded are used for control device. The zener diode acts as a variable instead of a fixed load.

The zener regulation circuit shown in Fig. 15-14 is a shunt type regulator (regulating device in parallel with the load). Basic design for the shunt zener regulator circuit requires that when no load is applied, the zener must pass the amount of current the load would require, and also an additional amount of current that would keep the zener regulating when the load is applied.

This total amount of current is determined by the value of R_S. If the value of R_S increased, the available current to the load and zener would decrease. The zener could possibly stop regulating, thus causing a deterioration in voltage regulation. Shunt regulation is the least efficient since there is always some current through the zener that performs no useful purpose.

A more efficient regulator system is the series type which has the regulating device in series with the load. A series regulator system requires a control device in series with the load like a power transistor shown in Fig. 15-15.

Basically, the circuit is the same except for the added transistor. The circuit actually is a series-shunt regulator circuit since the zener diode is still used as a shunt regulator. The only reason why the circuit is called a series regulator circuit is because the largest amount of current flows through the emitter-collector path of the transistor. Effectively, the transistor is the R_S. The advantage of this circuit is that a large change in current flow at the emitter shows up as a small change of current in the base circuit. This means that the amount of current required through the zener will be less. Because the transistor is used as an emitter follower, the emitter will remain very close to the zener rating, if the load does not draw too much current. Should the load circuit develop a short and excessive current be drawn, the transistor would be damaged, unless a limiting resistor (shown as R limiting) is used.

A +24 volt zener controlled regulator circuit is shown in Fig. 15-16. This circuit is that shown in Fig. 15-15. Note that C255 has a very large capacitance value. This is to maintain as stable a reference voltage for the zener as possible. Also note that C241, the output capacitance, is of a small value. Since an emitter follower is a low impedance device, and the base is

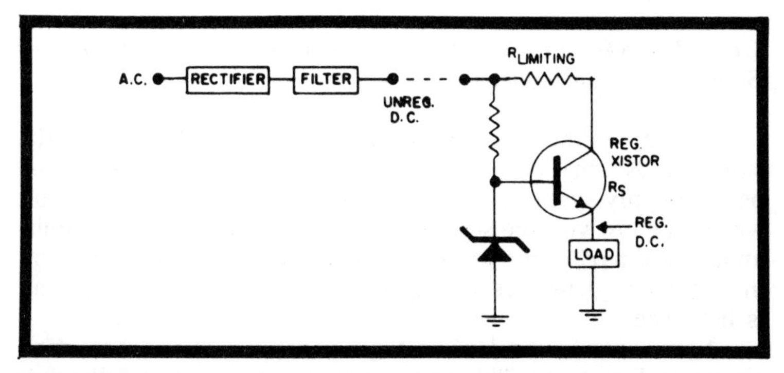

Fig. 15-15. Series power transistor regulator with shunt zener diode (Zenith).

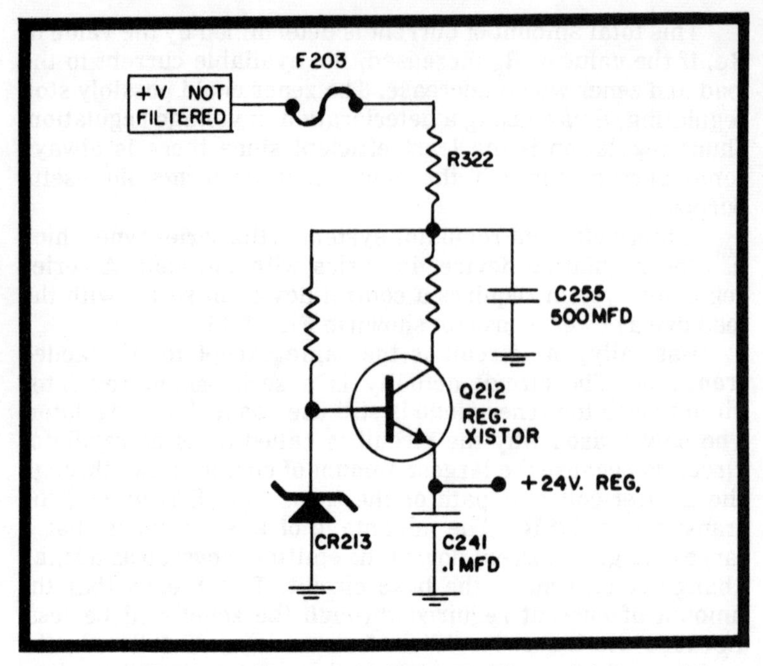

Fig. 15-16. Zener controlled +24-volt regulator.

clamped by the zener, there is no need for a large capacitor to be placed at the emitter of Q212. A larger capacitor may be located at another point in the chassis, the reasoning being is that a long wire, having sufficient current passing through it, can look like a resistor to high frequency information.

One way of increasing current gain, and maintaining voltage regulation, is to use a Darlington circuit like that shown in Fig. 15-17. The advantage of a Darlington regulator system is a greater amount of current can be handled with reasonable voltage regulation.

Error Voltage Sensing and Side Pincushion Regulated Power Supply Troubleshooting

In some of the new model solid-state color TV chassis, the side pincushion correction circuit is part of the regulated power supply system. A trouble symptom indication of a fault within the low voltage supply or the "side pincushion" circuit, an integral part of the supply performance, will be a variation in raster size, with line voltage changes, or side pincushioning is incorrect.

Now refer to the low voltage and pincushion correction circuit in Fig. 15-18. This power supply delivers the following voltages:

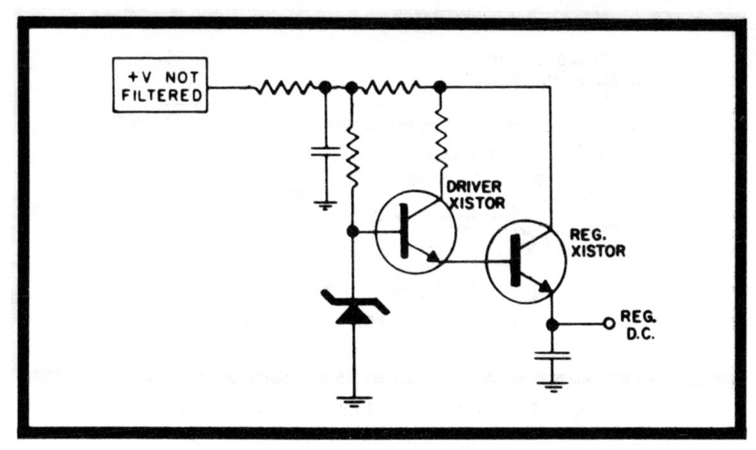

Fig. 15-17. A Darlington regulator circuit.

B+ 130V, unregulated, for audio circuits only.

B+ 128V, to the horizontal flyback transformer, regulated and containing a vertical parabolic curve.

B+ 125V, regulated, and filtered to remove the vertical parabola.

The following are some checks to determine normal operation of the regulating/pincushion system:

Emitter of Q214 "error sensing" transistor must hold at +24 volts. If not, check for defective zener CR216.

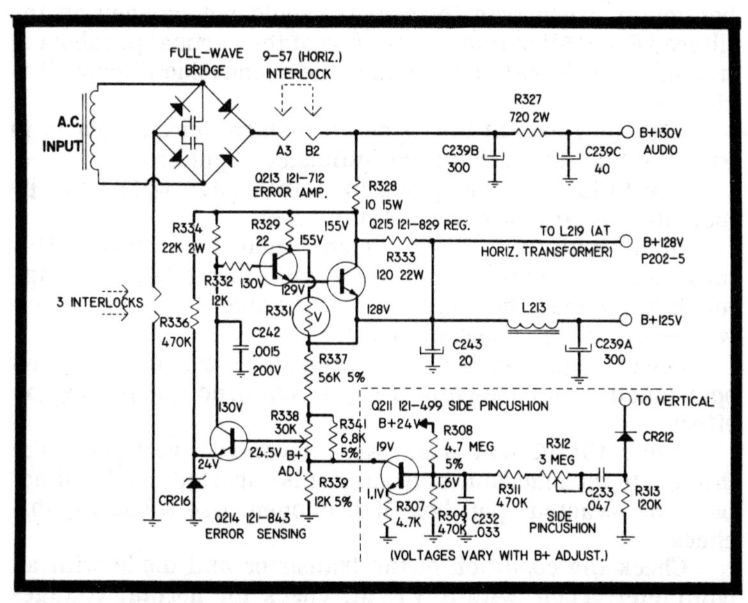

Fig. 15-18. Zenith 40 BC50 low voltage, regulated power supply with side pincushion circuitry.

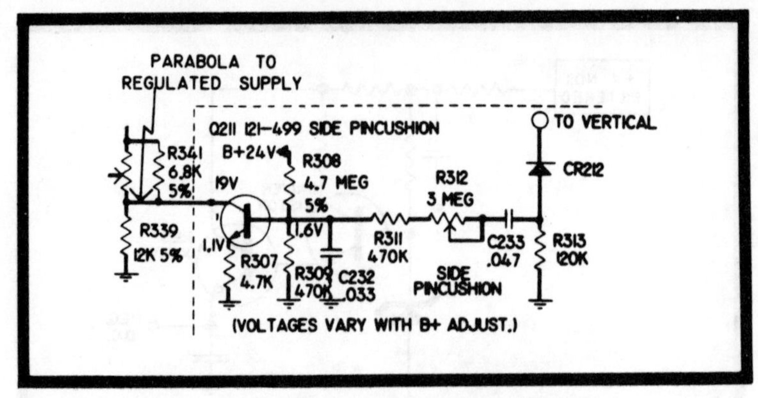

Fig. 15-19. Side pincushion circuit (Zenith).

Regulated B+ output voltages (+128V and +125V) will vary as the B+ adjust "pot" R338 is varied. A decrease in the "adjust" control arm voltage will cause an increase in the regulated B+ outputs.

The oscilloscope should display a 60 Hz parabolic curve with a pronounced vertical positive-going spike on the cathode of diode CR212. The diode should block the spike so only the parabola will appear on the anode. The parabola should also appear on the base of Q211, then it is amplified and inverted in polarity at the collector and through the regulator circuitry. Then the parabola should appear on the B+ 128V output to the horizontal sweep transformer. It should not be seen on the filtered B+ 125V output. Amplitude of the vertical parabola at transistor Q211 will vary as the "side pincushion" control is changed.

Let's now look at the "side pincushion" circuit shown in Fig. 15-19, as this circuit can influence regulation.

The failure of side pincushion transistor Q211 may be indicated by any of several symptoms:

High regulated voltage, which cannot be adjusted. This may be an emitter-collector short, with the collector voltage much lower than the indicated +19V value. Of course, there will be no side pincushion effect.

Lower regulated B+, but adjustable, which can be an open junction or base-emitter short. Again, no side pincushion effect.

Diode CR212 will pass the (unwanted) vertical spike through to the pincushion circuit if it is shorted and will not pass the (wanted) parabola if it is open. Use scope for this check.

Check the condition of the transistor and diode with an ohmmeter. Then, with a VTVM, check for normal voltages and transistor forward bias. Then with your scope look at the parabolic curve as it progresses through these circuits; this will help locate the faulty components.

When it is apparent the low voltage supply is not working correctly, and there is a hint of a fault in the "side pincushion" circuit as well, then the problems should be dealt with singly. First, remove Q211 and proceed with servicing the regulating system, then follow up by servicing the side pincushion system.

If the regulating system has been determined to be faulty, then a systematic analysis approach should be used.

The schematic in Fig. 15-20 shows that there is a voltage divider network from +155V to ground. The bridge rectifier output should be just over +155V, with a negligible drop across the 10 ohm, 15W resistor R328. There should be +155V on the collector of the regulator transistor Q215.

If you suspect the Q215 regulator is defective, remove it from the circuit. Full voltage drop from B+ 155V to the +128V and +125V outputs will be through R333, 120 ohm, 22 watt. Readings much lower than designated values will indicate excess current drain in either of these circuits and the

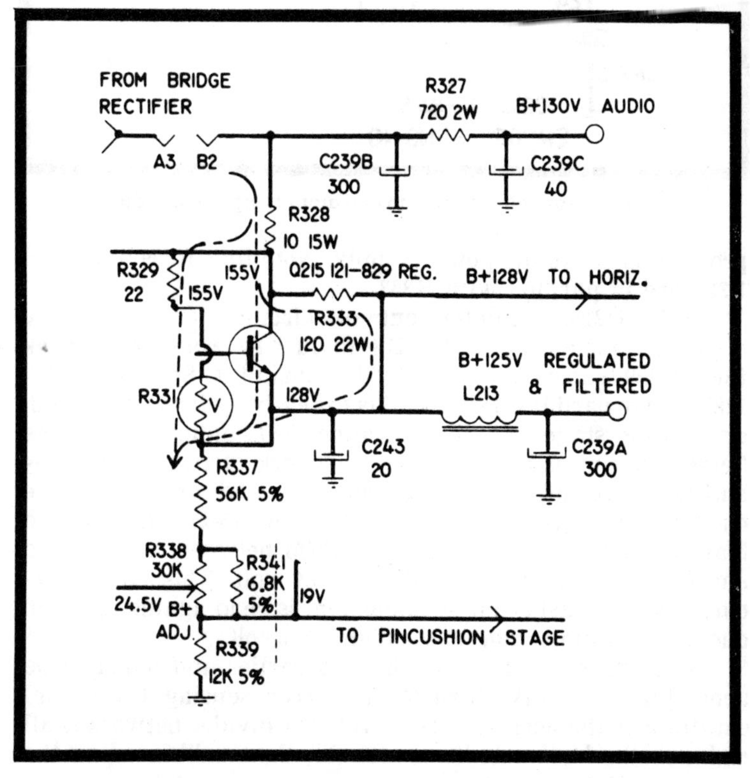

Fig. 15-20. Voltage divider from +155V to ground (Zenith).

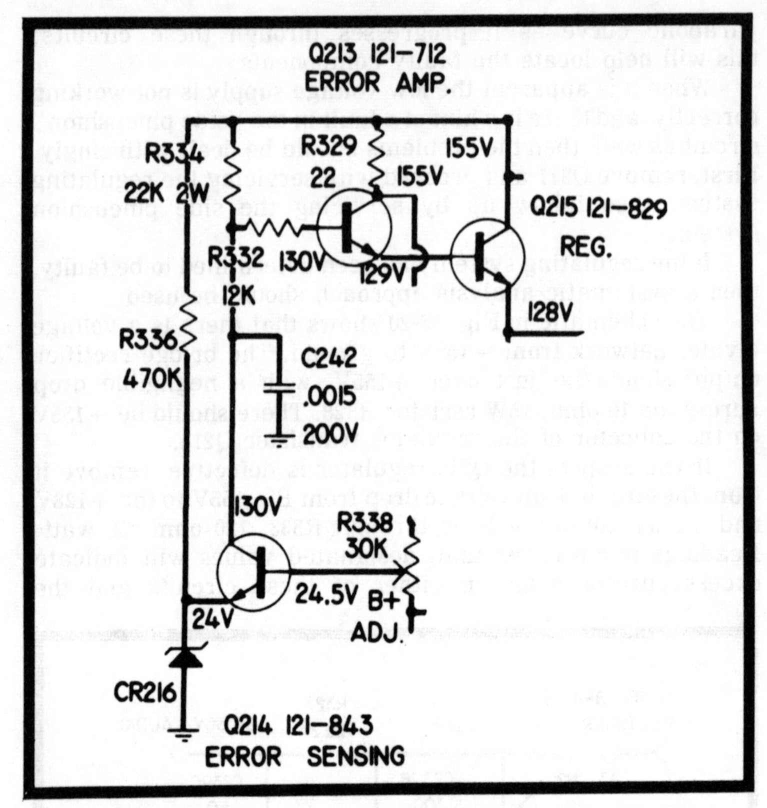

Fig. 15-21. Error voltage sensing and amplifying stages (Zenith).

problem is not in the power supply. Note that R329 and VDR R331 are in parallel with R333.

With Q215 regulator out of circuit, if the voltage distribution is normal at the B+ divider network, then check the 121-829 transistor and if good, reinstall it. Should the B+ 128V rise toward B+ 155V, Q215 is excessively forward-biased, driving the stage toward saturation. This indicates that the "error sensing and amplifying" network of transistors Q214 and Q213 are not functioning normally and will have to be checked out. This circuit is shown in Fig. 15-21. The forward bias of Q215 is measured with a VTVM between its base and emitter, and should read about +0.6 volt. This cannot accurately be measured if each element is read against ground, due to meter limitations at the high +128 volts.

Since zener CR216 was checked earlier and found to be good, that is, +24V, then Q214 (error sensing transistor) emitter is at the normal +24V level. The divider network is all right, so the base can be reasonably forward-biased on the order of +0.6V or so.

Next, check the forward bias of Q214. When the forward bias is all right, and can be adjusted through the normal range with the B+ adjust "pot," check the collector for +130V. A low value at this point may be caused by a leaky C242 (.0015 uf) capacitor at the collector. Higher voltage on the collector indicates either Q214 is not conducting or Q213 (error amplifier) is conducting too much.

Remove Q213, and check to see if Q214 can be adjusted normally. Removing Q213 will also disable regulator Q215 and the B+ dropping network voltages should return to normal, or slightly low at B+ 128 and 125V. Check out Q213 and be sure a good transistor is reinstalled.

As a final test, check the forward bias of Q213, error amplifier, for a reading on the order of +0.6V, base to emitter.

With the regulator system now locked back into normal operation, reinsert the side pincushion transistor, Q211, and proceed with tracking down with an oscilloscope any defect interfering with its ability to sample and invert the vertical parabola.

Symptoms and Troubleshooting Tips

If you suspect trouble in the power supply (or even if you don't) start by measuring the DC voltage test points to determine if these circuits are operating properly. Compare the voltages measured with those noted in the service information.

If the DC voltage is low, but the rectifier diodes are good and there are no short circuits, the first suspect would be an open input filter capacitor. This is easy to check by simply bridging a good capacitor across the input filter and noting a rise back to a normal DC voltage level. Watch polarities. The input filter is designated as C246 in Fig. 15-22. When this filter opens up you may find a picture on the screen as shown in Fig. 15-23.

Low Resistance or Shorts

Shorts in the TV chassis power supply can give many types of indications such as blown fuses, dead set, weak picture or sound, and puffs of smoke and bad odors. Make any resistance measurements with the power turned off and with all filter capacitors discharged before you zonk out the ohmmeter.

If the B+ line fuse blows, this indicates a short in the circuits fed from the power supply. Use an ohmmeter to locate the shorted part by following the B+ circuit path of lowest resistance. When this resistance check leads you to a junction

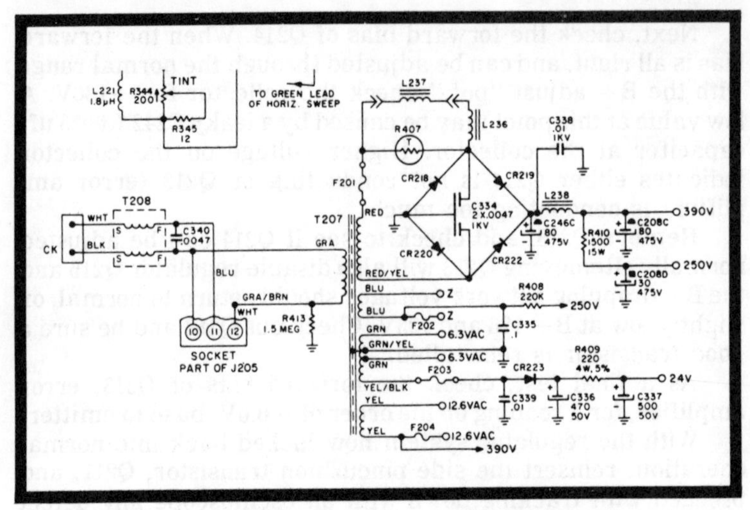

Fig. 15-22. Late model Zenith color TV power supply.

point with many branches, then isolate the short by disconnecting each lead wire, one at a time.

You may find a situation where resistors in the B+ line have become burnt or overheated, but can find no shorts in any of the associated circuitry. In a TV chassis this is usually caused by high amplitude horizontal frequency pulse voltage riding on the B+ line as shown in Fig. 15-24. Look, with the

Fig. 15-23. Effect of open input filter capacitor.

scope, for high amplitude pulses (this will occur when a filter opens up) on the B+ line as this will cause some resistors to burn up or overheat.

Here is another example to illustrate that trouble in the B+ power supply will give you misleading symptoms. A color set had no sound or video information, just a blank, milky-white screen. We first suspected trouble in the solid-state video IF amplifier stages. However, the trouble was a very low B+ voltage supply that was fed to the video IF stages. Note in Fig. 15-22 that +24 volts is used for all of the transistor stages. The failure of filter capacitor C337 (was very leaky) reduced this to about 5 volts and caused this set not to operate at all. Any time you come across a set with "odd-ball" symptoms or a multitude of troubles, always connect up the scope and check out the B+ filtering system.

OSCILLOSCOPE WAVEFORM CHECKS

The oscilloscope is the most accurate indicator for checking the ripple level of the various B+ points throughout the power supply system. Let's now look—with the scope—at the normal wave traces found for the input and output filter capacitors in a TV chassis. The hum level (shown in Fig. 15-25) at the input filter should be from 15 to 35 volts peak-to-peak, while that measured at the output filter (note Fig. 15-

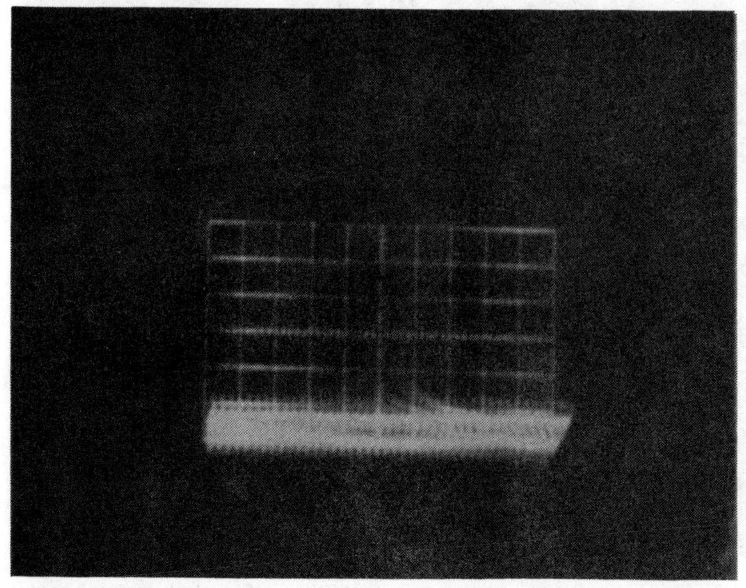

Fig. 15-24. High-amplitude horizontal pulse riding the B+ line can burn resistors in the B+ supply. (Scope at vertical rate.)

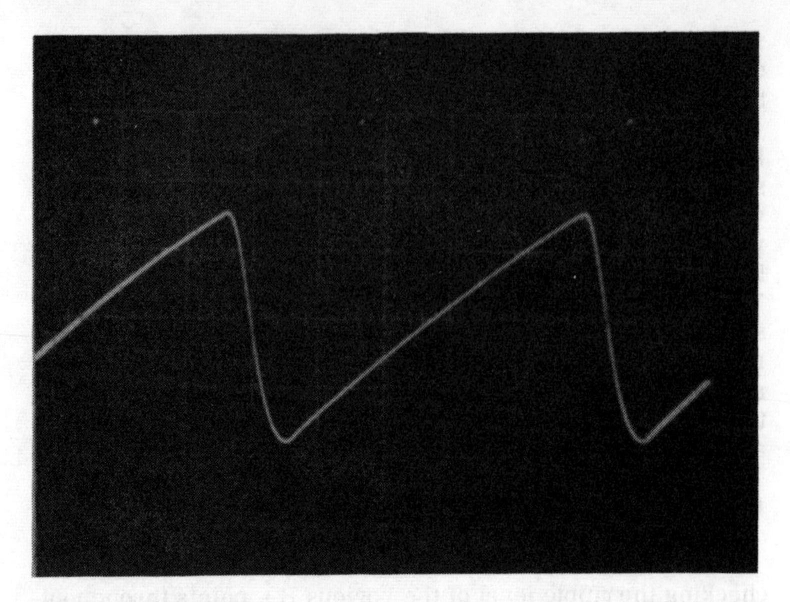

Fig. 15-25. Normal hum level at input filter. (Scope at vertical rate, amplitudes 35V P-P.)

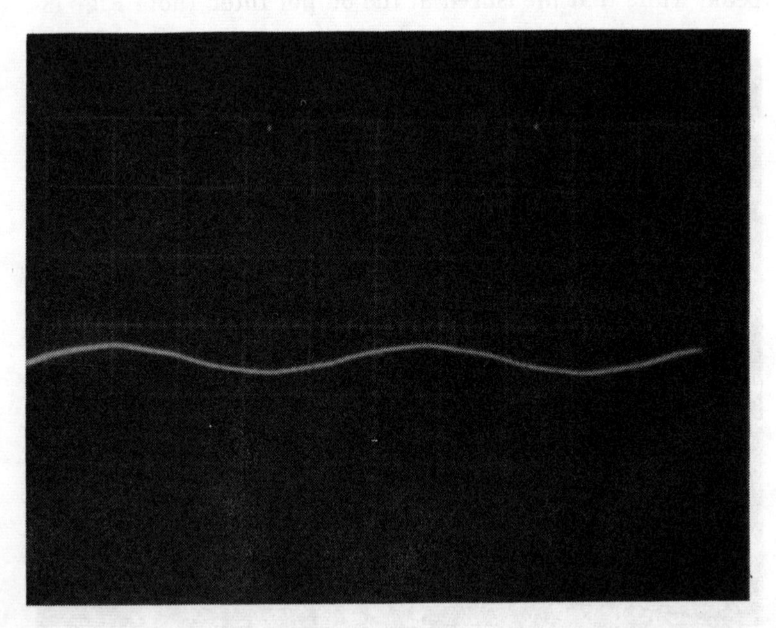

Fig. 15-26. Normal hum level at output filter. (Scope at vertical rate, amplitude 1.2V P-P.)

26) will usually have a level of 1 to 2 volts peak-to-peak. After some more filter units, the B+ line should appear as a nice smooth line across the scope screen. If you see any AC flickering around, start looking for an open capacitor.

When the oscilloscope test is made, use the direct probe as you are looking at a low-impedance circuit. Of course, with a high vertical scope amplifier gain, you may see some of the sweep signals that are fed back into the power supply. These should be no more than 0.2 volt P-P and if it's higher than this, again look for filter problems. The power supply B+ line in the solid-state TV receiver should have an almost zero ripple.

When using the scope to troubleshoot in the power supply, you actually look for signal that should not be there. Open filters will indicate trouble symptoms in circuits other than the B+ supply. Some of the circuits like the sweep, sync, video, and audio circuits will interact with each other due to a faulty filter or decoupling capacitors. This will show up as unstable sync, jitter, distorted picture, dark bars or sound in the picture. The photo in Fig. 15-27 shows the dark bands across the TV screen due to an open filter capacitor. Hum may be found in the picture or sound or both. And power supply hum may also affect the raster.

An easy way to check for an open filter is to bridge a good filter across the suspected one and watch the results either on

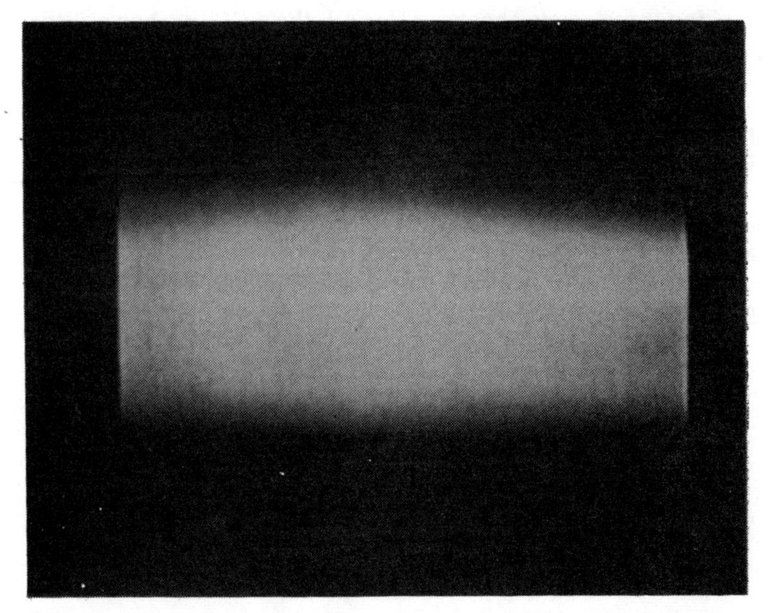

Fig. 15-27. Dark bands on TV screen caused by open filter capacitor. Note the pull-in at the sides.

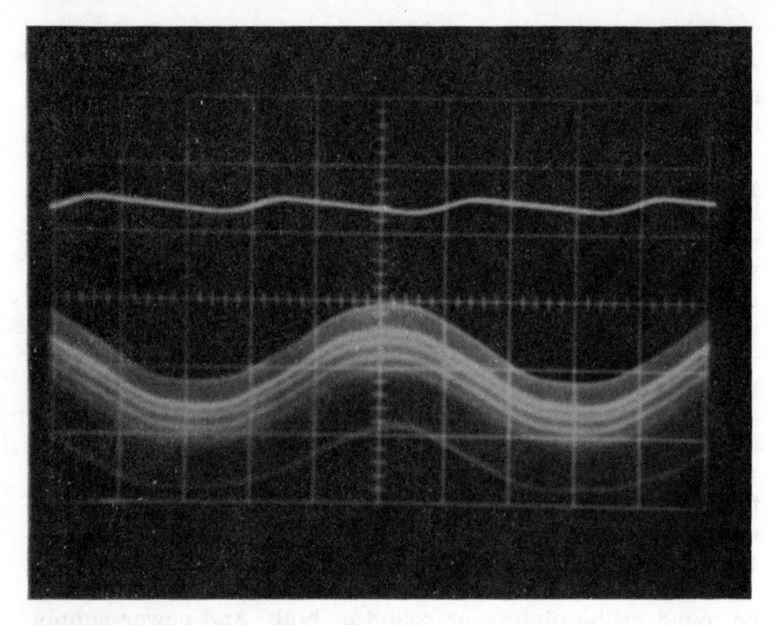

Fig. 15-28. Open AGC filter capacitor bottom trace. Top trace, normal AGC. (Scope at vertical rate, 6V P-P.)

the scope or TV screen. And don't overlook the AGC line while making these scope checks. After all, the AGC line has a filter system of its own. To cite another example, this set had an AGC problem that caused a picture bend. When the scope was connected to the AGC line the scope pattern shown in the bottom trace of Fig. 15-28 was seen. Note the 60 Hz and some video information of about 6V P-P. This was caused by an open AGC filter capacitor. The top trace shows what the line looked like after a good filter was bridged in. A scope check is ideal for locating any improperly operating voltage filtering system. A peak-to-peak AC waveform in excess of a few volts points to improper filtering.

INDEX

A

B

C

D

E

F

OTHER TAB BOOKS BY ROBERT GOODMAN

No. 436 Practical Color TV Servicing Techniques

No. 490 Advanced Techniques for Troubleshooting with Scope

No. 502 Zenith Color TV Service Manual, Volume 1

No. 522 Philco Color TV Service Manual

No. 536 General Electric Color TV Service Manual, Volume 1

No. 562 Zenith Color TV Service Manual, Volume 2

No. 577 How to Use Color TV Test Instruments

No. 595 199 Color TV Troubles and Solutions